TRANSACTIONS

OF THE

AMERICAN SOCIETY

OF

CIVIL ENGINEERS

(INSTITUTED 1852)

VOLUME 147

1982

NEW YORK
PUBLISHED BY THE SOCIETY
1983

NOTE—The Society is not responsible for any statement made or opinion expressed in its publications.

ISBN: 0-87262-355-6
Manufactured in the United States of America

CONTENTS, VOL. **147**

Foreword

Transactions of the American Society of Civil Engineers, Vol. 147, 1982 contains abstracts for all ASCE journal papers and technical notes, *Civil Engineering—ASCE* feature articles, special publications, and Manuals and Reports on Engineering Practice published by the Society during 1982. In addition, the president's annual address to the Society, abstracts for all papers published by ASCE that won 1982 awards or prizes, and memoirs of distinguished deceased members are presented. A keyword index and an author index are provided to help the user locate an abstract by subject or author.

The user will notice that abstracts for all 1982 Society journal papers are grouped together according to journal. For example, abstracts for all papers and technical notes published in the *Journal of the Construction Division of ASCE* are together, beginning with the first paper published in that journal in 1982 and proceeding according to the order in which they were published in that journal. The journals are in alphabetical order. Abstracts for all *Civil Engineering—ASCE* feature articles follow the journals. Abstracts for all special publications and Manuals and Reports appear next in alphabetical order according to title.

Each abstract entry includes an entry number, title of the paper, article, or book, author's name and affiliation, complete bibliographic information. Following each abstract is a list of discussions regarding that document which have appeared to date. Sample entries are shown below.

Sample entry for journal paper

[1]**0078** [2]**Photosynthesis-Light Mathematical Formulations**

[3]**Stephen D. Field**, [4](Asst. Prof., Dept. of Civil Engrg., Louisiana State Univ., Baton Rouge, La. 70803) and [5]**Steven W. Effler**, (Adjunct Prof., Dept. of Civil Engrg. Syracuse Univ. Syracuse, N.Y. 13210)

[6]*Journal of the Environmental Engineering Division,* Vol. 108, No. EE1, February, 1982, pp. [7]199-203

[8]Photosynthesis-light curves for photoplankton describe the characteristic dependence of the instantaneous rate of photosynthesis on incident light. Six different mathematical formulations are evaluated for the photosynthesis-light relationship, based on their ability to describe experimental data collected from nutrient saturated, hypereutrophic Onandaga Lake in New York State. The data are comprised of measurements of in situ gross primary productivity, and the available light from 115 experiments. [9]Errata: GT March '82, pp. 514-515.

1. Entry number. 2. Title. 3. Primary author (and membership grade, if applicable). 4. Author's affiliation. 5. Additional author and affiliation. 6. Journal title, volume number, issue number, month, year. 7. Pagination. 8. Abstract. 9. Discussions, errata, and/or closure.

Sample entry for special publication

[1]0974 [2]**Proceedings of the Conference on Construction of Power Generation Facilities**

[3]Experience with the Implementation of Construction Practices, Codes, Standards and Regulations

[4]**Jack H. Willenbrock**, (editor), (Dept. of Civ. Engrg., Pennsylvania State Univ., University Park, PA)

[5]*New York: ASCE,* 1982, [6]623pp.

[7]The impact of the present construction practice, codes, standards and regulations on the construction of power generation facilities is reviewed and examined in papers presented at the September, 1981 conference on construction of power generation facilities. The papers examine major problems related to codes and regulations, the difficulty of implementing some of the requirements, and the possible improvement recommendations that can be made. The major areas covered include: Nuclear standards, licensing, enforcement, and construction inspection programs; requirements and performance of concrete materials; quality assurance and quality control practices; concrete test requirements and process control practices; steel fabrication practices, concrete anchorage systems, plant modification and retrofit experience; concrete batch plant production transportation and field adjustment practices; concrete formwork, reinforcement and placement tolerances and practices; and concrete placement and post placement practices. The perspectives of the utilities industry, the architect-engineer, the government, and the concrete industry are also examined.

1. Entry number. 2. Title. 3. Subtitle. 4. Editor. (If this book was the work of a committee, the committee name would appear here.) 5. Publisher information. 6. Number of pages. 7. Abstract.

In the subject index the user will find under each keyword a list of titles pertaining to that subject. Each title is followed by the entry number for that abstract. In the author index, the document title and entry number are given if the author listed is the primary author of the document. For secondary authors, the primary author and entry number are given. Sample entries are shown below.

Sample subject index entry

[1]**Nuclear power plants**

[2]Proceedings of the Conference on Construction of
Power Generation Facilities, [3]0974

1. Keyword (subject heading). 2. Title. 3. Entry number.

Sample author index entry for primary author

[1]**Field, Stephen D.**

[2]Photosynthesis-Light Mathematical Formulations,
[3]0078

1. Author's name. 2. Title. 3. Entry number.

Sample author index entry for additional author(s)

[1]**Effler, Steven W.**

see [2]Field, Stephen D., [3]0078

1. Additional author's name. 2. Cross reference to primary author's name. 3. Entry number

PRESIDENT'S ANNUAL ADDRESS

Highlights of the Year

James R. Sims

It has been my honor to serve the Society during the past year as its 113th President, an experience that I will cherish for the balance of my life. This has been a year of change. Mr. Louis L. Meier, Jr., who has been in charge of our Washington office for the past 10 years, assumed the position of Executive Director and Secretary of the Society on April 1, and Mr. Don Buzzell of the staff assumed the position of Assistant Executive Director. Lou came to the Society after a distinguished career in the United States Navy and will retire from Society service at the Annual Meeting in Houston in October of 1983. A search committee, chaired by Vice President Al Grant, will recommend a successor to the Board of Direction during the coming year.

The Society rededicated itself during the past year to its primary reason for existence--service to membership. This rededication is a result of three and one-half days of deliberation by the Board of Direction at the time of the Las Vegas Convention during the past spring. Items discussed were in many instances generated as a result of preliminary reports from a Task Committee on Service to Membership, chaired by Past President Joe Ward. That committee determined that the principal goals of the Society continue to be education and continuing education or the maintenance of professional competence by the membership. The committee has studied the organizational structure of the Society and the staffing needed to improve services to membership. The Board will act on the recommendations of this committee during the coming months. The committee found strong support from society membership to study the location of headquarters to determine whether or not a relocation of all or a portion of headquarters might better serve Society function and the membership. A task committee, under the chairmanship of Vice President Paul Kuhn, is actively pursuing this study and will report to the Board in mid-January. This item is subject to much emotion and I am pleased to report that the committee has approached its deliberations in depth and its recommendations will be based upon facts as best they can be determined. We must all recognize that ASCE is one of the five Founder Societies and through its participation in the United Engineering Trustees has supported the present headquarters structure of the United Engineering Center. Decisions must give proper weight to our past association with the United Engineering Trustees and to our mutual relationships which have been developed over a long period of time. In addition to these ties, there is the prestige and influence in engineering matters that has been generated in concert with the other Founder Society organizations for benefit to our membership, to our profession, and to the well-being of our nation.

This has been a year in which the Branches, Sections and the General Membership have focused on membership growth in order to attract to the Society those civil engineers who might benefit from membership. The organized Membership Campaign, first in Society history, was aimed not at simply growth in numbers for numbers' sake, but at generating a foundation of strength for the Society. Such strengths manifest themselves in the programs of the Society and the finances to support those programs without depending solely on dues increases from an already-burdened membership. We only have to look at other societies which are not growing to recognize the consequences of static membership numbers in a time of economic inflation. On October 1 last year, we had a total membership of 78,104. On October 1 of this year, after all adjustments to membership had been applied, our membership stood at 86,183, for an increase of 8,079 members. This increase consists of 3,334 additional members in the grade of Associate Member and Member. In order to achieve that increase, it was necessary to recruit 7,300 new members to accommodate the loss of those deceased, resigned, or who ceased paying annual dues. The remainder of our increase is the 4,745 student chapter members who have affiliated with the Society as Student Members. Those Student Members whose new status became effective on January 1 receive full benefits of membership. The Society thanks all those who participated in bringing the new members and the student members into affiliation. It is hoped

---

Presented at the Annual Convention of the American Society of Civil Engineers, New Orleans, La., October 27,1982.

that with a body of over 86,000 members, we can accomplish professionally many things that could not have been done through individual efforts.

A direct result of the membership campaign is a year-end surplus of about 1 million dollars as related to a total budget of approximately 12 million dollars. This is about three times our budgeted surplus and indicates wise management of Society funds coupled with a growth in the dues-paying membership. Most certainly this surplus postpones the day of a dues increase.

This year ASCE has focused on the crisis conditions which exist on our overcrowded engineering campuses due to a shortage of competent faculty, deteriorating physical facilities and obsolete laboratory equipment. A task committee studying this subject reported to the Board at its annual meeting in St. Louis and provided the Society with an Educational Involvement Handbook, with an action plan and case studies to enable Sections and Branches to aid institutions on a local basis. One result has been to inform the membership that the crisis is not in a shortage of the number of graduates, but a shortage of faculty to man the classrooms and laboratories to insure that quality education of engineers can continue to meet our future national needs. A campaign of advertising by our Public Communications Committee funded from voluntary contributions from members and Industry has sought to inform the public of our education crisis in engineering.

With the change in federal funding patterns, it becomes even more important to mobilize all possible financial support from the employers of engineers, namely industry and the profession, to augment funds from private or government sources to support our system. An Industry/Founder Societies Forum during the year has provided a base for informing influential parties of their new roles in support of engineering education.

The continued challenge to the profession will be to reestablish the attractiveness of our university faculty positions as being one of the most prestigious of our professional activities. Through various conditions that have been generated over the past two decades, the attractiveness of academic appointments in certain professional areas has been tarnished. In some instances this is due to the development of personnel policies generated campus-wide that deter the practitioner-educator from participation in our educational programs. The current practice of requiring the Ph.D. (a research degree) as the minimum entry requirement for our faculties together with the nearly universal tenure policies which make final judgement on new faculty at the end of six years--applying in many instances research performance criteria alone--are the real reasons for much of our present circumstance. These are factors which have been discussed at length since the presidency of Walter Blessey, and will no doubt continue to be discussed until our crisis becomes appreciated so that representatives of education, government and the profession come to a common understanding of the jeopardy the current circumstance holds for our national welfare. I look forward to continued Society involvement with educational concerns during the presidencies of John Wiedeman and Russ Stearns.

The History and Heritage Program of the Society has been a dynamic part of the effort to provide recognition to projects which have contributed to the nation's quality of life. In the past year recognition has been given to the George Washington Bridge for its size and unique structural character; the Louisville Waterworks for its introduction of rapid sand filtration on a municipal scale; the Charles River Basin which converted an open sewer into a monumental recreation area; and the Chain of Rocks Water Purification Plant in St. Louis which provided that community safe drinking water when other cities on the Mississippi River suffered yellow fever epidemics.

Also recognized were Union Station in St. Louis which defined an important junction of a massive transportation network; the Bow-string Truss Bridge of Squire Whipple, which represented the first rationally proportioned simple bridge structure ever built; the George Washington Monument, constructed as a tribute to our first president, who was a practitioner of the surveying art; and the Kinzua Viaduct, a massively-proportioned structure built in a very short period of time in 1882 to open commerce in western Pennsylvania; the Courtland Street Drawbridge which was the first trunion bascule bridge opened to traffic, thus providing a model urban transportation structure; and the Carrollton Viaduct, the first masonry railroad viaduct in the United States which initiated rail traffic westward out of Baltimore.

ASCE recognition of Outstanding Civil Engineering Achievements (OCEA), included the transformation of a much-maligned sanitary land fill in Industry Hills, California, into a recreational paradise featuring a civic convention center, multiple championship golf courses, a tennis complex and Olympic swimming pool, all topped with a fine residential hotel.

Transportation was recognized in the development of the San Diego Trolley, connecting that metropolis with Tijuana, Mexico. The OCEA Program also helped provide recognition to our country's energy needs, via an award to the Louisiana Offshore Oil Port (LOOP), which will serve supertankers working that coastal area.

The Society is served by a dedicated headquarters staff. Executive Director Meier has instilled in them a renewal of dedication to member service, with the objective of maintaining the confidence and support of the members so essential to the continued well-being of the Society.

My wife Midge and I have met many hundreds of Society members as we have visited from coast to coast with as many Sections and Branches as physically possible during my year as President. We are grateful to all those individuals who are serving the Society so well for making this a memorable period indeed. The Society is fortunate to have a young, vigorous individual in the stature of John Wiedeman to direct its activities during the coming year. I pledge him my utmost support for his undertakings.

Finally, Midge and I would like to thank Rice University and the Brown Foundation for providing the resources which made possible our service to the Society, and the membership of the Society for the privilege of serving.

1982 ASCE AWARDS AND PRIZES

0001 A. P. Greensfelder Construction Prize

Dynamic Consolidation—Dramatic Way to Strengthen Soil

S. D. Ramaswamy, Fellow, ASCE, (Assoc. Prof., Dept. of Civ. Engrg., Natl. Univ. of Singapore), **S. L. Lee**, Fellow, ASCE, (Prof. and Head, Dept. of Civ. Engrg., Natl. Univ. of Singapore) and **I. U. Daulah**, Member, ASCE, (Regional Mgr., Techniques Louis Menard S.A., Singapore)

Civil Engineering--ASCE, Vol. 51, No. 4, April, 1981, pp. 70-73

The dynamic consolidation technique consists basically of dropping a heavy pounder of up to 4 x 10^4 kg from a height varying between 50 to 148 ft. The weight, height of fall, number of drops per pass and number of passes to achieve required compaction levels depend on thickness of ground to be compacted, nature of soil, groundwater level and consistency of material underlying soil to be compacted. This technique has been successfully used in Singapore to improve properties of loose ground by deep compaction for different types of soils (such as silty clay, peaty clay and dredged granular fill) up to about 26 ft in depth. From experience gained from various compaction jobs, the dynamic consolidation method can provide a quick and economical solution to a variety of deep compaction problems.

0002 Arthur M. Wellington Prize

St. Lucie No. 2 Foundation Construction

James P. Burket, (Principal Civ. Engr., Ebasco Services, Inc., New York, N.Y.) and **Hung Jen Kuo**, (Chf. Civ. Engr., Ebasco Services, Inc., Newport Beach, Calif. 92660)

Journal of the Construction Division, Vol. 107, No. CO2, June, 1981, pp. 393-404

St. Lucie Unit No. 2 is an extension and a duplication of the St. Lucie Unit No. 1 which is a nuclear power generating station of 790 MW capacity utilizing a two loop PWR nuclear steam supply system. The construction of the Unit 2 Reactor Building (RB) foundation (adjacent to the existing Unit 1) required special provisions to assure the safe operation of Unit 1. The circular cofferdam that was designed and constructed to meet all seismic category 1 requirements for the construction of Unit no. 2 RB foundation is considered. Restrictions and tests required for sheet pile driving are also presented.

0003

ASCE State-of-the-Art of Civil Engineering Award

Remote Sensing for Tunnel Siting Studies

Orville R. Russell, (Sr. Geologist, Earth Satellite Corp., Wash., D.C.), **Dennis T. Stanczuk**, (Sr. Geologist, Science Applications, Inc., McLean, Va.) and **John R. Everett**, (Chf. Geologist, Earth Satellite Corp., Wash., D.C.)

Transportation Engineering Journal, Vol. 106, No. TE5, September, 1980, pp. 523-537

Data evaluated included Landsat, Skylab, black-and-white panchromatic, color, color-infrared, and low-sun-angle black-and-white infrared photography. Also used were thermal infrared and multispectral scanner imagery and aerial electromagnetic and magnetometer data. No single array of sensors is optimal for all terrains and environments. There is, however, a suite of remotely sensed data, including space imagery, black-and-white and color aerial photography, low-sun-angle photography, and side-looking-airborne radar (if it already exists), that should be analyzed for most sites because of high information content and relatively low cost. Other systems can provide unique information under particular sets of conditions, but are deemed too expensive for inclusion in all site studies. The results of the evaluation demonstrate that remote sensing can reduce the cost of tunnel site selection.

0004

Rickey Medal

James Bay Powerhouses—Layout and Design Criteria

Klaus R. Ludwig, Member, ASCE, (Chef de Service, Structures, Direction Ingenierie, La Societe d'energie de la Baie James, Bechtel Quebec Limitee, Montreal, Quebec, Canada), and **Richard T. Olive**, Member, ASCE, (Specialiste en structures, Direction Ingenierie, La Societe d'energie de la Baie James, Bechtel Quebec Limitee, Montreal, Quebec, Canada)

Journal of the Energy Division, Vol. 106, No. EY2, October, 1980. pp. 235-255

The La' Grande River is the first of several flowing into the Eastern side of James Bay to be developed for its energy potential. The structural criteria and layout guidelines developed during the design of the four La Grande powerhouses will be used to advantage in subsequent projects in the James Bay region. Layouts of unit blocks, service bay, superstructure and penstocks are considered. A summary of cases to be studied for the structural stability of the powerhouse is presented, showing the various loading conditions to be considered, along with the corresponding factors of safety. Criteria for the detailed design of structural elements, including loadings, are also presented.

0005

J. C. Stevens Award

Discussion of: Theory of Minimum Rate of Energy Dissipation

Gary Parker, Member, ASCE, (Assoc. Prof., Dept. of Civ. Engrg., Univ. of Alberta, Edmonton, Alberta, Canada)

Journal of the Hydraulics Division, Vol. 106, No. HY7, July, 1980, pp. 1260-1262

In the original paper ("Theory of Minimum Rate of Energy Dissipation," by Chih Ted Yang and Charles C. S. Song, *Journal of the Hydraulics Division*, Vol. 105, No. HY7, July, 1979, pp. 769-784), a general theory of minimum rate of energy dissipation for a class of open channel flows with or without the movement of sediment is proposed. In this discussion several contentious points are listed. Equations are presented which clarify or disprove several of the equations presented in the original paper.

0006 J. James R. Croes Medal

Seismic Behavior of Structural Subassemblies

Egor P. Popov, Fellow, ASCE, (Prof., Dept. of Civ. Engrg., Univ. of California, Berkeley, Calif.)

Journal of the Structural Division, Vol. 106, No. ST7, July, 1980, pp. 1451-1474

The behavior of both structural steel and reinforced concrete specimens subjected to quasi-static cyclic loadings simulating severe seismic effects are examined. A large number of experimentally determined hysteretic loops for different structural arrangements based on the work mainly done at the University of California, Berkeley, serve as illustrations. The important distinction among the material, the member and the subassemblage ductilites is emphasized. The significance of the different character of the hysteretic behavior on the overall seismic structural performance is discussed by relating it to the widely-used response spectrum approach for seismic analysis. The relationship of the latter approach to the conventional code design procedures is also given.

0007 Karl Emil Hilgard Hydraulics Prize

Surface Hydrology: I—Explanation of Phenomena

Dan Zaslavsky, (Prof. of Soil and Water Engrg., Faculty of Agricultural Engrg., Soils and Fertilizers Div., Technion—Israel Inst. of Tech., Technion City, Haifa 32000, Israel) and **Gideon Sinai**, (Lect., Faculty of Agricultural Engrg., Soils and Fertilizers Div., Technion—Israel Inst. of Tech., Technion City, Haifa, 32000, Israel)

Journal of the Hydraulics Division, Vol. 107, No. HY1, January, 1981, pp. 1-16

Nonisotropic sloping soil leads to a horizontal flow component during rain and infiltration. Splashing of raindrops and flow in the transition zone between the soil bulk and the air also lead to lateral flow. The result is concentration of moisture in concave spots to the point of saturation and even seepage out of the soil. The same occurs at steep soil cuts. Such moisture concentration can explain the formation of runoff during rainfall at low rates that do not exceed infiltration capacity, partial area contribution to runoff, net recharge of groundwater by low-rate rainfall, the formation of rills and gullies in a relatively dry area, delay of runoff after a certain total amount of rain irrespective of rain intensity, variations in soil formation in different parts of the landscape, poor distribution of irrigation water in nonplane fields, and leaching of the soil surface by water that was assumed to run only above the surface.

0008

Moisseiff Award

Diagnostic Testing of a Bridge

Baider Bakht, (Sr. Research Officer, Research and Development Div., Ministry of Transportation and Communications, Downsview, Ontario, Canada) and **Paul F. Csagoly**, (Principal Research Officer, Research and Development Div., Ministry of Transportation and Communications, Downsview, Ontario, Canada)

Journal of the Structural Division, Vol. 106, No. ST7, July, 1980, pp. 1515-1529

Diagnostic load testing is an effective means of locating the sources of weakness and establishing the true load distribution pattern of a bridge. In support of this argument test data is given from an actual load test on a bridge which had experienced some distress in the past, and was not considered safe for carrying normal traffic according to conventional analyses. The sources of distress that were located by the test could be removed by simple remedial measures. By utilizing the enhanced load distribution pattern of the bridge that was also established by the test, the bridge, after minor repairs, was opened to normal traffic.

0009

Norman Medal

Prediction of Movements for Braced Cuts in Clay

Abdulaziz I. Mana, Assoc. Member, ASCE, (Asst. Prof., Dept. of Civ. Engrg., Univ. of Petroleum and Minerals, Dhahran, Saudi Arabia) and **Wayne G. Clough**, Member, ASCE, (Prof., Dept. of Civ. Engrg., Stanford Univ., Stanford, Calif. 94305)

Journal of the Geotechnical Engineering Division, Vol. 106, No. GT6, June, 1981, pp. 759-777

A simplified method is presented for the prediction of wall and soil movements for braced excavations in soft to medium clay deposits. The method is developed by combining the results of field data performance and finite element analyses. A careful review of 130 published and unpublished case histories was carried out to select well-documented projects with similar support systems and free of unusual construction effects. The finite element analyses were directed toward the evaluation of effects of variation of soil shear strength, soil stiffness, support stiffness and system geometry. Both the field data and the finite element results show that a strong correlation exists between system movements and the factor of safety against basal heave. Based on the established behavior trends, nondimensional plots are developed for predicting maximum magnitudes and distributions of wall deflections and ground settlements as a function of the factor of safety against basal heave and the other important parameters of the soil and structural system.

0010

Raymond C. Reese Research Prize

Vertical Vibration Analysis of Suspension Bridges

Ahmed M. Abdel-Ghaffer, Assoc. Member, ASCE, (Asst. Prof., Dept. of Civ. Engrg., Princeton Univ., Princeton, NJ.)

Journal of the Structural Division, Vol. 106, No. ST10, October, 1980, pp. 2053-2075

A method of analysis is outlined for free vertical vibrations of suspension bridges. The method employs a digital computer and a finite element approach, and uses a linearized theory which restricts the amplitudes of vibrations to be small. The analysis is designed to determine the natural frequencies, modes of vibration, and energy storage capacities of the different members of the bridge. Both symmetric and antisymmetric modes of vibrations are considered. For the symmetric modes (which include additional vibrational cable tension), it is shown that it is imperative to include the elasticity of the cables if correct solutions are to be obtained. A detailed numerical example, which includes the effect of the extensibility of the cables and the elasticity of the towers, is presented and the results are examined by a comparsion between measured and computed natural frequencies and mode shapes of a real bridge; the comparison shows good agreement.

0011 Samuel Arnold Greeley Award

Port Valdez Environmental Studies: 1976-1978

Joseph M. Colonell, Member, ASCE, (Sr. Project Engr., Woodward-Clyde Consultants, Anchorage, Alaska)

Journal of the Environmental Engineering Division, Vol. 106, No. EE5, October, 1980, pp. 907-917

All oily ballast water from the holds of tankers arriving at the Trans-Alaska Pipeline terminus in Valdez is pumped ashore for treatment before being discharged into Port Valdez. The Institute of Marine Science (IMS) of the University of Alaska conducted a three-year multidisciplinary program of studies to monitor the effect, if any, that this discharge would have upon the water quality and biota of Port Valdez. Consisting of eight separate projects and involving seven investigators, the IMS program (1) Extended substantially the baseline data for selected physical, chemical, and biological parameters; and (2) developed monitoring procedures as required to detect changes in the Port Valdez marine environment. The conclusion of these studies was that there had been no detectable impact during the term of this program on those elements of the environment that had been monitored prior to and following initiation of the treated ballast water discharge.

0012 Thomas A. Middlebrooks Award

Vibration Tests of Full-Scale Earth Dam

Ahmed M. Abdel-Ghaffar, Assoc. Member, ASCE, (Asst. Prof., Civ. Engrg. Dept., Princeton Univ., Princeton, N.J. 08544) and **Ronald F. Scott**, Member, ASCE, (Prof. of Civ. Engrg., California Inst. of Tech., Pasadena, Calif.)

Journal of the Geotechnical Engineering Division, Vol. 106, No. GT3, March, 1981, pp. 241-269

Extensive full-scale dynamic tests, including ambient (unique in earth dam research) and forced vibration as well as popper tests were conducted on the Santa Felicia earth dam in Southern California. For the forced vibration tests, the dam was excited into resonance in both the upstream-downstream and longitudinal directions. During the ambient test, the naturally occurring vibrations of the dam were measured. The popper tests involved the use of pressure waves originating from a controlled, submerged release of gas. During the tests, three-dimensional measurements of the motions of approximately 25 stations along the crest and seven stations on the downstream slope were recorded and then analyzed in both time and frequency domains. Modes of vibrations and associated natural frequencies as well as damping ratios were determined.

0013

Thomas Fitch Rowland Prize

Mixed Face Tunneling on Melbourne Underground

Alfred M. Petrofsky, Fellow, ASCE, (Vice Pres., Jacobs Assoc., San Francisco, Calif.)

Journal of the Construction Division, Vol. 106, No. CO3, September, 1980, pp. 409-425

Adequate preliminary investigation, continued monitoring of ground conditions and construction procedure by the engineer, and a resourceful contractor combined to result in the economical design and construction of a four-track (double-under-double-over) rapid transit railway project in Melbourne, Australia. One tunnel contract required a variety of heading-and-benching-type operations to overcome mixed face conditions, with the soft ground being either on top, underneath, or in the middle of the face. Close monitoring of the tunnel construction by the engineer, skillful execution of the work by the contractor, and a good specification, permitted several variations of the original design resulting in economies.

0014

Wesley W. Horner Award

Ground-Water Quality at Rapid Infiltration Site

James V. Olson, Assoc. Member, ASCE, (Sr. Environmental Engr., Creegan & D'Angelo, San Jose, Calif.), **Ronald W. Crites**, Assoc. Member, ASCE, (Project Mgr., Metcalf & Eddy, Inc., Sacramento, Calif.) and **Paul E. Levine**, (Geologist, Ecology and Environment, San Francisco, Calif.)

Journal of the Environmental Engineering Division, Vol. 106, No. EE5, October, 1980, pp. 885-899

The ground water quality beneath a 30-yr old rapid infiltration was studied over a 12-month period. The system receives primary effluent and treats it to a very high quality through the land treatment process. The flow of treated water passes through 10 m to 20 m of permeable sands and then moves laterally off the site. This shallow ground water is not used as a water supply and does not affect the deeper, separate aquifers. A 96% BOD and TOC removal was achieved. Because of the high BOD to nitrogen ratio, the relatively low loading rate of 15.4 m/yr, and the moderate climate, nitrogen was removed by nitrification and denitrification in the soil. The average nitrogen removal was 93%. Phosphorus removal was 30%. Trace element removal varied, but shallow ground-water levels were generally lower than guidelines established for unrestricted irrigation.

1982 ASCE PUBLICATIONS

0015 Compression Overlapping Precedence Networks

Srilal Perera, (Lect. of Construction Engrg. & Management, Dept. of Civ. Engrg., Univ. of West Indies, St. Augustine, Trinidad, W.I.)

Journal of the Construction Division, Vol. 108, No. CO1, March, 1982, pp. 1-12

A study is presented, which is directed at finding a solution to the problem of compressing networks having overlapping activities. The method involves the use of cost duration charts, network diagrams and linear programming. The computer is used to solve the linear programming model. The results give information pertaining to earliest start and finish times of activities, what activities to expedite in order to complete the project within the specified time limit and by how many days, how much money should be allocated to each activity and what additional costs are incurred in expediting the work.

0016 Games People Play with Work Sampling

H. Randolph Thomas, Jr., Member, ASCE, (Assoc. Prof. of Civ. Engrg., Pennsylvania State Univ., University Park, Pa.), **Mason P. Holland**, (Sr. Field Engr., Stone & Webster Engrg. Corp., Lycoming, N.Y.) and **Carl T. Gustenhoven**, Assoc. Member, ASCE, (Engr., Stone & Webster Engrg. Corp., Boston, Mass.)

Journal of the Construction Division, Vol. 108, No. CO1, March, 1982, pp. 13-22

Guidelines and principles are presented for planning and implementing a work sampling program. Principles to be followed include careful planning to fulfill objectives, the need to be responsive to changing needs and conditions, the requirement for strong program leadership, and the need to implement changes. It is suggested that work sampling should be used as a means of communicating information much in the same way as cost account data are used. Planning and implementation as a joint venture between owner, construction manager, contractor, and labor is suggested. The role of the project director in establishing and maintaining acceptable levels of performance by the observers is also presented. Sources of errors in the data are discussed along with the problems of superficial data analysis. Several difficulties with continuous sampling programs are also noted.

0017 Underground Transit Station Construction in Japan

Boyd C. Paulson, Member, ASCE, (Assoc. Prof. of Civ. Engrg., Stanford Univ., Stanford, Calif.)

Journal of the Construction Division, Vol. 108, No. CO1, March, 1982, pp. 23-37

Technologies used in Japan for constructing underground stations for heavy-rail urban rapid transit systems are described. Two example projects have been chosen that illustrate different approaches which incorporate several advanced engineering features. These examples illustrate the top-down, or reverse construction method, secant-pile walls, and underground mining of stations in soft ground. With higher direct costs and longer durations, the methods discussed here will not offer solutions to all U.S. station construction projects. Nevertheless, there is a lot to be learned from the Japanese about working on congested sites in urban areas, particularly under the tough geotechnical conditions in Japan. Considering total costs, including disruptions to commerce and traffic in urban areas, the bad publicity, loss of citizen and political support, litigation and other factors that are increasingly encroaching on United States construction methods, it is argued that Japanese construction methods might provide attractive alternatives.

0018 Technical Aspects of Metric Construction

Edward R. Lewandowski, Member, ASCE, (Asst. Chf., Div. of Research, Water and Power Resources Services, Denver, Colo.)

Journal of the Construction Division, Vol. 108, No. CO1, March, 1982, pp. 39-45

The Water and Power Resources Service, formerly the Bureau of Reclamation, has begun construction of a number of major projects using SI metric units. Technical aspects of the design and construction activities are discussed. Experiences of both Government staff and contractor's staff are presented, with emphasis on metric equipment used. Recommendations are given for techniques and procedures to be used when future projects are constructed in metric.

0019 The Effect of Contract Wording on Inspector Liability

Lawrence J. Green, Assoc. Member, ASCE, (Vice Pres., Construction Management, J.C. Zimmerman Engrg. Corp., 858 E. 120th Ave., Denver, Colo. 80233)

Journal of the Construction Division, Vol. 108, No. CO1, March, 1982, pp. 47-53

Design professionals should endeavor *not* to write burdensome or inaccurate contracts; i.e., those that inaccurately describe duties. Inspector liability is not relieved by ambiguity in the wording of a contract. The answer to reduced inspector liability lies not in the type of contract or wording of the contract but in the quality and quantity of inspection. It is argued that while it is necessary to write more accurate and descriptive contracts, words are not to be depended upon as protection from the losing end of a lawsuit. It is stated that better inspection is a better protection from such losses.

0020 Quality Control During Construction of Power Plants

Robert F. Hartstern, (Mgr., Inspection Services, Quality Services Dept., Gilber/Commonwealth, Jackson, Mich.)

Journal of the Construction Division, Vol. 108, No. CO1, March, 1982, pp. 55-62

The need for quality control during the construction of power plants has always been recognized as beneficial. Federal law requires a comprehensive program be provided to control and assure quality during the construction of nuclear power plants. Many methods to control quality have been developed to meet the mandatory requirements set forth by the Nuclear Regulatory Commission. However, many in the industry have criticized these as expensive, unnecessary, and uncompromising. Making quality control cost effective and supportive of construction, requires more than just meeting federal regulations and many other aspects have to be considered. It should be considered another management tool.

0021 Slipforming of Reinforced Concrete Shield Building

M. C. Hsieh, (Principal Civ. Engr., Ebasco Services, Inc., New York, N.Y. 10048) and **Jerome R. King**, Member, ASCE, (Chf. Engr., Fegles-Power Services Corp., Minneapolis, Minn.)

Journal of the Construction Division, Vol. 108, No. CO1, March, 1982, pp. 63-73

The unique design and construction features of slipforming the heavily reinforced concrete cylindrical shield walls at the Satsop nuclear plant site are presented. It proved that, with proper planning, design, and coordination, the slipform construction technique can be successfully applied to large and heavily reinforced concrete structures. The shield walls were designed in compliance with seismic requirements which resulted in the need for reinforcing steel averaging 550 pounds per cubic yard (326 kg/m^3). A 25 ft (7.6 m) high, three-deck moving platform was designed to permit easy installation of the reinforcing steel, embedments, and blockouts, and to facilitate concrete placement and finishing. Two circular box trusses, one on each side of the shield wall, were used in combination with a spider truss to meet both the tolerance and strength requirements for the slipform assembly. Use of preassembled, lifted-in-place sleeve tower racks that combined diagonal and vertical bars for the openings with two vertically aligned pipe sleeves greatly eased the installation. Specific recommendations for future slipforming are also presented.

0022 Scheduling: Effective Methods and Techniques

Elise Hosten McGough, Member, ASCE, (Sr. Schedule Engr., Barton-Malow Col, P.O. Box 5200, Detroit, Mich. 48235)

Journal of the Construction Division, Vol. 108, No. CO1, March, 1982, pp. 75-84

Network planning and analysis (CPM) is a proven means to outline the many activities to be performed and to investigate the feasibility of a given construction project. Commencing with a Master Project Schedule, developed in a low level of detail, the criteria is set for engineering and document development together with the logistics of construction. Costs, determined by conceptual estimates, added at this time forecast financial requirements. Schedule development continues to expand in detail as engineering information proliferates, and alternative construction methods and procedures are studied. With contractor input, a detailed Procurement and Construction Schedule is generated. Ultimate success of the project depends upon careful implementation and surveillance of the plan and schedule. Regular monitoring and comparing of actual performance with the plan and schedule enables maintenance of effective time and cost control.

0023 Contract Administration: A Resident Engineer's View

Stephen B. Quinn, Member, ASCE, (Project Mgr., Howard, Needles, Tammen & Bergendoff, Houston, Tex.)

Journal of the Construction Division, Vol. 108, No. CO1, March, 1982, pp. 85-91

Professional liability related to field work is a major concern of the consulting profession. Field personnel should review the plans and specification in the pre-construction period and systemize the plans. Revisions are a major source of error and must be thoroughly reviewed and documented. Extensive documentation consisting of field reports, project reports, memoranda, and letters must be kept. The resident must be aware of all phases of the contractor's operations, the projected work schedule and the capability of his own staff. Client involvement before, during, and after construction in relation to cost changes is necessary. The paper points out procedures that should be utilized in order to limit exposure to liability to the client and to the consultant.

0024 Modernized Metric System—SI—The American Choice

Cornelius Wandmacher, Honorary Member, ASCE, (Prof. of Engrg. Education and Dean Emeritus, Univ. of Cincinnati, Cincinnati, Ohio)

Journal of the Construction Division, Vol. 108, No. CO1, March, 1982, pp. 93-104

Americans have played a major role in the development of the new International System of Measurement Units (SI). The world is moving rapidly in the adoption of this modernized metric system. SI is an absolute, coherent, simplified system. SI is particularly well suited for all aspects of the space age, the nuclear age, the electronic age, and the computer age. SI will bring the measurements of all branches of engineering, science, and commerce together under one system of units. Planning ahead on the basis of a clear knowledge of SI will give an optimum change-over with maximum benefits and minimum costs. Opportunities will abound for significant advances in modular construction, dimensional coordination, and rationalization of the sizes of components, especially in engineered construction. SI is a significantly evolutionary advancement in measurement systems.

0025 Aggregate Gradings for Nuclear Concrete

Kurtz L. Paulson, Member, ASCE, (Chf. Civ. Construction Engr., Bechtel Power Corp., San Francisco, Calif.) and **R. Harlan Wright**, (Construction Specialist, Bechtel Power Corp., San Francisco, Calif.)

Journal of the Construction Division, Vol. 108, No. CO1, March, 1982, pp. 105-120

Evaluation of concrete shows that segregation, inherent in rehandling of coarse aggregate, is significantly reduced in the large mixers used in concrete production. The investigation also shows that relatively broad ranges in coarse aggregate gradation as shown by the on site test data had no measureable effect on the strength of the inplace concrete. On nuclear work the dependence on data resulting from single increment samples taken from rehandled material results in misleading information as to the actual gradation of the material in use. Application of statistical techniques used to evaluate data received from the aggregate manufacturer, from the rehandled material, and the freshly mixed concrete yields reliable information by which product control can be established and maintained.

0026 Planning, Scheduling, and Communicating on Crew Productivity

R. Court Olson, Assoc. Member, ASCE, (Consultant, Leas Crutcher Construction Co., P.O. Box 817, Redwood, Wash. 98502)

Journal of the Construction Division, Vol. 108, No. CO1, March, 1982, pp. 121-127

Construction project planning and scheduling along with communication of the plan have major impact on the productivity of field crews. Though the size of the impact is practically impossible to document, the cause-effect relationship has been revealed in recent productivity studies. It appears that construction managers are not recognizing their full responsibilities in establishing a framework for project teamwork in job planning and are not following through with good communication to foremen and tradesmen to promote true team spirit. Ways of improving these management shortfalls are suggested.

0027 Construction Operation and Microcomputers

Jose Lluch, Assoc. Member, ASCE, (PhD Candidate, Georgis Inst. of Tech., Atlanta, Ga. 30332) and **Daniel W. Halpin**, Member, ASCE, (Prof., School of Civ. Engrg., Georgia Inst. of Tech., Atlanta, Ga. 30332)

Journal of the Construction Division, Vol. 108, No. CO1, March, 1982, pp. 129-145

The availability of small reasonably priced computers—micro-computers—has made computer based techniques and management control accessible for the analysis of construction processes. A microcomputer implementation of the CYCLONE method for the analysis of a contruction operation is presented. The CYCLONE method is an analytical technique for the analysis of construction processes. The system, as discussed, is implemented on the TRS-80 and Apple II plus microcomputers. Two complete examples analyzing a paving process and a precast process are presented and discussed. The implementation of CYCLONE on a microcomputer provides a cost effective means by which construction decision makers can compare various construction methodologies and conduct sensitivity analysis of a selected methodology to determine the optimal or best resource mix.

0028 Managerial Decisions Under Inflation

Abraham Warszawski, Member, ASCE, (Assoc. Prof., Faculty of Civ. Engrg. Technion—Israel Inst. of Tech., Haifa, Israel)

Journal of the Construction Division, Vol. 108, No. CO1, March, 1982, pp. 147-157

The problems of managerial decision making under inflation are discussed, and some solutions are offered. The risk of estimating under inflation is explained, and a method to improve the prevailing compensation procedure is presenteed. Later, the influence of erosion of the depreciation allowance on the economy of investment in equipment is examined. Then the inadequacy of the conventional cost control methods under conditions of inflation is shown. Finally, the limitations of financial reporting based on historic values are described.

0029 Construction Industries and Metrication

A. Fattah Chalabi, Member, ASCE, (George I. Alden Prof. of Engrg., The Worcester Polytechnic Inst., Worcester, Mass.) and **Frank E. Vanzler**, Member, ASCE, (Construction Engr., The Turner Construction Co., Inc., Cleveland, Ohio)

Journal of the Construction Division, Vol. 108, No. CO1, March, 1982, pp. 159-174

A national survey of architects, engineers, contractors, material suppliers and manufacturers is conducted to determine what they feel about conversion to the SI system of measurement. The respondents perceptions of the problems in planning and implementation, their estimated costs and schedules for conversion and their comments and solutions are summarized. No strong opposition to metrication was prevalent. Apprehension about conversion is caused by fear of added cost and difficulties in training labor, increased errors and psychological factors. Many felt however that these problems are short term. The designers and design/construct sectors are the most optimistic in their expectations. Smaller companies have no present plans for conversion while some large companies have significant experience in using the SI system. The majority expect conversion by 1990.

0030 Construction Risk: Multivariate Approach

C. William Ibbs, Jr., Assoc. Member, ASCE, (Asst. Prof., Dept. of Civ. Engrg., Univ. of Illinois, Urbana, Ill. 61801) and **Keith C. Crandall**, Member, ASCE, (Prof., Dept. of Civ. Engrg., Univ. of California, Berkeley, Calif. 94720)

Journal of the Construction Division, Vol. 108, No. CO2, June, 1982, pp. 187-200

Numerous mathematical decision models have been formulated in an attempt to analyze construction risk. Most such efforts have failed because of the limited scope of their objective functions. This study presents field interview results substantiating the hypothesis that decision-making is multiattributed in nature, contrary to the implicit assumption of these other efforts. A theory is introduced which facilitates analysis of complex decisions and selection of consistent courses of action. An illustrative decision problem is solved, not only to explain the mechanics of the theory, but to demonstrate that construction risk is a function of competitive economics as well as project-related characteristics.

0031 Turnkey Construction in the United States

Seymour S. Greenfield, (Chmn. of the Board, Parsons, Brinckerhoff, Quade & Douglas, Inc., One Penn Plaza, New York, N.Y. 10001)

Journal of the Construction Division, Vol. 108, No. CO2, June, 1982, pp. 201-210

Turnkey construction has been in practice in the United States since the early 1900s, and has been used most successfully in private capital projects, particularly those involving proprietary process systems. In Europe, South America, and developing nations (where the legal/institutional climate differs significantly from that of the U.S.) turnkey is more universal in public works projects. U.S. government agencies have made limited attempts at implementing turnkey projects in public works. Generally the major impediments to the use of turnkey construction contracts in public works projects are current competitive bidding laws, the diffuse decision-making process in government, restrictions on contract negotiations, traditional owner-engineer-contractor relationships, lack of shared risk between owner and contractor, and substantial likelihood of downstream regulatory roadblocks, all of which serve to make the potential risks to the contractor (and cost to the owner) too great in comparison to the potential gains.

0032 Graphical Method of Optimization: A Short Cut

Sandor Popovics, Fellow, ASCE, (Prof., Dept. of Civ. Engrg., Drexel Univ., Philadelphia, Pa. 19104)

Journal of the Construction Division, Vol. 108, No. CO2, June, 1982, pp. 211-218

Although optimization problems are usually solved by computer, argument is presented that simpler cases can be handled by non-computer methods. A new, graphical method based on triangular systems of coordinates is presented. The triangular system described is usually considered an equilateral triangle. Each of the three sides of the triangle can be used as a coordinate axis for a variable in a suitable scale. Each combination of the three variables has a correspondent point in the triangle. The coordinates of a point inside the triangle represent the magnitude of each of the three variables, and can be read off of the corresponding scales. Two examples illustrate the application of this method. The method is compared with the Simplex method to show advantages and limitations.

0033 Managing Employee Conflict in the Middle East

Orfan Rabbat, (Professional Engr., Vartan Assocs., Inc., Harrisburg, Pa.) and **Robert B. Harris**, Fellow, ASCE, (Prof. of Civ. Engrg. and Program Advisor, Construction Engrg. and Management, Univ. of Michigan, Ann Arbor, Mich.)

Journal of the Construction Division, Vol. 108, No. CO2, June, 1982, pp. 219-225

Management of the employee by an international construction company operating in the Middle East involves many interrelated factors that can lead to conflict. Management must recognize the motivations and personalities of individual workers, use straight forward communication practices, have good company organization, and do quality planning if conflict is to be avoided. Conflicts arising from cultural differences stem from the pervasive Islamic religion that determines the values, attitudes, and social traditions of the worker. Language differences and advanced technology are additional sources of conflict. Management must adapt to the culture if conflict is to be minimized.

0034 Application of Cost-Flow Forecasting Models

Shlomo Peer, (Assoc. Prof., Technion - Israel Inst. of Tech., Technion City, Haifa, 32 000, Israel)

Journal of the Construction Division, Vol. 108, No. CO2, June, 1982, pp. 226-232

Cost-flow models are formulated for forecasting the cumulative cost-flow in housing and public buildings, as a function of total cost and construction period. This permits to estimate the cost increment for any interval even before any schedule is available. Detailed cost-flows are compiled for all individual projects, and brought to a common denominator by relating cost to time in percentage. The prediction curves are obtained by polynomial regression. Considering the dynamic characters of a company's financial plan, and the need for permanent updating, the deviations obtained by this procedure between the predicted and actual cost of the individual project are insignificant for practical use. These deviations also tend to balance out in compiling the company's overall picture that management is concerned with. The implementation of such models for all major categories of projects engaged in by a construction company can thus greatly simplify one of tha most time and cost consuming steps in establishing and updating a financial plan.

0035 Protecting the Environment During Construction

Enno Koehn, Member, ASCE, (Assoc. Prof. of Construction Engrg. and Management, Purdue Univ., West Lafayette, Ind.) and **James A. Rispoli**, Member, ASCE, (Lieutenant Commander, Civ. Engr. Corps., U.S. Navy, Northern Div., Naval Facilities Engrg. Command, Philadelphia, Pa.)

Journal of the Construction Division, Vol. 108, No. CO2, June, 1982, pp. 233-246

Present practice appears to indicate that nonpoint source pollution will be controlled most effectively by a combination of fitting the development to the particular site, reducing, if possible, the grading and exposure of bare soils, and by applying control measures at the proper times. Methods to reduce the amount of erosion and the resulting pollution due to the transportation of sediments are presented and an expression designed to measure soil loss is outlined. Examples and construction cases are presented. The importance of air quality and control of miscellaneous wastes such as pesticides and fertilizers is emphasized. The control of storm water during construction and the life of the facilities being built is also discussed.

0036 Modeling Project Milestones: Dependent Activities

David Morris, Member, ASCE, (Prof., Dept. of Civ. Engrg., School of Engrg. and Applied Sci., Univ. of Virginia, Charlotesville, Va.)

Journal of the Construction Division, Vol. 108, No. CO2, June, 1982, pp. 247-259

The current methods of project planning by network analyses and the assumed relationships between activity duration, cost, start date and risk elements are reviewed. The concept of modeling the interdependencies between activities by means of milestone data achievement is presented. The Graphical Evaluation and Review Technique (GERT) of structuring conditional branching and discrete probabilistic outcomes in scheduling networks are used to develop probable frequency distributions of project duration and cost. The role of the milestone model in project planning is illustrated by an example of a small highway construction project with weather sensitive operations.

0037 Reinforced Earth Case Histories in New York State

Bernard E. Butler, Member, ASCE, (New York Metropolitan Regional Mgr., The Reinforced Earth Co., Englewood Cliffs, N.J.)

Journal of the Construction Division, Vol. 108, No. CO2, June, 1982, pp. 260-276

Three projects which used earth reinforcement concepts are described from design through construction. Reinforced earth (RE) was used for two of the projects. One of these was a long high expressway ramp contained within two RE walls which crossed an alluvial flood plain. Geotechnical studies were necessary for pile drag on a nearby existing structure as well as ramp settlement and stability. The other RE project involved stabilizing a very old unsafe dam with a downstream buttress treatment consisting of RE slope panels and rock fill. Both RE projects were cost effective and are performing well. The third project discusses two retaining walls built of granular material and geotextile to stabilize sidehill shear failures. Construction went well and performance, thusfar, is satisfactory based on a limited monitoring program. However, the cost for this type of wall was much higher than expected.

0038 Marketing Design and Construction Services

L. William Murray, Jr., (Prof. of Business, Univ. of San Francisco, San Francisco, Calif.)

Journal of the Construction Division, Vol. 108, No. CO2, June, 1982, pp. 277-288

A survey of approximately ninety of the largest domestic professional firms engaged in process engineering design and construction, and of eighty of the largest clients of those services, was undertaken to ascertain current marketing practices and preferences. Forty-three of the firms surveyed and thirty-eight of the clients responded. The client and firm respondents agreed that the single most important marketing asset is the project manager. Firm respondents tend to overestimate the importance of proposed project costs in the client's evaluation of the proposal. Client respondents heavily weight the firm's scope of work statement in their determination of the quality of services offered by the firm. As the complexity of the services increase, clients tend to use more and more pre-qualification of firms. This results in a two-stage marketing problem for the firm: The first is to be initially pre-selected, and the second is to win the contract award over other qualified firms.

0039 Inflation and Equipment Replacement Economics

C. J. Schexnayder, Jr., Member, ASCE, (Assoc. Prof., Dept. of Civ. Engrg., Louisiana Tech. Univ., Ruston, La.) and **Donn E. Hancher**, Member, ASCE, (Assoc. Prof., Dept. of Civ. Engrg., Purdue Univ., W. Lafayette, Ind.)

Journal of the Construction Division, Vol. 108, No. CO2, June, 1982, pp. 289-298

The proposed analysis procedure enables construction contractors to correctly account for the costs associated with machine price increases caused by inflation. Methods for acquiring the necessary input data for the procedure are presented. An example set of calculations based on actual contractor cost files is worked out and the results discussed. The dynamic nature of equipment economics is also stressed.

0040 Managerial Planning and Control in Precast Industry

Abraham Warszawski, Member, ASCE, (Head, Building Research Station, Technion-Israel Inst. of Tech., Haifa, Israel)

Journal of the Construction Division, Vol. 108, No. CO2, June, 1982, pp. 299-313

The production process in prefabrication plant for concrete building elements differs in several important aspects from a production process in typical manufacturing plants. The paper describes the more important elements of the production system in a prefabrication plant. The main problems of production planning and control, cost and quality control in prefabrication, are examined. The main features of its information system are also discussed.

0041 Impact of Inspector's Experience on Owner's Liability

Merlin D. Kirschenman, Member, ASCE, (Assoc. Prof. and Chmn., Dept. of Construction Management and Engrg., North Dakota State Univ., Fargo, N. Dak.)

Journal of the Construction Division, Vol. 108, No. CO2, June, 1982, pp. 314-320

Inspection of construction methods and procedures is a very important part of the total process of constructing a facility. This inspection is accomplished by the total efforts of the resident engineer's staff and all the people involved with quality control and quality assurance programs. The lack of proper field experience and training of key individuals has had a significant impact on the liability and increased costs to the owners. Many of these additional costs such as extended overhead and delay costs to contractors can be directly traced to the lack of adequate experience of the inspectors. The industry has not put proper significance on adequate education, training and experience of the total inspection team. This is reflected by the salary structures and lack of accountability of the individual as it relates to their decision and actions.

0042 Schedule Development under Stochastic Scheduling

Keith C. Crandall, Member, ASCE, (Prof. of Civ. Engrg., Univ. of California, Berkeley, Calif.) and **John C. Woolery**, Member, ASCE, (Superintendent, Turner Construction Co., San Francisco, Calif.)

Journal of the Construction Division, Vol. 108, No. CO2, June, 1982, pp. 321-329

A most effective use of Monte-Carlo schedule generation is to create a control schedule of the planning, design and construction of a major project. The normal Monte-Carlo process will yield a distribution for the time required for project completion, activity criticality and list potential critical paths. A methodology is described whereby milestone dates can be created at selected node points within the project without the necessity to maintain an entire record of each possible outcome at these nodes. The methodology utilizes the first four moments about the mean and the Edgeworth Series Method. Examples are presented to demonstrate the methodology.

0043 Current Aspects of Construction Rehabilitation

Enno Koehn, Member, ASCE, (Assoc. Prof. of Construction Engrg. and Management, Purdue Univ., West Lafayette, Ind.) and **Steven E. Tower**, (Lieutenant, Civ. Engr. Corps, U.S. Navy)

Journal of the Construction Division, Vol. 108, No. CO2, June, 1982, pp. 330-340

Rehabilitation work suffers from wasteful attitudes, social conditions, and regulatory barriers that prevent it from reaching its potential. There are four primary ways in which rehabilitation projects are adversely affected by codes which are intended for use in new construction. Such codes tend to increase project costs 10-20%, add unnecessary project approval time, discourage feasible rehabilitation projects, and encourage payoffs to building officials. The Federal Government however, has issued draft rehabilitation guidelines designed to alleviate the negative impact building codes have on rehabilitation projects. The study conducted for this presentation shows that the management of rehab work is difficult and that many construction companies tend to shun rehabilitation. The investigation also indicates that producing plans and specifications is more time consuming, and that rehabilitation projects require more inspection and change orders compared with that of new construction. In addition, there may be a productivity decrease in rehabilitation that could possibly affect cost estimates and decrease profitability. Rehabilitation's positive aspects, however, include conservation of resources, providing contractors with interim work during slack periods, and helping to alleviate the current housing shortage.

0044 Financial Analysis under Inflation in Construction

Abraham Warszawski, Member, ASCE, (Head, Building Research Station, Technion, Israel Inst. of Tech., Haifa, Israel) and **Yechiel Rosenfeld**, (Instr., Dept. of Civ. Engrg., Technion, Israel Inst. of Tech., Haifa, Israel)

Journal of the Construction Division, Vol. 108, No. CO2, June, 1982, pp. 341-354

The stated value of main items in the financial reports of a construction company does not reflect correctly under inflation their actual value. This introduces a considerable difficulty into the process of financial analysis. The paper presents a method for appropriate adjustment of the main financial reports - the Balance Sheet and the Income Statement, to eliminate the distorting effects of inflation. The procedure is illustrated by an example of application.

0045 Integrated Project and Process Management

Majed A. A. Dabbas, (Construction Management Consultant, P.O. Box 1554, Amman, Jordan) and **Daniel W. Halpin**, Member, ASCE, (Prof., School of Civ. Engrg., Georgia Inst. of Tech., Atlanta, Ga. 30332)

Journal of the Construction Division, Vol. 108, No. CO3, September, 1982, pp. 361-374

This paper presents a computerized system, called PROMAX, for integrating project and process level planning and management. The system integrates the project management level capabilities of the PROJECT I system with those of the CYCLONE process level modeling system. Simulation is used at the process or technological level to determine the durations of project level activities. Variations in resource allocation and cost and their impact on schedule and cost at project level can be studied using the system.

0046 **The Resident Engineer: Intermediary Between Owner and Contractor**

Robert Del Re, Fellow, ASCE, (Sr. Assoc., Greeley and Hansen, Tampa, Fla.)

Journal of the Construction Division, Vol. 108, No. CO3, September, 1982, pp. 375-378

The resident engineer must possess significant talents other than the necessary technical abilities that qualify him for his position. The tactful and diplomatic handling of project interrelationships not only can resolve construction problems but can forestall them as well. Forming successful on-the-job relationships are covered together with a discussion of personnel introductions, meetings, and other techniques for developing project harmony. The resident engineer's role as an intermediary, promoting the orderly and expedient conduct of the work, is highlighted.

0047 **Coping with Uncertainty in Unit Price Contracting**

James E. Diekmann, Member, ASCE, (Asst. Prof. of Civ. Engrg., Univ. of Colorado, Boulder, Colo.), **Robert H. Mayer, Jr.**, (Asst. Prof. of Civ. Engrg., Univ. of Colorado, Boulder, Colo.) and **Robert M. Stark**, Member, ASCE, (Prof. of Civ. Engrg., Univ. of Delaware, Newark, Del.)

Journal of the Construction Division, Vol. 108, No. CO3, September, 1982, pp. 379-389

Unbalancing of Unit Price Proposals is widely used to improve cash flows and the competitiveness of a tender. Such practices can alter a contractor's risk favorably or otherwise. This paper is concerned with selecting unit prices which achieve a balance between maximizing the present worth of profit and minimizing potential contractor loss due to quantity misestimates. A quadratic programming model is devised to determine optimal unit prices and an example illustrates features of implementation.

0048 **Prevention and Resolution of Construction Claims**

Roy L. Wilson, Member, ASCE, (Pres., Wilson, Ma & Assocs., Inc., 125 Willis Ave., Roslyn Heights, N. Y.)

Journal of the Construction Division, Vol. 108, No. CO3, September, 1982, pp. 390-405

The owner and architect/engineer's roles in the prevention and resolution of construction claims are examined. A carefully written contract is invaluable in the prevention of a claim. Important aspects of construction contracts are presented, including scheduling, submission of shop drawings, payment terms, rates for time and material work and notification of delays and extra work. The owner and architect/engineer must be aware that his or her actions should be avoided whenever possible. Causes of valid construction claims include extra work, project delays and acceleration, lack of management, limited site access and change in work schedule. The importance of record keeping is stressed. Detailed records help to establish facts if a dispute later develops. Records that should be kept include minutes of meetings, correspondence, progress reports, status logs and photographs.

0049 Supervisory Problems in International Construction

William F. Maloney, Member, ASCE, (Asst. Prof., Dept. of Civ. Engrg., The Univ. of Michigan, Ann Arbor, Mich. 48109)

Journal of the Construction Division, Vol. 108, No. CO3, September, 1982, pp. 406-418

The success of supervisors on construction projects is, to a large extent dependent upon their ability to develop effective interpersonal relationships with their subordinates. The task of developing such relationships on international construction projects is more difficult because of the multicultural work environment. The impact of culture on the development of interpersonal relations is examined. Three factors subject to cultural influences that greatly determine the interpersonal relationship between two people in an employment context were examined: managerial style, motivation, and communication. Very few firms provide training for their prospective overseas supervisors. A case for training is made and supported by the results of a reported study.

0050 General Contractor Project Control Practices and MIS

Alan D. Russell, Member, ASCE, (Assoc. Prof. and Assoc. Dir., Centre for Building Studies, Concordia Univ., Montreal, Canada) and **Emmanuel Triassi**, (Pres., Les Entreprises TTS, Montreal, Canada)

Journal of the Construction Division, Vol. 108, No. CO3, September, 1982, pp. 419-437

Building contractor project control, information documentation, flow practices, and the attributes of project management personnel provide the focus for this paper. The methodology employed involves case studies of three successful general contracting firms whose annual work volumes range from $10 to 40 million. Key findings include: (1) Little written information is collected and processed pertaining to actual job progress; (2) few contractor personnel think of construction as a process which can be formally modelled; and (3) areas in need of greatest attention include subcontractor and change-order control, and forecasting time and cost to complete. Analysis of the case studies provides a set of requirements that should be met by MIS users and designers working both separately and jointly.

0051 Guidelines for Producing Better Specifications

Eli T. Abdallah, Member, ASCE, (Mgr., Specifications Dept., Parsons, Brinckerhoff, Quade & Douglas, Inc., One Penn Plaza, New York, N.Y. 10119)

Journal of the Construction Division, Vol. 108, No. CO3, September, 1982, pp. 438-444

Careful preparation, procedures, and methods to produce better specifications will minimize errors and reduce costs of delays and claims for extras. Specifications should be reviewed at preliminary, intermediate, and prefinal submittal stages by all disciplines of the design team. Specifications should be coordinated with the drawings to delete conflicts. The specifications should be checked for constructibility by reviewing existing conditions, restrictions as to access, utility availability, work areas, disposal areas, sequencing of work, time of completion, availability of materials, submittal requirements, quality assurance, and quality control. Many clauses should be discussed with the owner such as changed conditions, escalation, liquidated damages, retained percentages, mobilization, and insurance requirements. An important aspect is to have feedback from the field on how the specifications held up under actual conditions. Owners give an implied warranty to the contractor that the specifications are constructible, otherwise the contractor deserves additional compensation with time extensions to make them constructible.

0052 Time Space Scheduling Method

Dr. Oldrich Stradal, (Dr. Sci. Prof., Inst. of Construction Management, Swiss Federal Inst. of Tech., Zurich-8093 Zurich-Honggerberg, Switzerland) and **Josef Cacha**, (Civ. Engrg. Master of Science Research Fellow, Inst. of Construction Management, Swiss Federal Inst. of Tech., Zurich--8093 Zurich-Honggerberg, Switzerland)

Journal of the Construction Division, Vol. 108, No. CO3, September, 1982, pp. 445-457

A construction scheduling technique is presented which is, in many cases, more advantageous than bar charts and network schedules, particularly for repetitive activities and for scheduling the work in sections of a project (floors, bridge spans, road and sewage sections, etc.). The technique is labeled the time space scheduling technique because it shows clearly the connections of activities, their duration times, and sections (space) where they take place in a given time. The principal organization of the construction process is described. Examples of the application of time space scheduling are given for the pump foundation in a pipeline, construction of an apartment house, a multistoried concrete building, a road section, and a railway bridge. The time space scheduling technique is compared to bar charts and network scheduling techniques. Advantages and limitations are examined.

0053 Construction of Su-Ao Harbor Project, Taiwan

Shae-Chee Wang, Member, ASCE, (Deputy Chf. Engr. and District Mgr., Ret-Ser Engrg. Agency, Chang-Yuan Univ., Republic of China)

Journal of the Construction Division, Vol. 108, No. CO3, September, 1982, pp. 458-480

Su-Ao Harbor is surrounded on three sides by mountains, and the opening, approximately 6550 ft (2000) wide, is directly facing the Pacific Ocean. Besides the monsoon season, during which the wave height in the harbor is usually as high as 6.5 ft (2 M) to 13 ft (4 M), it suffers from typhoons for a period about three months every year. According to the planning and design, there are a number of caissons, piles, and shaped concrete blocks to be placed and piled in the sea. For overcoming geographical and oceanographical disadvantages, floating docks and slip forms were used to construct the huge caissons in floating state, tug boats to tow the huge caissons in rough water were designed, and methods were arranged for protecting various caissons under construction from typhoons. Special machines for launching small caissons, and a long-distance belt-conveyor for earth-moving were also designed.

0054 Contractor Quality Control: An Evaluation

Merlyn Isaak, Member, ASCE, (Pres., Testing Engineers, Inc., P.O. Box 24075, Oakland, Calif. 94623)

Journal of the Construction Division, Vol. 108, No. CO4, December, 1982, pp. 481-484

CQC (Contractor Quality Control) is a system mandated by certain government agencies to place the entire responsibility for quality control on a project upon the contractor. The contractor is required to submit a quality control plan, the extent of which depends on complexity of the project. Extensive experience has shown serious deficiencies in the system which can result in inferior quality, early failures and major repair expense. CQC encourages contractors to select inspectors based on lowest prices rather than professional qualifications. Quality control is too important to be left to low price bid shopping! It is argued that the system invites abuses due to lack of oversight, and many specified tests are never accomplished and often failing test results are screened out of the submittals. Quality control should be the responsibility of the design professional.

0055 Predicting Driving Stresses in Piles

Balbir S. Sandhu, Member, ASCE, (Consulting Engr., Kansas City, Mi.)

Journal of the Construction Division, Vol. 108, No. CO4, December, 1982, pp. 485-503

Reliable methods for the prediction of pile stresses are necessary because of the ever increasing damage and financial loss incurred on many piling projects due to improper selection of driving hammer. There is a delicate balance between pile stress, driving resistance, and hammer-to-pile weight ratio. Too large a hammer may destroy the pile whereas too small may not be adequate to drive the piles effectively. Existing methods of pile dynamic analysis have been found to give misleading results for certain driving conditions. This paper presents rigorous analytical procedures based on the theory of longitudinal wave transmission in conjunction with the equations of motion of pile-hammer system. The various cases of pile resistance distribution considered are: (1) 100% friction with uniform distribution; (2) 100% friction with triangular distribution; (3) 50% tip resistance and 50% uniformly distributed friction; (4) 50% tip resistance and 50% friction with triangular distribution; and (5) 100% point resistance. Charts are developed for steel, concrete, and wood piles. A method for selection of the most appropriate driving hammer is presented.

0056 Construction of Offshore Platforms

Stephen A. Will, Member, ASCE, (Mgr., Advanced Structures Dept., McDermott Marine Construction Unit, Research & Development Div., McDermott Inc., New Orleans, LA 70160)

Journal of the Construction Division, Vol. 108, No. CO4, December, 1982, pp. 504-519

The size and weight of offshore platforms have increased steadily as the search for oil and gas moves into deeper water and more hostile environments. This growth is portrayed in terms of water depth record setting projects. The capabilities and limitations of conventional construction techniques and equipment are described along with their impact on the fabrication and erection phases at construction. In the past five years, three offshore platforms have been installed in water depths approaching 1000 ft (305 m) and more with each project utilizing a different installation concept. The relationship between each of these concepts and the capacities of available equipment are explored. Future developments in water depths greatly exceeding 1000 ft (305 m) will utilize the compliant structural concepts, which require radically different installation procedures and more sophisticated equipment.

0057 Bid Markup Calculation by Crew-Day Method

Richard C. Ringwald, (Asst. Prof., Construction Engrg., Iowa State Univ., Ames, Iowa 50010)

Journal of the Construction Division, Vol. 108, No. CO4, December, 1982, pp. 520-530

A method for the calculation of bid markup to be used by contractors is developed which relates to the capacity of a firm during a given time period and to its particular financial goals. Capacity is measured for these purposes by normally available crew-time in a calendar period rather than in dollars, thus acknowledging the variation in dollar production achievable by one crew doing various types of work. It is demonstrated that setting markup by straight percent methods can fail to produce desired financial goals.

0058 Motivating Construction Workers

Alexander Laufer, (Asst. Prof., Civ. Engrg. Dept., Texas A & M Univ., College Station, Tex.) and **G. Douglas Jenkins, Jr.**, (Asst. Prof., Management Dept., Univ. of Texas at Austin, Austin, Tex.)

Journal of the Construction Division, Vol. 108, No. CO4, December, 1982, pp. 531-545

The paper opens with a review of previous construction motivational studies and an examination of their dated theoretical basis which demonstrate their limited usefulness. The contemporary approaches to motivation are then presented. The massive experience that has accumulated regarding the application of motivational programs in other industries is discriminately used to develop a new approach to motivate construction workers. The proposed guidelines are developed along the three variables that affect motivation: the individual, the job, and the work environment. The major implications for construction managers included in the findings are: (1) Management should play an active and continuous role in managing on-site motivational processes; (2) employees' desired outcomes should be tied to performance; and (3) on large projects, management should focus on eliminating performance obstacles. The paper also examines various managerial approaches to motivate construction workers and concludes that construction management would benefit from a general move toward a more participative decision-making style of leadership.

0059 Inflation and the Construction Industry

George Stukhart, Member, ASCE, (Assoc. Prof. of Civ. Engrg., Dept. of Civ. Engrg., Texas A & M Univ., College Station, Tex.)

Journal of the Construction Division, Vol. 108, No. CO4, December, 1982, pp. 546-562

Inflation has become a chronic problem whose effects permeate the entire construction industry. Owners are not only paying for the increased costs of facilities and capital but also for premiums on construction prices because of the uncertainties of inflation and its side effects. Contractors are faced with severe uncertainty in bidding and financing work on projects.

Productivity is affected because contractors cannot accurately forecast long-term returns on their investments and are required to divert necessary capital to meet resource costs. Owners and contractors must plan for these effects and attempt to reduce the risks entailed. In particular, the proper assignment of economic risks in contracting should reduce costs in the long term, although this would entail considerable change in construction industry operations.

0060 Realistic Tolerances for Double-Tee Construction

Richard L. Tucker, Fellow, ASCE, (Prof., Dept. of Civ. Engrg., The Univ. of Texas, Austin, Tex.), **Luh-Maah Chang**, (Grad. Student, Dept. of Civ. Engrg., The Univ. of Texas, Austin, Tex.) and **Ned H. Burns**, Member, ASCE, (Prof., Dept. of Architectural Engrg., The Univ. of Texas, Austin, Tex.)

Journal of the Construction Division, Vol. 108, No. CO4, December, 1982, pp. 563-576

This paper is mainly concerned with the problems of tolerance specification for precast concrete construction. Some articles have reported that constructed tolerances are much larger than those commonly specified and that tolerances recommended in published standards were unrealistic. In order to confirm the aforementioned statements, an investigation was made by directly measuring the existing variation of bearing lengths in the connections of the beams and their supports at five precast parking structures in Austin, Texas. The results of the investigation illustrate the problem of tolerance specification. A realistic tolerance standard should be established. More effort should be dedicated to measuring the variations in existing satisfactory structures to obtain more realistic data for establishing tolerance standards.

0061 Implementation of Foreman-Delay Surveys

Richard L. Tucker, Fellow, ASCE, (Prof. of Construction Engrg. and Project Management, Univ. of Texas, Austin, Tex. 78712), **David F. Rogge**, Member, ASCE, (Asst. Prof. of Construction Engrg. and Project Management, Oregon State Univ., Corvallis, Oreg.), **William R. Hayes**, (Mgr., Labor Relations, Monsanto Co., St. Louis, Mo.) and **Frank P. Hendrickson**, (Mgr., Engrg. - Construction, Monsanto Co., St. Louis, Mo.)

Journal of the Construction Division, Vol. 108, No. CO4, December, 1982, pp. 577-591

The practice of using surveys of construction foremen to make a qualitative and quantitative determination of the job factors resulting in lost time, and the subsequent use of this information as a management tool for the reduction of the magnitude of these factors and the improvement of worker morale is investigated. The development of this process and the current state-of-the-art are considered. Comparative data from eight construction sites are presented. A comparison of costs with potential benefits is provided. Recommendations are made for the effective use of foreman-delay surveys. Specific instructions for performing foreman-delay surveys are given.

0062 Foreman-Delay Surveys: Work Sampling and Output

David F. Rogge, Member, ASCE, (Asst. Prof. of Construction Engineering and Management, Oregon State Univ., Corvallis, Oreg.) and **Richard L. Tucker**, Fellow, ASCE, (Prof. of Construction Engrg. and Project Management, Univ. of Texas, Austin, Tex. 78712)

Journal of the Construction Division, Vol. 108, No. CO4, December, 1982, pp. 592-604

The usefulness of foreman-delay surveys (FDS) for performance measurement and productivity improvement at construction sites is evaluated by comparison with production rates and work sampling observations. A comparison of measurements obtained with concurrent work sampling and FDS at a nuclear plant construction site is presented. This comparison is followed by an examination of the changes in key site labor performance factors during the implementation of the work sampling and FDS programs at this site. An examination of relationships between FDS reported values, performance factor, and weld production for the major mechanical contractor of a refinery construction project is also presented. Both projects utilize union labor.

0063 Construction Education: Past, Present, and Future

Clarkson H. Oglesby, Fellow, ASCE, (Silas Palmer Prof. of Civ. Engrg., Emeritus, Stanford Univ., Stanford, Calif.)

Journal of the Construction Division, Vol. 108, No. CO4, December, 1982, pp. 605-616

Higher education for construction managers demands increased attention by the colleges and universities, and the construction industry. The present-day programs have developed only recently because of constraints such as the industry's diversity and divisions, and differences in the backgrounds and traditions of the participants. These programs have followed three main streams: (1) Four-year programs in civil engineering; (2) five-year programs in civil engineering leading to the M.S. degree; and (3) four-year programs outside engineering. University research on construction problems is minimal. Neither of the four-year programs, which produce most of the graduates, cover the subject matter the industry says it wants, but the transition to, or development of, more five-year programs will be slow if present conditions continue. To maintain and improve construction education and launch the universities into research and in-service education will be impossible without industry attention and support far beyond the present low levels. Fortunately, there are strong signs that this attention and support is developing so that the future seems bright.

0064 Owner's Guidelines for Selecting Safe Contractors

Nancy M. Samelson, Assoc. Member, ASCE, (Sr. Research Assoc., Dept. of Civ. Engrg., Stanford Univ., Stanford, Calif. 94305) and **Raymond E. Levitt**, Member, ASCE, (Assoc. Prof., Dept. of Civ. Engrg., Stanford Univ., Stanford, Calif. 94305)

Journal of the Construction Division, Vol. 108, No. CO4, December, 1982, pp. 617-623

A study of practical means by which owners can reduce construction costs was undertaken as part of a larger investigation of construction cost control. The study involved three parts: 1) documenting the costs of accidents, 2) selecting safe contractors and 3) monitoring for construction safety. The major costs of accidents are found to be workers' compensation insurance costs and hidden costs, conservatively estimated from the research to total 4% of project costs on industrial construction. The size of construction accident costs combined with the fact that research shows that they can be reduced very substantially by management methods makes selection and monitoring by owners cost effective techniques. Criteria for selection of safe contractors are described in detail: Experience modification ratings, OSHA incidence rates, and management safety accountability. Effective monitoring methods are also listed.

0065 Risk Element in Profit Planning under Inflation

A. Warszawski, Member, ASCE, (Head, Bldg. Res. Station, Technion - Israel Inst. of Tech., Haifa, Israel)

Journal of the Construction Division, Vol. 108, No. CO4, December, 1982, pp. 624-638

The paper deals with the influence of inflation on profit from construction ventures, and on contractors price determination. First the effect of price escalation on profit is examined under two types of contracts: a fixed price contract and a contract with a compensation clause. Then, the influence of inflation risk on construction price of contractors with different attitudes towards risk taking is explored. The risk is represented by a range of possible inflation rates with different statistical distributions. Finally, the influence of inflation on return from long range investment in equipment and facilities is examined.

0066 General Bidding Model

Robert I. Carr, Member, ASCE, (Prof., Dept. of Civ. Engrg., Univ. of Michigan, Ann Arbor, Mich. 48109)

Journal of the Construction Division, Vol. 108, No. CO4, December, 1982, pp. 639-650

The general model of competitive bidding is not limited by the assumptions on which Friedman's and Gates' models depend. It is applicable to competition for which a contractor's cost distribution and opponents' bid distributions can be estimated. Historical data of contractor's cost and competitors' bids on different projects produce a distribution for the ratio between them, the bid/cost ratio. Standardized distributions for contractors' cost and competitors' bids, estimated to have respective means of one and the mean bid/cost ratio and to have equal variance, are inserted into the general model. Results are compared with Friedman's and Gates' models for competition against average competitors. Because markup adjustments are counterbalanced by shifts in probability of winning, contractor expected value is not very sensitive to markup, or its method of selection. However, the models vary considerably in their estimates of expected value. Gates' model is always more accurate than Friedman's model where variance of bids is due to variance of costs.

0067 Nuclear Waste Management

Task Committee on Nuclear Effects of the Committee on Air, Noise and Radiation Management of the Environmental Engineering Division

Journal of the Environmental Engineering Division, Vol. 108, No. EE1, February, 1982, pp. 1-50

This Task Committee report to the members of ASCE reviews the current technology and many of the key technical issues concerned with the long-term storage and disposal of high-level and intermediate-level nuclear wastes from the commercial nuclear fuel cycle. Although national policy is not firm, major efforts are focused on permanent storage of nuclear wastes in engineered waste repositories with multiple barriers to transport to the biosphere. Many of the issues focus on the desire for assurances that the wastes can be isolated from the biosphere for geologic time periods in excess of thousands of years. The report reviews the origin of nuclear wastes from the total nuclear fuel cycle, the biological and environmental significance of the wastes, and technical options for waste repositories, both for shallow land burial and federal repositories. The report disucsses pressing issues on spent fuel reprocessing and concludes with recommendations that could accelerate activities to relieving national need for nuclear waste disposal facilities.

0068 Dynamic Estuary Model Performance

Robert B. Ambrose, Jr., (Environmental Engr., Athens Environmental Research Lab., Office of Research and Development, Environmental Protection Agency, Athens, Ga. 30613) and **Stephen E. Roesch**, (Environmental Engr., Central Regional Lab., Region III, Environmental Protection Agency, Annapolis, Md. 21401)

Journal of the Environmental Engineering Division, Vol. 108, No. EE1, February, 1982, pp. 51-71

Applications of the Dynamic Estuary model (DEM) to both the Delaware and Potomac Estuaries by the Environmental Protection Agency during the 1970s are summarized and evaluated. Methods for calibrating, refining, and validating this model, and statistics for evaluating its performance are discussed. Following a brief description of DEM, observed and predicted hydrodynamic, mass transport, and water quality variables are statistically analyzed for the Delaware and Potomac applicatins. Weaknesses in the model and in these applications are identified. Model performance is evaluated for its usefulness to investigations of wastewater management alternatives in tidal rivers and the upper reaches of estuaries. Results indicate that the DEM can describe real-time longitudinal tidal hydraulics with good accuracy for relatively steady flow periods and can predict longitudinal mass circulation with sufficient accuracy for well-mixed, far-field conditions. The DEM can adequately predict relative differences in water quality to be expected from incremental wastewater management alternatives, but more realistic kinetic structures are needed to predict water quality response to large changes in waste load characteristics.

0069 Urban Impacts of Water Supply Reservoir

Ernesto Baca, (Research Sci., Dept. of Environmental Sci. and Engrg., Rice Univ., Houston, Tex. 77001), **Philip B. Bedient**, Assoc. Member, ASCE, (Assoc. Prof., Dept. of Environmental Sci. and Engrg., Rice Univ., Houston, Tex. 77001) and **Richard Olsen**, (Research Sci., Dept. of Environmental Sci. and Engrg., Rice Univ., Houston Tex. 77001)

Journal of the Environmental Engineering Division, Vol. 108, No. EE1, February, 1982, pp. 73-87

Water quality of Lake Houston, a rapidly urbanizing water supply reservoir near Houston, Texas, was characterized in terms of sediments, nutrients, and algal population. Data was collected at five in-lake stations and all the tributaries, on a weekly basis during 1979. Analyses showed major water quality differences between the urbanized tributaries to the lake and the forested tributaries. Frequent storms and silty soils cause very high turbidities in the lake, which in turn limit the algal population by storm wash-out and light restriction. Algal productivity rate is highly controlled by seasonal factors, such as light and temperature. Two equations were developed relating physical factors (i.e., flow rate, turbidity, and water temperature) to biological factors (i.e., chlorophyll a and algal productivity rate). A suggested set of management strategies was also developed.

Errata: EE Aug. '82, pp. 818.

0071 Computer Design of Landfill Methane Migration Control

Charles A. Moore, Member, ASCE, (Prof. of Civ. Engrg., Ohio State Univ., Columbus, Ohio), **Iqbal S. Rai**, Member, ASCE, (Sr. Research Engr., Goodyear Tire & Rubber Co., Akron, Ohio) and **John Lynch**, Assoc. Member, ASCE, (Staff Engr., Benedict, Bowman, Craig, and Moos, Columbus, Ohio)

Journal of the Environmental Engineering Division, Vol. 108, No. EE1, February, 1982, pp. 89-106

Computer programs for the design of methane migration control systems around sanitary landfills are described. These programs allow the effectiveness of individual designs to be evaluated and allow comparison of alternative designs. The effects of varying pumping rates on active systems can be assessed. In addition to describing the mathematical principles used in the computer programs, results for typical control installations are shown in a manner that allows for comparing their effectiveness.

0072 DO Model for Discharges from Deep Impoundments

John M. Higgins, Member, ASCE, (Environmental Engr., Tennessee Valley Authority, 248 401 Bldg., Chattanooga, Tenn. 37401) and **Byung R. Kim**, Assoc. Member, ASCE, (Asst. Prof., Georgia Inst. of Technology, Atlanta, Ga 30332)

Journal of the Environmental Engineering Division, Vol. 108, No. EE1, February, 1982, pp. 107-122

A simple model is presented for the annual dissolved oxygen (DO) cycle of releases from the hypolimnion of deep impoundments. The model includes mathematical formulations and rate constants for the processes of BOD decay, atmospheric reaeration, and thermal stratification. It was used to characterize the DO cycle of releases from 17 Tennessee Valley Authority reservoirs, using three years of weekly DO and temperature data. Rate constants were estimated using the quasilinearizatin technique for estimating unknown parameters in a system of ordinary differential equations. The results indicate that the model can reasonably simulate annual DO variations of turbine discharges from hypolimnetic waters. Since the model is a simple formulation of complex reservoir processes, it is best suited for preliminary analyses.

0073 Pressure Distribution Design for Septic Tank Systems

Richard J. Otis, (Specialist, Dept. of Civ. and Environmental Engrg., Univ. of Wisconsin, Madison, Wisc.)

Journal of the Environmental Engineering Division, Vol. 108, No. EE1, February, 1982, pp. 123-140

Pressure distribution of septic tank effluent in subsurface soil absorption systems is gaining in popularity. This method of effluent distribution has the advantage over other methods of providing dosing in addition to more uniform distribution which reduces the rate of soil clogging and prevents localized overloading. However, the application of these networks has been limited because of the complexity of their design. To overcome this difficulty, the design criteria are discussed and a simplified design procedure presented.

0074 Heat Loss from Rapid Infiltration Basin in Winter

Heinz G. Stefan, Member, ASCE, (Prof. and Assoc. Dir., St. Falls Hydraulic Lab., Dept. of Civ. and Mineral Engrg., Univ. of Minnesota, Minnneapolis, Minn. 55455)

Journal of the Environmental Engineering Division, Vol. 108, No. EE1, February, 1982, pp. 141-158

A method to estimate heat losses from a rapid wastewater infiltration basin under winter operation is developed. The analysis is theoretical and considers two modes of operation, one with a permanent ice cover on the basin and one without. The loading of the infiltration basin is intermittent. The analysis provides a complete heat budget. The equations are expressed in a form which permits estimation of frequency of loading of a basin as a function of loading depth. Heat losses to the atmosphere, formation of an ice cover and cyclic freezing, and thawing of the soil are included in the analysis.

0075 Evaluation of Selected Initial Dilution Models

Abu M. Z. Alam, Member, ASCE, (Project Manager, Metcalf & Eddy, Inc., Boston, Mass.), **Donald R. F. Harleman**, Member, ASCE, (Ford Prof. and Dir., Ralph M. Parsins Lab. for Water Resources and Hydrodynamics, Massachusetts Inst. of Tech., Cambridge, Mass.) and **Joseph M. Colonell**, Member, ASCE, (Project Engr., Woodward-Clyde Consultants, Anchorage, Alaska)

Journal of the Environmental Engineering Division, Vol. 108, No. EE1, February, 1982, pp. 159-186

A number of mathematical and computer models are currently available to calculate initial dilution of treated wastewater effluent discharges into the ocean. This paper presents a comparative evaluation of six initial dilution models using data from Boston Metropolitan District Commission's proposed ocean discharge. To determine the reliability of application of these models, field data and results of dye studies conducted for the Alyeska Marine Terminal outfall in Port Valdez, Alaska were used to check against calculated dilutions from these models. It was found that some of the models provided dependable results.

0076 Calorific Value of Wastewater Plant Sludges

Alphonse E. Zanoni, Member, ASCE, (Prof. of Civ. Engrg., Marguette Univ., Milwaukee, Wis. 53233) and **David L. Mueller**, (Regional Engr., Zimpro, Inc., Levittown, Pa., 19054)

Journal of the Environmental Engineering Division, Vol. 108, No. EE1, February, 1982, pp. 187-195

Calorific values of wastewater plant sludges are best determined using an oxygen bomb calorimeter, an apparatus which is not readily available in many plants. Using a variety of sludges from thirteen municipal wastewater treatment plants in southeastern Wisconsin, an excellent correlation is established between the calorific value and chemical oxygen demand of the sludges. A well-tested recommended procedure for conducting the chemical oxygen demand of wet sludges and semi-dry cakes is also presented.
Errata: EE Dec. '82, pp. 1393.

0078 Photosynthesis-Light Mathematical Formulations

Stephen D. Field, (Asst. Prof., Dept. of Civil Engrg., Louisiana State Univ., Baton Rouge, La. 70803) and **Steven W. Effler**, (Adjunct Prof., Dept. of Civil Engrg. Syracuse Univ. Syracuse, N.Y. 13210)

Journal of the Environmental Engineering Division, Vol. 108, No. EE1, February, 1982, pp. 199-203

Photosynthesis-light curves for photoplankton describe the characteristic dependence of the instantaneous rate of photosynthesis on incident light. Six different mathematical formulations are evaluated for the photosynthesis-light relationship, based on their ability to describe experimental data collected from nutrient saturated, hypereutrophic Onandaga Lake in New York State. The data are comprised of measurements of in situ gross primary productivity, and the available light from 115 experiments.
Errata: GT March '82, pp. 514-515.

0080 Leachate Collection Design for Containment Landfills

Edward A. McBean, (Prof., Dept. of Civ. Engrg., Univ. of Waterloo, Waterloo, Ontario, Canada), **Ronald Poland**, (Dist. Engr., Waste Management of Canada, Maple, Ontario, Canada), **Frank A. Rovers**, (Principal, Conestoga-Rovers & Assoc. Ltd., Waterloo, Ontario, Canada) and **Anthony J. Crutcher**, (Assoc., Conestoga-Rovers & Assoc. Ltd., Waterloo, Ontario, Canada)

Journal of the Environmental Engineering Division, Vol. 108, No. EE1, February, 1982, pp. 204-209

A mathematical model is proposed for use in designing the leachate collection system for containment landfills, specifically with the respect to the spacings of drains, and to ground-water mounding. It is shown that parameter sensitivities can be easily explored for specific applications of the model to determine the need for more accurate definition of input data to the model.
Errata: EE Oct. '82, pp. 1088.

0082 Models of Landfill Leaching: Moisture Flow and Inorganic Strength

William A. Straub, (Project Engr., Wright-Pierce Engrs., 25 Vaughn Mall, Portsmouth, N.H. 03801) and **Daniel R. Lynch**, Member, ASCE, (Asst. Prof., Thayer School of Engrg., Dartmouth Coll., Hanover, N.H. 03755)

Journal of the Environmental Engineering Division, Vol. 108, No. EE2, April, 1982, pp. 231-250

Mathematical models are applied to the movement and inorganic contamination of moisture in sanitary landfills situated above the water table. The models are based on simple well-mixed reactor concepts, and on unsaturated flow and transport in porous media. Computer simulations are obtained for laboratory-scale experimental landfills reported by others. General agreement between simulated and observed results indicates that leachate behavior is explainable in terms of fundamental transport processes. The roles of moisture retention in the landfill and dilution by infiltrating water are shown to be of primary importance in understanding and predicting leachate quantity and quality.

0083 Models of Landfill Leaching: Organic Strength

William A. Straub, (Project Engr., Wright-Pierce Engrs., 25 Vaughan Mall, Portsmouth, N.H.) and **Daniel R. Lynch**, Member, ASCE, (Asst. Prof., Thayer School of Engrg., Dartmouth Coll., Hanover, N.H. 03755)

Journal of the Environmental Engineering Division, Vol. 108, No. EE2, April, 1982, pp. 251-268

Mathematical models are applied to the dissolution, transport, and decay of organic materials in unsaturated sanitary landfills. The models are based on cascaded reactor concepts and on unsaturated moisture and contaminant transport in porous media. The roles of aerobic and anaerobic bacteria are simulated using conventional kinetic formulations. Computer simulations are obtained for landfill experiments reported by others, and good agreement with observed behavior is demonstrated. The results indicate that aerobic activity is negligible over the leaching life of a landfill, and that anerobic activity is significantly inhibited from that observed in coventional municipal wastewater treatment. Further, the processes of dilution of landfill moisture by infiltration of rainwater, and convection of materials from the landfill via leaching, are significant and in some cases the predominant influences on overall organic behavior. The retention of cells within the landfill system during leachate recycling can be a significant contributor to accelerated rates of biostabilization. The landfill models are useful in the analysis of empirical data and in the identification of important leaching processes.

0084 Role of Surface Active Media in Anaerobic Filters

Khalique A. Khan, (Sanitary Engr., Burk and Assoc., Inc., 4176 Canal St., New Orleans, La. 70119), **Makram T. Suidan**, (Assoc. Prof. of Civ. Engrg., Dept. of Civ. Engrg., Univ. of Illinois, Urbana, Ill. 61801) and **Wendall H. Cross**, (Research Sci., School of Civ. Engrg., Georgia Inst. of Tech., Atlanta, Ga. 30332)

Journal of the Environmental Engineering Division, Vol. 108, No. EE2, April, 1982, pp. 269-285

Equal-sized activated carbon and anthracite coal were compared when used as packing media in completely mixed anaerobic filters. Synthetic glucose and phenol bearing solutions and a diluted paint stripping bath wastewater were used as the test substrated. Each wastewater was fed to the two anaerobic filters until stable operating conditions were reached. The granular activated carbon packed filter resulted in better chemical oxygen demand conversion, higher rates of methane production, and lower biomass production than the anthracite packed filter for all the conditions tested.

0085 Hydraulic Regime and Activated Sludge Performance

Robert M. Sykes, Member, ASCE, (Assoc. Prof., Dept. of Civ. Engrg., 470 Hitchock Hall, 2070 Neil Ave., Columbus, Ohio 43210)

Journal of the Environmental Engineering Division, Vol. 108, No. EE2, April, 1982, pp. 286-296

The standard bio-kinetic theory used to describe the activated sludge process predicts that the soluble organic matter (SOM) concentration in the effluent of plug flow aeration tanks will be less than the SOM concentration in the effluent of completely mixed aeration tanks, if both contain the same total biomass and are operated at the same solids' retention time (SRT). Large-scale field studies indicate that this prediction is false. An alternate model of the process is presented that identifies the effluent SOM as a product synthesized and excreted by the microbes of the activated sludge. This model predicts that all aeration tanks treating a given wastewater will produce the same effluent SOM concentration if they contain the same total biomass and are operated at the same SRT. The alternate model also predicts that the effluent SOM concentration will be proportional to the influent organic matter concentration, that the apparent microbial decay rate will vary as different nutrients are considered, and that plots of the specific uptake rate vs. the effluent SOM concentration will show a positive SOM intercept. All these predictions are supported by experiment.

0086 1980 Needs Survey: Combined Sewer Site Studies

James E. Scholl, Assoc. Member, ASCE, (Water Resources Engr., CH2M Hill, Inc., 7201 N.W. 11th Place, P.O. Box 1647, Gainesville, Fla. 32602) and **Ronald L. Wycoff**, Assoc. Member, ASCE, (Water Resources Engr., CH2M Hill, Inc., 7201 N.W. 11th Place, P.O. Box 1647, Gainesville, Fla. 32602)

Journal of the Environmental Engineering Division, Vol. 108, No. EE2, April, 1982, pp. 315-326

The U.S. Environmental Protection Agency is mandated by the Clean Water Act to prepare and report to the Congress a biennial estimate of the construction cost for publicly owned wastewater treatment facilities necessary to achieve the goals of the Act. This required biannual report is known as the U.S. EPA Needs Survey. Ten combined sewer site studies were conducted as part of the 1980 Needs Survey to develop transferable criteria, principles, and relationships which could be applied nationwide to estimate the cost to construct publicly owned wet-weather pollution control facilities. The site studies were performed using a receiving water simulation program called the "Continuous Stormwater Pollution Simulation System" (CSPSS) and an economic optimization program called the "Computer Optimzied Stormwater Treatment" (COST) program. It was found that wet-weather pollution control requirements are very site-specific and that CSPSS and COST were valuable tools for quantifying removal requirements and optimal control costs.

0087 Sewers and Cities: France and the United States Compared

Gabriel Dupuy, (Prof., Directeur de L'Enseignement, Ecole Nationale des Ponts et Chaussees, France) and **Carnegie-Mellon Univ., Pittsburgh, Pa. 15213 Tarr,. Joel A.**, (Prof. of History and Public Policy, Dept.of Social Science and School or Urban Public Affairs)

Journal of the Environmental Engineering Division, Vol. 108, No. EE2, April, 1982, pp. 327-338

The development and impacts of wastewater systems in France and the United States and their application for developing countries are discussed. Watercarriage technology was adopted in both nations because of the public health problems created by cesspools and privy vaults in growing urban areas and the nuisances generated by the development of piped-in water systems. Both nations experienced controversy over questions of sewerage system design and sewage disposal. French cities had fewer difficulties with drinking water pollution than American because they seldom utilized adjacent rivers for drinking water supplies. A major problem today in France is that of insufficient stormwater capacity. In the United States problems of water quality are of most concern. In terms of impacts, in both nations questions of waste disposal resulted in a larger role for the central government in local affairs, creation of a new sanitary ethos, and reliance upon technology for problem solutions.

0088 Rotating Biodisks Recarbonate Alkaline Wastewater

Charles I. Noss, (Research Engr., U.S. Army Medical Bioengineering Research and Development Lab., Ft. Detrick, Frederick, Md. 21701), **Roy D. Miller**, Member, ASCE, (Chf., Environmental Health Engrg. Branch, U.S. Army Environmental Hygiene Agency, Regional Division-North, Fort Meade, Md. 20755) and **Edgar D. Smith**, Member, ASCE, (Environmental Engr., U.S. Army Construction Engrg. Research Lab., Champaign, Ill. 61820)

Journal of the Environmental Engineering Division, Vol. 108, No. EE2, April, 1982, pp. 339-348

Lime addition to raw wastewater as an upgrading technique can significantly aid a rotating biological contactor (RBC) plant in producing effluents in compliance with federal discharge permit limitations for biochemical oxygen demand (BOD_5), amonia nitrogen and phosphorus. Low-level lime addition does not require recarbonation and does not produce the sludges typical of high pH lime treatment schemes. The pilot RBC process provided BOD_5 removal when subjected to an influent pH of 9.5 for hydraulic loading rates of 2.0, 3.0, and 4.0 gpd/sq ft (0.08, 0.12, and $0.6 m^3/m^2 \cdot$ day). In addition to the removal of phosphorus, lime pretreatment reduced the organic loading on the RBC process and allowed for recarbonation by microbial populations which produced carbon dioxide, thereby forming carbonate alkalinity necessity for nitrification after the initial BOD_5 had been removed. The resultant pH after recarbonation was also in the optimal range for nitrification.

0089 Algorithm for Optimal Activated Sludge Operation

Patrick W. Hughey, (Grad. Student, The Lewis and Clark Law School, Portland, Oregon), **Dale D. Meredith**, Member, ASCE, (Prof., Dept. of Civ. Engrg., State Univ. of New York at Buffalo, Buffalo, N.Y.) and **Andrew C. Middleton**, Assoc. Member, ASCE, (Vice Pres., Environmental Resources, Koppers Co., Inc., Pittsburgh, Pa.)

Journal of the Environmental Engineering Division, Vol. 108, No. EE2, April, 1982, pp. 349-366

A method for determining the minimum cost operation of an activated sludge wastewater treatment plant is developed and applied to a plant in Erie County, New York. The plant includes an aerated grit chamber, two aeration basins, two secondary clarifiers, two solids contact clarifiers, a flotation thickener, sand-drying beds, centrifuge, an aerobic digester, two gravity sand filters, and chlorination-reaeration facilities. The complex method of constrained optimization is modified for discrete-valued variables and used with nonlinear cost and process equations to determine the minimum cost operation of the plant for a specified 12-month flow sequence. The results from the optimization are compared with actual operation of the plant.

0090 Anerobic Treatment in Novel Contact Digester

Jan A. Oleszkiewicz, (Duncan, Lagnese, and Assocs., Inc., 3185 Babcock Blvd., Pittsburgh, Pa. 15237.) and **Szymon Koziarski**, (Section Head, Research Inst. for Environmental Development, Rosenbergow 28, 51-616 Wroclaw, Poland)

Journal of the Environmental Engineering Division, Vol. 108, No. EE2, April, 1982, pp. 367-378

Laboratory studies are presented on a combined jet gas mixed and sludge blanket anaerobic contact digester with continuous sludge recycle. Raw piggery waste substrate is used as feed with a COD_{nf} contents of 13-14 g/L. The aims of the study are to establish lower limits of hydraulic retention and upper limits of COD loading, in conditions of continuous sludge recycle without intentional sludge wasting, in order to formulate design criteria for the process operated either as a roughing treatment or a combined primary-secondary treatment of strong industrial wastewaters. The studies show the feasibility of obtaining methanogenic conditions at hydraulic retention (HRT) as low as 0.5 d and corresponding organic load of 28 kg COD/m^{3}d. Satisfactory secondary treatment is attained at HRT = 3.5d and a load of 4 kg COD_{nf}/m^{3}d, yielding steady state COD_{nf} removals in excess of 85% and filtered effluent COD_f and BOD_{5f} of 1,200 and 500 mg $O_2/^{L3}$, respectively. The biogas production at that loading is 3.15 m^3/m^{3}d (75% CH_4) which corresponds to 0.17 m^3/kg COD_{nf} removed.

0091 Improvement of Swirl Concentrator

Michel Alquier, (Docteur es Sciences, Institut de Mechanique des Fluides, Toulouse, France), **Daniel Delmas**, (Docteur Ingenieur, Ecole Polytechnique de Montreal, Quebec, Canada) and **Michel Pellerej**, (Chercheur, Institut de Mechanique des Fluides, Toulouse, France)

Journal of the Environmental Engineering Division, Vol. 108, No. EE2, April, 1982, pp. 379-390

The analysis of a treatment device for pluvial water, proposed by the American Public Works Association (APWA) and called a swirl concentrator, shows that its efficiency greatly depends on a large eddies structure generated by an obstacle placed to obtain a usable hydraulic load-discharge characteristic. The characterization of the convective-diffusive process is done by measuring the transit-time distribution of fluid and liquid particles. The study of similarity law done on two hydraulic models observes the critical velocity of solid particles as similarity parameter. In order to optimize the concentrator efficiency, a configuration of vertical bars is proposed to obtain a flow with a minimum of large eddies structure. A pattern of bar an be added to an already built concentrator with a gain for the removal efficiency.

0092 Total Suspended Solids in Highway Runoff in Washington State

Randy Asplund, (Water Systems Engr., RH2 Engrg., Seattle, Wash. 98195), **Brian W. Mar**, Member, ASCE, (Prof., Dept. of Civ. Engrg., Univ. of Washington, Seattle, Wash. 98195) and **John F. Ferguson**, Member, ASCE, (Prof., Dept. of Civ. Engrg., Univ. of Washington, Seattle, Wash. 98195)

Journal of the Environmental Engineering Division, Vol. 108, No. EE2, April, 1982, pp. 391-404

During the first year of full-time operation of ten highway runoff monitoring sites in the State of Washington, a total of 241 storms were sampled with a composite sampling system developed by Clark. Vehicle traffic during the storm was found to be the principal factor influencing the mass of suspended solids in the runoff. During winter ice conditions a major fraction of solids loadings in runoff can be traced to sand used. A predictive model for total suspended solids loadings was derived as a function of cumulative traffic during storms. Traffic during the dry periods appeared to remove pollutants from the highways and is not significant in predicting pollutant loadings in the State of Washington.

0093 Pharmaceutical Waste Treatment by Anaerobic Filter

Elliot F. Sachs, Assoc. Member, ASCE, (Project Engr., Camp, Dresser, & McKee, New York, N.Y.), **J. Charles Jennett**, Member, ASCE, (Dean of Coll. of Engrg. and Prof. of Environmental Systems Engrg., Clemson Univ., Clemson, S.C.) and **Myrton C. Rand**, (Prof. of Civ. Engrg. and Environmental Sci., Univ. of Alabama at Huntsville, Huntsville, Ala. 35899)

Journal of the Environmental Engineering Division, Vol. 108, No. EE2, April, 1982, pp. 397-314

The anaerobic filter has been used on a laboratory scale to treat a variety of industrial wastewaters. The process was known to be effective at treating fermentation pharmaceutical wastes but synthesized organic chemical wastes were considered to be less amenable to treatment. Preliminary studies did indicate that the raw waste exhibited toxicity to both aerobic and anerobic organisms. This investigation dealt with analyzing the physical and chemical variations in the constituents of the pharmaceutical waste. By using laboratory scale models under conditions of controlled temperature, hydraulic loading and organic strength and using a waste feed composed of varying percentages of methanol and waste at a constant COD, it was possible to investigate the toxic effects of the waste on anaerobic digestion. The waste was found to be toxic but the filter could acclimate to it. Final results indicate that the anaerobic filter would treat the waste faster than the existing aerobic extended aeration system producing an effluent as good or better, and because of the methane generated would be an energy producer rather than a consumer.

0094 Microbial Activity in Activated Carbon Filters

John G. den Blanken, (Sr. Scientific Officer, Dept. of Civ. Engrg., Lab. of Sanitary Engrg., Delft Univ. of Tech., Stevinweg 4, 2628 CN Delft, The Netherlands)

Journal of the Environmental Engineering Division, Vol. 108, No. EE2, April, 1982, pp. 405-425

To obtain information on the adsorption, bacteriological oxidation and desorption processes in filters, the oxygen consumption (ΔO_2) and the removal of organic carbon (ΔTOC) as well as their relation to one another ($\Delta C/\Delta O_2$) were studied. The results show that bacteriological activity in activated carbon filters can make an important contribution to the removal of organic substances both in the water and on the activated carbon. This activity was dependent on various factors such as the water temperature, contact time, type of activated carbon, the running time of the filter, number of attached bacteria, and composition of the influent water. At steady state 0.010 g of carbon was oxidized per hour by 10^{11} attached bacteria at 20°C and pH 8.0, which was measured in filter and in batch experiments. It is important that water treatment plants with activated carbon filters make use of the bacteriological activity and benefit from it as a contribution to water treatment. In terms of bacteriological activity the activity of the slow sand filter (0.4 m/hour) was comparable with the activated carbon filter (4 m/hour).

0095 Heat Treatment and Anaerobic Digestion of Refuse

James M. Gossett, Assoc. Member, ASCE, (Asst. Prof., Environmental Engrg. Dept., Cornell Univ., Ithaca, N.Y. 14853), **David C. Stuckey**, (Civ. Engrg. Dept., Stanford Univ., Stanford, Calif. 94305), **William F. Owen**, Member, ASCE, (Culp, Wesner, and Culp Consulting Engrg., P.O. Box 518, Cameron Park, Calif. 95682) and **Perry L. McCarty**, Member, ASCE, (Chmn. and Prof., Civ. Engrg. Dept., Stanford Univ., Stanford, Calif. 94305)

Journal of the Environmental Engineering Division, Vol. 108, No. EE3, June, 1982, pp. 437-454

Alkaline heat treatment was evaluated as a pretreatment process for increasing the anaerobic digestibility of municipal refuse. The goal was an increased methane yield from refuse lignocellulose. Batch digestibility assays were employed to ascertain the separate effects of solids concentration (0.79 - 10.0 percent), sodium hydroxide concentration (75 - 300 meq/l), and temperature (25 - 250°C) on the methane yield from refuse. Consumption of added alkali was observed and found to be relatively constant per unit mass of refuse solids. Various semi-continuous reactor configurations were evaluated. These studies resulted in investigation of a two-stage digestion system, with the second digester fed alkaline-heat-treated (180 meq/l NaOH, 200°C, 2.1 percent solids, 1 hour) effluent from the first. Such a system operated on newsprint was capable of increasing the methane yield by 80 percent, compared with the yield from a single-stage digester fed raw newsprint. A major limitation appears to be digester inhibition noticeable whenever the concentration of heat-treated lignin products exceeds 1 g/l.

0096 Acid Rain Model: Hydrologic Module

Carl W. Chen, Member, ASCE, (Vice Pres., Environmental Systems Engrg., Tetra Tech., Inc., Lafayette, Calif.), **J. David Dean**, (Agricultural Engr., Environmental Systems Engrg., Tetra Tech., Inc., Lafayette, Calif.), **Steven A. Gherini**, (Chf. Engr., Environmental Systems Engrg., Tetra Tech., Inc., Lafayette, Calif.) and **Robert A. Goldstein**, (Project Mgr., Environmental Assessment and Analysis, Electric Power Research Inst., Palo Alto, Calif.)

Journal of the Environmental Engineering Division, Vol. 108, No. EE3, June, 1982, pp. 455-472

The acid rain model has been developed to simulate lake-watershed acidification processes. It provides a quantitative linkage between atmospheric deposition and lake water chemistry. The continuous simulation model is comprised of several modules including watershed hydrology, stream and lake hydraulics, canopy, snow, soil, stream and lake chemistry. The hydrology module simulates the processes of interception, throughfall, evapotranspiration, freezing and thawing of soil, and snow accumulation and melting. The model segregates the watershed horizontally into subcatchments and vertically into soil layers. Flows are routed through a layered soil profile to the streams and ultimately to the lake. Incoming precipitation and deep groundwater flow. Application of the model to Panther and Woods Lake basins in the Adirondack mountains of New York indicates the importance of flowpaths in determining lake water acidity.

0097 Nitrification Kinetics and Its RBC Application

Ching-San Huang, (Environmental Engr., Air Force Regional Civ. Engr., M-X, Norton Air Force Base, Calif.)

Journal of the Environmental Engineering Division, Vol. 108, No. EE3, June, 1982, pp. 473-487

The surface reaction model for the mechanism of mass transfer-metabolism in the fixed-film nitrification process is investigated and discussed for the design of trickling filter systems and rotating biological contactor (RBC) systems. Two experimental studies are performed. The first study, using a stationary fixed-film reactor to simulate the trickling filter process, reveals that the surface reaction kinetics follow a pseudo-homogeneous model. The second study, using a bench-scale RBC unit, indicates that the pseudo-homogeneous model is also applicable to the rotating fixed-film process. The effective slime thickness of an RBC system can be estimated from this model by locating the optimum NH3-N removal rotating speed and finding the corresponding liquid film thickness at that rotating speed.

0098 Mean of Autocorrelated Air Quality Measurements

Cynthia S. Hirtzel, Assoc. Member, ASCE, (Asst. Prof., Dept. of Chemical and Environmental Engrg., Rensselaer Polytechnic Inst., Troy, N.Y. 12181), **Jimmie E. Quon**, Fellow, ASCE, (Former Prof. and Chmn., Dept. of Civ. Engrg., Northwestern Univ., Evanston, Ill. 60201) and **Ross B. Corotis**, Member, ASCE, (Prof., Dept. of Civ. Engrg./Material Science and Engrg., Johns Hopkins Univ., Baltimore, Md. 21218)

Journal of the Environmental Engineering Division, Vol. 108, No. EE3, June, 1982, pp. 488-501

The high degree of autocorrelation present among the observations in a continuous record of air quality has significant implications with respect to the estimation of the mean value of air pollutant concentration and to the determination of data collection requirements for estimating air quality levels. Based on the characterization of the form of the autocorrelation function, equations are derived for the equivalent number of independent observations in a sample of given size. An expression for the variance of the sample mean of an autocorrelated series is derived; and equations to estimate data collection requirements for a sequence of air quality measurements are developed. These equations yield both the number of equivalent independent measurements and the number of measurements in a continuous record needed to estimate the mean value at given levels of precision and significance. The derived results are applied to observed ambient carbon monoxide measurements.

0099 Dispersion in Mountainous Natural Stream

Hasa Bajraktarevic-Dobran, (Independent Research Engr., Inst. of Water Resources Engrg., Civ. Engrg. Faculty, Univ. of Sarajevo, Sarajevo, Yugoslavia)

Journal of the Environmental Engineering Division, Vol. 108, No. EE3, June, 1982, pp. 502-514

A chronological and critical review is made of existing methods for predicting coefficients and the area of the longitudinal turbulent dispersion, which were chosen by their importance and applicability to open natural streams. For natural mountain streams, a suitable method for calculating dispersion coefficient on the basis of flow parameters is accepted. A new method is proposed for determining characteristic mixing lengths l'_{sr} for nonuniform winding streams. There is a description of investigations carried out in a mountain, expressly nonuniform flow, which were used to check the given assumptions.

0100 Modeling Toxicity in Methane Fermentation Systems

Gene F. Parkin, Assoc. Member, ASCE, (Asst. Prof. of Civ. Engrg., Drexel Univ., Philadelphia, Pa.) and **Richard E. Speece**, Member, ASCE, (Betz Prof. of Environmental Engrg., Drexel Univ., Philadelphia, Pa.)

Journal of the Environmental Engineering Division, Vol. 108, No. EE3, June, 1982, pp. 515-531

Wastewater treatment using anaerobic methane fermentation offers several significant advantages over aerobic methods, but is considered unreliable when treating wastewaters containing toxicants. Three models describing toxicity phenomena are given to help engineers develop more rational design and operational strategies. The first model describes the recovery pattern of methane fermentation systems from slug addition of toxicants as a function of toxicant type and concentration, and time. This model allows for prediction of periods of zero gas production and threshold toxicant concentrations. The second two models incorporate the effect of toxicity on the fundamental Monod-type expressions. Unsteady-state behavior is effectively described by both models and the importance of providing proper biological solids retention time is emphasized. The primary effect of toxicant addition is to alter process kinetics by temporarily or permanently increasing bacterial washout time. Therefore, to minimize the severity of both transient and chronic toxicity, a sufficiently large solids retention time is required.

0101 Design Approach to Dual-Power Aerated Lagoons

Linvil G. Rich, Fellow, ASCE, (Prof., Coll. of Engrg., Dept. of Environmental Systems Engrg., Clemson Univ., Clemson, S.C. 29631)

Journal of the Environmental Engineering Division, Vol. 108, No. EE3, June, 1982, pp. 532-548

Completely- and partially-suspended lagoon cells are defined and their use in dual-power level, multicellular systems is discussed. A procedure is presented for the design of such systems on a rational basis using biokinetic coefficient data. Factors relating to the performance of these systems with respect to soluble BOD5 and suspended solids in the effluent are also discussed. Special attention is paid to cell size and number, power requirements both for solids suspension and oxygen demand, benthal solids decay, and sludge accumulation. A design example is given. It is concluded that further research and performance data derived from systems designed by the procedure are needed for a better evaluation of the parameters used and to test the assumptions made in applying the procedure.

0102 Performance of Modified Chlorine Contact Chamber

Frederick L. Hart, (Assoc. Prof. of Environmental Engrg., Dept. of Civ. Engrg., Worcester Polytechnic Inst., Worcester, Mass.) and **Zinovios Vogiatzis**, (Grad. Student, Dept. of Civ. Engrg., Worcester Polytechnic Inst., Worcester, Mass.)

Journal of the Environmental Engineering Division, Vol. 108, No. EE3, June, 1982, pp. 549-561

A field study on the effectiveness of perforated baffles in an existing cross-baffled serpentine flow chlorine contact chamber is reported. The object of the method studied was to increase the hydraulic efficiency and lower the required chlorine dose. A cross-baffled serpentine flow chamber was chosen because it is a common design. Tracer curves and disinfection efficiency were monitored on a modified and unmodified chlorine contact chamber simultaneously operating under equal conditions. Data generated from these field tests and batch reactor experiments were used to determine the efficiency increase achieved from these baffles, and to project the expected benefit of these modifications as an interim solution at municipal treatment plants.

0103 Managing Potomac Water Quality: Evolving Approaches

Austan S. Librach, (Dir., Metropolitan Washington Council of Government's, Dept. of Environmental Programs, 1875 Eye St., NW, Washington, D.C. 20006), **Cameron Wiegand**, (Chf., Metropolitan Washington Council of Government's, Water Programs, Dept. of Environmental Programs, 1875 Eye St., NW, Washington, D.C. 20006), **Stuart Freudberg**, (Water Resources Engr., Metropolitan Washington Council of Government's, Dept. of Environmental Programs, 1875 Eye St., NW, Washington, D.C. 20006), **Michael Sullivan**, (Water Resources Engr., Metropolitan Washington Council of Government's, Dept. of Environmental Programs, 1875 Eye St., NW, Washington, D.C. 20006) and **Robert N. Magill**, (Environmental Planner, Metropolitan Washington Council of Government's, Dept. of Environmental Programs, 1875 Eye St., NW, Washington, D.C. 20006)

Journal of the Environmental Engineering Division, Vol. 108, No. EE3, June, 1982, pp. 562-577

While an ambitious point source control program has yielded measurable improvements in water quality of the Upper Potomac River Estuary, the degree to which additionally strict point source controls are needed is the subject of current investigations by Washington, D.C. metropolitan area local jurisdictions. Indications are strong that eutrophication trends and periodic events of low dissolved oxygen will continue in the Estuary despite public expenditures for advanced wastewater treatment. Contributing to poor water quality conditions are significant uncontrolled loadings of oxygen demanding material and nutrients estimated to come from natural sources, unregulated agricultural activities, urban stormwater and other nonpoint pollution sources generated locally and upstream of the Washington region. Area jurisdictions are beginning to reappraise the region's water quality control program with the intent of developing a more comprehensive program that considers pollutant loadings from all sources. Investigations will involve an assessment of the costs of achieving desired water quality levels, and a study of possible mixes of point source, nonpoint source and combined sewer overflow controls. Studies will involve use of refined prediction models of the Estuary which employ new information on the magnitude and fluctuation of pollution loadings.

0104 Rotating Biological Contactor System Design

Yeun C. Wu, Member, ASCE, (Assoc. Prof., 944 Benedum Hall, Dept. of Civ. Engrg., Univ. of Pittsburgh, Pittsburgh, Pa. 15261) and **Ed D. Smith**, Member, ASCE, (Environmental Engr., Construction Engrg. Research Lab., Dept. of the Army, Champaign, Ill. 61820)

Journal of the Environmental Engineering Division, Vol. 108, No. EE3, June, 1982, pp. 578-588

Practical application of Wu's model for the prediction of soluble BOD removal in RBC systems is discussed extensively for various operating conditions. The model was tested by the use of more than 80 data sets obtained from the operation of six full-scale RBC plants. Also, nomograms were constructed to demonstrate a method for selecting the design surface hydraulic loadings. More importantly, the hydraulic loading rates predicted/calculated on the basis of the criteria employed for four full-scale RBC plant designs were compared with the actual design hydraulic loadings. No significant difference between the predicted and the designed hydraulic loading was found.

0105 Modeling and Interpreting Oxygen Transfer Data

Linfield C. Brown, (Dept. of Civ. Engrg., Tufts Univ., Medford, Mass. 02155) and **C. Robert Baillod**, Member, ASCE, (Dept. of Civ. Engrg., Michigan Technological Univ., Houghton, Mich. 49931)

Journal of the Environmental Engineering Division, Vol. 108, No. EE4, August, 1982, pp. 607-628

Recommendations for modeling, parameter estimation, and experimental design in unsteady state clean water aeration tests are presented. The basic model for both surface and subsurface aeration systems is the exponential form of the first order mass transfer equation. A nonlinear least squares regression procedure is recommended for estimating the oxygen transfer coefficient and equilibrium saturation concentration. Techniques for selecting the number and spacing of data values and for assessing model adequacy are suggested. Methods for applying clean water test results to field water conditions are described and an example problem is presented.

0106 Temperature Effects: Methane Generation from Landfill Samples

Kenneth E. Hartz, (Assoc. Prof., Dept. of Civ. and Environmental Engrg., Washington State Univ., Pullman, Wash. 99164), **Robert E. Klink**, (Project Engr., Donahue and Assocs., Sheboygan, Wisc. 53081) and **Robert K. Ham**, (Prof., Dept. of Civ. and Environmental Engrg., Univ. of Wisconsin, Madison, Wisc. 53706)

Journal of the Environmental Engineering Division, Vol. 108, No. EE4, August, 1982, pp. 629-638

The objective of this investigation was to study the impact of temperature variations on the rate of methane generation from solid waste. The temperatures investigated ranged from 21 °C to 48 °C. Two approaches were applied. These were short term residence at seven different temperatures and intermediate term residence at two different temperatures. For the short term studies, samples were obtained from the Freshkills landfill (M.Y.) and the Operating Industries landfill (Calif.). Three samples were used in the intermediate term studies, and were from Palos Verdes landfill and Menlo Park landfill, both in California. From the short term results, energy of activation values of 22.4 kilo calories per mole to 23.7 kilo calories per mole were calculated. The intermediate term results produced values ranging from 18.7 to 21.8 kilo calories per mole. From the results it was concluded that some minor population shifts occurred with minor temperature changes but all of the energy of activation values were higher than any previous reportings. In addition, the temperature of 41°C was found to be the optimum for methane generation on a short term basis.

0107 Distribution of Toxic Substances in Rivers

Peter R. Jaffe, (Grad. Student, Research Assoc., Dept. of Civ. Engrg., Princeton Univ., Princeton, N.J.), **Frank L. Parker**, Member, ASCE, (Prof., Dept. of Civ. and Environmental Engrg., Vanderbilt Univ., Nashville, Tenn.) and **David J. Wilson**, (Prof., Dept. of Chemistry and Environmental Chemistry, Vanderbilt Univ., Nashville, Tenn.)

Journal of the Environmental Engineering Division, Vol. 108, No. EE4, August, 1982, pp. 639-649

A statistical technique was used to determine if hydrophobic organic substances have vertically a uniform distribution in a river. At a river polluted with chlorinated organic pesticides, several water samples were collected at different depths to perform an analysis of variance on a nested design. To check if the data had to be normalized for the analysis of variance and hypothesis testing, the probability distribution of a grab sample was studied. The lognormal distribution gave in most cases a better description of the grab sample distribution than the normal distribution. Water and bottom sediment samples were taken at several points of the polluted river reach. Although statistically no attenuation in the pesticide concentration over distance was observed, upper limits for a first order decay at which decay would have been observed are reported, as well as the observed partition coefficients. The pesticides appeared to be fractioned in three categories; adsorbed on suspended sediments, adsorbed on fine sediments, and in solution.

0108 Treatment of Saline Domestic Wastewater Using RBCs

Nancy E. Kinner, (Grad. Student, Dept. of Civ. Engrg., Univ. of New Hampshire, Durham, N.H.) and **Paul L. Bishop**, Member, ASCE, (Assoc. Prof., Dept. of Civ. Engrg., Univ. of New Hampshire, Durham, N.H.)

Journal of the Environmental Engineering Division, Vol. 108, No. EE4, August, 1982, pp. 650-664

The treatment of high salinity wastewaters from small island communities by use of rotating biological contactors was evaluated. Domestic waterwaters containing up to full strength seawater were utilized. The effects of salinity, loading rate and disk media composition on organic removal rate were examined. It was shown that RBCs can treat high salinity wastewaters to the same degree as non-saline wastewaters. The resulting effluent organic content was adequate for discharge into either marine or freshwater receiving waters. Mean COD removals of 61% and 64% were achieved using hydraulic loading rates of 0.04 and 0.08 m^3/m^2•d (1.0 and 2.0 gpd/sq ft), respectively. At higher loading rates, up to 0.33 m^3/m^2•d (8.0 gpd/sq ft), organic removal efficiency was reduced. No difference in treatment efficiency was observed using either plastic or Masonite^R as the disk media.

0109 Cost-Effective Pretreatment Programs

Lewis A. Rossman, Member, ASCE, (Operations Research Analyst, U.S. Environmental Protection Agency, Cincinnati, Ohio 45268)

Journal of the Environmental Engineering Division, Vol. 108, No. EE4, August, 1982, pp. 665-679

A method is developed for planning cost-effective control programs for toxic pollutants in municipal treatment systems. It chooses the levels of industrial pretreatment and type of municipal sewage treatment a city can use so that all applicable standards and criteria on sewer discharges, receiving water quality, and municipal sludge quality are met at minimum cost. The problem is formulated as a nonlinear 0-1 integer programming model. An implicit enumeration procedure is developed for its solution. The method is applied to control of five heavy metals in Kokomo, Ind. The economic impact of EPA's national pretreatment policies and alternative water quality goals is studied.

0110 Ozone and Weather Patterns in Japan and the U.S.

Richard A. Wadden, (Prof., Univ. of Illinois, School of Public Health, P.O. Box 6998, Chicago, Ill. 60680), **Yasushi Ogawa**, (Sr. Environmental Sci., Air Environment Div., National Inst. for Environmental Studies, Tsukuba, Ibaraki 300-21, Japan), **Michio Okuda**, (Former Div. Head, Air Environment Division, National Inst. for Environmental Studies, Tsukuba, Ibaraki 300-21, Japan) and **Kentaro Murano**, (Environmental Scientist, Air Environment Div., National Inst. for Environmental Studies, Tsukuba, Ibaraki 300-21, Japan)

Journal of the Environmental Engineering Division, Vol. 108, No. EE4, August, 1982, pp. 680-687

Air pollution and meteorological data were collected in April-June, 1979 for 49 consecutive days in the Tsukuba area about 60 km northeast of Tokyo. Hourly O_3, total oxidant, CO, NO and NO_2 were collected and the ozone-forming potential, a simple characterization of chemical reactivity, was measured on a regular basis. In addition, 5 ½ months (April-September, 1976) of hourly pollution and meteorological data for 4 stations in the immediate Tokyo area were evaluated with respect to synoptic weather patterns. Ozone levels were strongly correlated with temperature ($r = 0.66$) but not with NO, NO_2, CO, UV radiation, wind speed or direction. Analysis on the basis of synoptic weather patterns revealed maximum values of maximum hourly NO_2, ozone-forming potential, ozone and oxidant occurring 3, 3, 4 and 4 days, respectively, after passage of a front. This trend was repeated for NO_2 and oxidant for the Tokyo Fujimi station in 1976 and has also been observed in the U.S. The analysis suggests that air mass age is an important factor for ozone development.

0111 Predicting Phosphorus Levels in a Future Reservoir

Steven W. Effler, (Engr., Environmental Engrg. and Adjunct Prof., Dept. of Civ. Engrg., Syracuse Univ., 149 Hinds Hall, Syracuse, N.Y. 13210), **Stephen D. Field**, (Asst. Prof., Dept. of Civ. Engrg., Louisiana State Univ., Baton Rouge, La.), **Joseph V. DePinto**, (Assoc. Prof., Dept. of Civ. and Environmental Engrg., Clarkson Coll., Potsdam, N.Y. 13676), **Emmet M. Owens, Jr.**, Assoc. Member, ASCE, (Asst. Engr., Stearns & Wheler Consulting Engrs., Cazenovia, N.Y. 13035), **John S. Dobi**, (Study Technical Mgr., Public Service Electric and Gas Co., Newark, N.J. 07101) and **Mihai A. Preda**, (Lead Engr., Public Service Electric and Gas Co., Newark, N.J. 07101)

Journal of the Environmental Engineering Division, Vol. 108, No. EE4, August, 1982, pp. 688-705

The concentration of total phosphorus at spring turnover in a proposed flow-augmentation reservoir is predicted, with a steady-state and two dynamic models, for a wide, but representative, range of reservoir operation. Simulations were performed utilizing hypothetical data sets which correspond to the hydraulic regime of the Delaware River from 1929 through 1970. The level of total phosphorus to be expected in the reservoir is shown to be highly dependent on its hydraulic operation. Phosphorus enrichment of the reservoir watercolumn will be coupled to the reservoir's primary function of augmentation of Delaware River flows during low runoff periods. The best estimate simulations of the dynamic system with the dynamic models indicate the total phosphorus concentrations at spring turnover will equal or exceed 0.020 mg/l continuously; levels which are generally associated with a eutrophic condition. There are some indications that mesotrophy may prevail for years when augmentation releases to the Delaware River are not made.

0112 Stormwater Pollution Control: Structural Measures

E. John Finnemore, Member, ASCE, (Assoc. Prof., Dept. of Civ. Engrg., Univ. of Santa Clara, Santa Clara, Calif.)

Journal of the Environmental Engineering Division, Vol. 108, No. EE4, August, 1982, pp. 706-721

Combinations of storage and wet-weather treatment, integrated with dry-weather treatment facilities and nonstructural control measures on an area-wide basis, are found to provide the greatest and most cost-effective control of storm-water pollution. Case studies of three very promising, constructed and operating structural measures for controlling combined sewer overflow pollution demonstrate these concepts. Information on the design, operation, performance and costs of these facilities is provided and compared to guide future planners and designers. In Seattle, Washington, computerized control of in-line storage offers operational flexibility and inexpensive storage. At Saginaw, Michigan, off-line storage/treatment integrated with in-line storage and dry-weather treatment has achieved high performance at moderate costs. At Mount Clemens, Michigan, physical and biological wet-weather treatment processes promise very high treatment levels for combined sewage, if needed.

0113 Performance of some Empirical Phosphorus Models

Dintie S. Mahamah, (Grad. Student, Dept. of Civ. and Environmental Engrg., Washington State Univ., Pullman, Wash. 99164) and **Surinder K. Bhagat**, Member, ASCE, (Prof. and Chmn., Dept. of Civ. and Environmental Engrg., Washington State Univ., Pullman, Wash. 99164)

Journal of the Environmental Engineering Division, Vol. 108, No. EE4, August, 1982, pp. 722-729

Applicability of four empirical phosphorus models to 22 western lakes is investigated. The models by Jones and Bachman, and Walker show significant agreement with observed phosphorus concentrations as indicated by the Spearman correlation of 0.809 and 0.812 respectively. The values calculated by the other two models by Dillon and Kirchner, show poor correlation with observed data. Vollenweider type classification charts are also derived for the four models. Although all the models show some degree of disagreement with the EPA National Eutrophication Survey trophic classifications, the models by Walker, and Jones and Bachman can be used for predicting a first cut estimate of total phosphorus for lakes low in nutrients.

0114 Maximizing Algal Yield in High-Rate Oxidation Ponds

Gideon Oron, (Research Engr., Inst. for Desert Research, Ben-Gurion Univ. of the Negev, Kiryat Sde Boker, Israel 84990) and **Gedaliah Shelef**, (Prof. of Environmental and Water Resources Engrg., Technion-Israel Inst. of Tech., Haifa, Israel 32000)

Journal of the Environmental Engineering Division, Vol. 108, No. EE4, August, 1982, pp. 730-738

One of the major problems in operating high-rate oxidation ponds (HROP) so as to give the maximal yield of algae is the determination of the best combination of retention time and effluent depth under varying environmental conditions. The criteria for choosing these variables were determined by applying the results previously obtained from pilot and field ponds operated in Israel and via computer simulation techniques. Guidelines for operating the HROP under different climatic conditions are also discussed.

0115 Localized Destratification at Lake Texoma

Kerry M. Robinson, (Research Assoc., Agricultural Engrg. Dept., Oklahoma State Univ., Stillwater, Okla.), **James E. Garton**, (Prof., Agricultural Engrg. Dept., Oklahoma State Univ., Stillwater, Okla.) and **Richard E. Punnett**, (Hydr. Engr., U.S. Army Corps of Engrs., Huntington, W.Va.)

Journal of the Environmental Engineering Division, Vol. 108, No. EE4, August, 1982, pp. 739-749

An axial flow Garton Pump was used to improve release water quality from Lake Texoma during thermal stratification. An 8-ft (2.44m) diam propeller pump, mounted on a floating support platform, was operated directly over the intake structure. The propeller, suspended 5.6 ft (1.7m) below the water surface, pumped high quality epilimnion water down to the outlet. Water quality parameters were monitored with and without pumping at several release rates to determine pump effectiveness. A maximum dilution factor of 32.5% was obtained at a pumping to release rate ratio of 0.48. The maximum observed improvements for turbidity, sulfides, manganese, total iron, phosphorous, and ammonia nitrogen were 83, 68, 49, 21, 53, and 37% respectively.

0116 Temperature/Sediment Model for a Shallow Lake

H. G. Stefan, (Prof., Dept. of Civ. and Mineral Engrg., St. Anthony Falls Hydr. Lab., Univ. of Minnesota, Minneapolis, Minn. 55455), **S. Dhamotharan**, Member, ASCE, (Mgr., Hydrological Analysis, Morrison-Knudsen Co., Boise, Idaho 83729) and **F. R. Schiebe**, Member, ASCE, (Research Hydr. Engr., U.S. Dept. of Agri., Sci. and Education Administration, Southern Plains Watershed and Water Quality Lab., Durant, Okla. 74701)

Journal of the Environmental Engineering Division, Vol. 108, No. EE4, August, 1982, pp. 750-765

The effects of turbid water inflow into a shallow reservoir (Lake Chicot, Ark.) are described, analyzed, and modeled. Relationships between turbidity, solar radiation reflection and solar radiation attenuation in the water are given for a reservoir with suspended sediment (clay) content up to 1,000 ppm. The stratification dynamics and the formation of underflows and overflows are illustrated using field data. A dynamic computer model (RESQUAL) for the prediction of stratification and water quality in a shallow reservoir is developed and described. Dynamic effects are simulated on a daily timescale and include mixing by wind and convection, inflow, and outflow. Predicted and measured distributions of temperature and suspended sediments are compared.

0117 Activated Carbon: Sorption of DOC from Wastewater

Frederick S. Cannon, Member, ASCE, (Engr., Brown and Caldwell, Consulting Engr., Inc., Pasadena, Calif. 91109) and **Paul V. Roberts**, Member, ASCE, (Adjunct Prof., Dept. of Civ. Engrg., Stanford Univ., Stanford, Calif. 94305)

Journal of the Environmental Engineering Division, Vol. 108, No. EE4, August, 1982, pp. 766-783

Activated carbons were prepared from a lignocellulosic material (prune pits) by pyrolysis to 900°C, and then activation in CO_2 at 900°C for activation times ranging from 15 to 60 minutes. These were compared with each other and to Filtrasorb 400 and Aquanuchar A with regard to specific surface area (N_2, 77K), pore volume distribution, and the rate and extent of adsorption of dissolved organic carbon (DOC) from filtered wastewater secondary effluent. It was found that the magnitudes of the above variables increased with increased time of activation, and that the magnitudes were greater for extensively activated fruit pits than for Filtrasorb 400 and Aquanuchar A. Moreover, the uptake of DOC onto these powdered activated carbons in batch reactors behaved according to a model based on linear equilibrium and rate controlled by pore diffusion. The apparent diffusivities estimated by this model were in the range of 1×10^{-10} to 3×10^{-10} m^2/s, which conform to expected values.

0118 Scrubbing Odors from Wastewater Treatment

Abul Basher M. Shahalam, Member, ASCE, (Prof., Civ. Engrg. Dept., Yarmouk Univ., Irbid, Jordan)

Journal of the Environmental Engineering Division, Vol. 108, No. EE4, August, 1982, pp. 785-799

A pilot scrubber system using sodium hypochlorite and sodium hydroxide solutions was tested for its effectiveness and the optimal operation conditions when used for removing odors from the air of a sewage treatment plant. Test equipments such as scrubbers and olfactometers were used for the collection of data. The data were utilized to determine the optimum concentrations of chemicals necessary for the most effective reduction of the sewage odor. From the data, the odor intensity was found to be a function of sewage temperature and ORP (Oxidation-Reduction Potential). The concentrations of chemicals in the treated air were determined by performing GC/MS analyses on the scrubber effluent air.

0119 Equal Volume Stack Sampling Methodology

Philip D. Baumann, Member, ASCE, (Research Engr., Texas Utilities Services Inc., 2001 Bryan Tower, Dallas, Tex. 75201), **Jerry W. Crowder**, Member, ASCE, (Asst. Prof. of Air Resources Engrg., Graduate Pgram of Environmental Sciences, Univ. of Texas at Dallas, P.O. Box 688, Mail Station BE 2.2, Richardson, Tex.) and **Ervin J. Fenyves**, (Prof. of Energy, Grad. Program of Environmental Sciences, Univ. of Texas at Dallas, P.O. Box 688, Mail Station BE 2.2, Richardson, Tex.)

Journal of the Environmental Engineering Division, Vol. 108, No. EE4, August, 1982, pp. 803-807

An air pollution sampling methodology, involving varying point locations determined to be representative of equal annular volumes, is developed for the case of uniform flow profiles. Results from the proposed method are compared to results from equal area methodologies. It is concluded that the equal area methodology is most representative under very high turbulent conditions and least representative as the transition region is approached.

0120 Characteristics of Fill-and-Draw Reactor

Reidar Zapf-Gilje, (Design Engr., SIGMA Resource Consultants Ltd., Vancouver, B.C., Canada) and **Donald S. Mavinic**, (Assoc. Prof., Dept. of Civ. Engrg., Univ. of British Columbia, Vancouver, B.C., Canada)

Journal of the Environmental Engineering Division, Vol. 108, No. EE4, August, 1982, pp. 808-812

Characteristics of a fill-and-draw (biological batch) reactor of a previously reported landfill leachate biotreatment study are examined. The operation of the reactor and the monitoring of the leachate are described. The settling characteristics of the mixed-liqour were evaluated during the study. Results and analysis of the study are presented.

0121 Cost Estimating for Conventional Water Treatment

Robert M. Clark, Member, ASCE, (Engrg. Systems Analyst, Drinking Water Research Div., Municipal Environmental Research Lab., U.S. Environmental Protection Agency, Cincinnati, Ohio 45268)

Journal of the Environmental Engineering Division, Vol. 108, No. EE5, October, 1982, pp. 819-834

The Safe Drinking Water Act of 1974 emphasizes the need to take costs into consideration when promulgating regulations. This paper presents cost equations that can be used for water treatment unit process cost estimating. The need for pre-design estimates is discussed, and the factors affecting the reliability of cost estimates is examined. These factors include availability of basic data, stage of development, definition of scope, the time expended on analysis, and experience of the analyst. Data for these equations were derived from a carefully designed study to establish water supply unit process costs on a consistent and understandable basis. The basis for derivation of these curves is presented. Examples of the application of these equations are developed based on a standard layout for a water supply treatment plant and assumed standard values for the input factors for the cost equations.

0122 Stormwater Pollution Control: Best Management Practices

E. John Finnemore, Member, ASCE, (Assoc. Prof., Dept. of Civ. Engrg., Univ. of Santa Clara, Santa Clara, Calif.)

Journal of the Environmental Engineering Division, Vol. 108, No. EE5, October, 1982, pp. 835-851

Nonstructural and elementary structural stormwater pollution control measures, known as best management practices (BMPs), are found capable of excellent urban runoff pollution control. Besides costing less than structural measures, which to date have received much attention, BMPs often provide multiple benefits. Case studies describe four successful and very

Promising BMPs. Their approaches, implementation, and resulting performance, costs, and impacts are provided for future planners and designers. In Bellevue, Washington, an innovative stormwater utility has effectively administered Bellevue's stormwater management program. In a regional approach at Montgomery County, Maryland, many source controls have minimized erosion and removed most pollution from urban runoff at moderate costs. Near Lake Tahoe, in California, planning for stormwater management aided by erosion controls has prevented tremendous increases in sediment yields. At San Jose, California, extensive studies revealed that improved street cleaning will effectively control total solids and heavy metals in runoff.

0123 Lysimeters for Simulating Sanitary Landfills

Hans-Jurgen Collins, (Prof., Leichtweiss-Inst. for Water Research, Technical Univ., Braunschweig, Federal Republic of Germany) and **Peter Spillmann**, (Research Asst., Leichtweiss-Inst. for Water Research, Technical Univ., Braunschweig, Federal Republic of Germany)

Journal of the Environmental Engineering Division, Vol. 108, No. EE5, October, 1982, pp. 852-863

An interdisciplinary team is investigating the various decomposing processes inside sanitary landfills. Eight different ways of operation of sanitary landfills are simulated by special lysimeters. These lysimeters consist of flexible walls, which allow settlements similar to that of nature. Thermal insulation of the lysimeters protect the biological and chemical processes against outside temperatures. The leachate is collected separately from different compartments in the base, and samples of seepage are drawn from different levels. The lysimeters allow taking samples of refuse or the inserting of material at different places and levels. The density of the refuse in the lysimeter, filled according to today's practice, is in this range; while mixing of the refuse or refuse and sludge enlarges the density considerably. After aerobic decomposing, sieving and compacting the required deposition volume was one-third compared to the present common practice. It was shown, by five years of observation, that leachate occurs before the material was saturated. As a consequence thereof and by means of interpretation of the accumulated outflow of the lysimeter one can state that the quantity of outflow will increase after the operation of a sanitary landfill has ended.

0124 Lake Photoplankton Model with Destratification

John S. Gulliver, Member, ASCE, (Asst. Prof., St. Anthony Falls Hydr. Lab., Dept. of Civ. and Mineral Engrg., Univ. of Minnesota, Minneapolis, Minn. 55455) and **Heinz G. Stefan**, Member, ASCE, (Prof. and Assoc. Dir., St. Anthony Falls Hydr. Lab., Dept. of Civ. and Mineral Engrg., Univ. of Minnesota, Minneapolis, Minn. 55455)

Journal of the Environmental Engineering Division, Vol. 108, No. EE5, October, 1982, pp. 864-882

The phytoplankton mass and available phosphorus during a destratification experiment, at Lake Calhoun, Minnesota, in 1972, are simulated by a numerical, multilayered, unsteady numerical model. A timescale of one day is used. The effects of surface mixed layer dynamics, PO_4-P inflows from surface runoff, recycling of PO_4-P from the sediments available light, temperature, and mixing in the hypolimnion on phytoplankton productivity are analyzed. The model is used to hindcast the events of 1972, and explore the effects of other lake management alternatives such as the reduction of PO_4-P from runoff, the elimination of phosphorus recycling, and the maintenance of an artifical deepened surface layer.

0125 Interaction of Hydrolyzed Al and Polyelectrolyte Coagulants

Raymond D. Letterman, Assoc. Member, ASCE, (Assoc. Prof., Dept. of Civ. Engrg., Syracuse Univ., Syracuse, N.Y.) and **Prasit Sricharoenchaikit**, (Grad. Student, Dept. of Civ. Engrg., Syracuse Univ., Syracuse, N.Y.)

Journal of the Environmental Engineering Division, Vol. 108, No. EE5, October, 1982, pp. 883-899

A laboratory jar test study was conducted to determine the mode of action of commercial cationic polyelectrolytes used in combination with aluminum salt coagulants. Electrophoretic mobility and residual turbidity data were collected as a function of coagulant dosage, solution chemistry, and the particle size distribution and concentrations of the silica particle suspension. Two general types of results were obtained depending on the tendency of the hydrolyzed aluminum to form a flocculent precipitate. This tendency is a function of the solution chemistry, including pH, concentrations of adsorbing anions such as sulfate and the ionic strength. With dilute solutions and low pH the aluminum hydrolysis products tend to be stable and under these conditions the cationic polyelectrolyte and the adsorbed hydrolyzed Al have similar effects on particle stability. Maximum turbidity removal is observed near an electrophoretic mobility of zero. A simple trade-off relationship exists between the amounts of Al and polyelectrolyte required to react this point. When the aluminum hydrolysis products are flocculent a coating of aluminum hydroxide precipitate is apparently formed on the particles which, as the aluminum concentration is increased, eventually controls particle stability and the efficiency of flocculation.

0126 Anaerobic-Filter Pretreatment Kinetics

Bruce E. Rittmann, Assoc. Member, ASCE, (Asst. Prof., Dept. of Civ. Engrg., Univ. of Illinois at Urbana-Champaign, Urbana, Ill.), **Charles E. Strubler**, (Research Asst., Dept. of Civ. Engrg., Univ. of Illinois at Urbana-Champaign, Urbana, Ill. 618O1) and **Thomas Ruzicka**, (Asst. Plant Mgr., Meadow Gold Milk, Champaign, Ill. 618020)

Journal of the Environmental Engineering Division, Vol. 108, No. EE5, October, 1982, pp. 900-912

Pilot studies with synthetic and actual dairy wastewater showed that anaerobic-filter pretreatment at ambient temperature (20-24°C) was capable of reducing total BOD_5 by up to 96%. Analysis of the reactor kinetics showed that the reactions were complete in about two days detention time, which corresponded to a volumetric loading of 1.7 kg COD/m^3-day. Because of declining marginal removals with increased detention time, simple kinetic expressions were not appropriate representations of the reactor performance. Fermentative reactions dominated during the first 0.75 days of treatment, while methanogenic reactions dominated thereafter. Although effluent quality deteriorated, the anaerobic filters could handle a severe shock load as long as sufficient alkalinity had been added to keep the pH above 6.8 throughout the reactor.

0127 Adsorption Isotherm Parameter Estimation

Michael W. Sweeney, (Process Control Engr., Liquid Waste Div., Dept. of Public Works, Indianapolis, Ind. 46221), **William A. Melville**, (David Ross Fellow, Environmental Engrg., School of Civ. Engrg., Purdue Univ., West Lafayette, Ind. 47907), **Boris Trgovcich**, (Water Resources Control Engr., California State Water Resources Control Board, Sacramento, Calif.) and **C. P. Leslie Grady, Jr.**, (Prof., Dept. of Environmental Systems Engrg., 501 Rhodes Engrg. Research Center, Clemson Univ., Clemson, S.C. 29631)

Journal of the Environmental Engineering Division, Vol. 108, No. EE5, October, 1982, pp. 913-922

Determination of the parameter values for the Langmuir and Freundlich adsorption isotherm equations for water and wastewater treatment samples often requires estimation of the concentration of nonadsorbable material. Present techniques for estimating the concentration of that material can introduce bias into the data. Furthermore, linearization of the isotherm equations usually leads to application of least-square analysis to variables other than the true dependent and independent variables. A nonlinear programming technique is presented which allows simultaneous evaluation of the concentration of nonadsorbable material and the isotherm parameters. The technique is based upon the flexible polyhedron concept and achieves minimiation of the sum of squares error term associated with the solute concentration when expressed as a function of the adsorbent concentration.

0128 Verification of Water Quality Models

Robert V. Thomann, Member, ASCE, (Prof., Environmental Engrg. & Science Program, Manhattan Coll., Bronx, N.Y.)

Journal of the Environmental Engineering Division, Vol. 108, No. EE5, October, 1982, pp. 923-940

The basic concepts of water quality models are reviewed and the need to recognize calibration and verification of models with observed data is stressed. Post auditing of models after environmental control procedures are implemented is necessary to determine true model prediction ability. The direct questions in the decision-making arena related to model credibility are examined by several quantitative measures of model performance: regression analysis between observed and computed values, relative and root mean square errors and comparison of means. An analysis of 15 DO water quality models indicated median relative error of 10%. For a eutrophication model of Lake Ontario, application of quantitative verification measures indicated the credibility of the calculations decreased as the model was applied to smaller scale phenomena. Inclusion of more complex kinetics reduced median relative error across five variables and over a 10 year modeling framework from 32% to 22%. It is suggested that quantitative measures of water quality model credibility be integrated into the modeling analysis whenever possible.

0129 Prediction of Sulfide Buildup in Sanitary Sewers

Karl E. Kienow, Assoc. Member, ASCE, (Project Engr., Buck Lewis Engrg., Inc., Tucson, Ariz.), **Richard D. Pomeroy**, Fellow, ASCE, (Principal Engr., James M. Montgomery Consulting Engrg., Pasadena, Calif.) and **Kenneth K. Kienow**, Member, ASCE, (Grp. Engr., ARCA Precast Products Grp., Corona, Calif.)

Journal of the Environmental Engineering Division, Vol. 108, No. EE5, October, 1982, pp. 941-956

A graphical solution is developed for determining sulfide buildup in gravity sanitary sewer pipe. The method utilizes the Pomeroy-Parkhurst equations for sulfide buildup first published in 1977. The Pomeroy-Parkhurst equations are based on studies conducted over the past 20 years in the Los Angeles Sanitation District's system. An extensive bibliography on sulfide generation, effects, and case histories is included.

0130 Decarbonation of Limestone-Treated Mine Drainage

Frank Pearson, Member, ASCE, (Asst. Dir., Sanitary Engrg. and Environmental Health Research Lab., Univ. of California, Berkeley, Calif.), **Stephen McBride**, Member, ASCE, (Project Engr., GEO-Technical Servicms, Inc., Harrisburg, Pa.) and **Morman Batcheler**, Member, ASCE, (Project Monitor, Pennsylvania Dept. of Environmental Resources, Harrisburg, Pa.)

Journal of the Environmental Engineering Division, Vol. 108, No. EE5, October, 1982, pp. 957-972

Crushed limestone is used in various processes to neutralize acid mine drainage in a 2.5 mgd (9500 m^3/day) facility near Hazelton, Pennsylvania. A 0.5 mgd (1900 m^3/day) stream was treated by two limestone packed tumbling drums in series to increase pH from 3.7 to 5.4, followed by a lagoon for carbon dioxide removal that further increased pH to 6.1, to meet the pH 6 effluent quality target. Decarbonation process theory and design procedures are proposed. Graphs show pH increase versus carbon dioxide removal and carbon dioxide removal versus time of treatment and the overall film transfer coefficient. Nomographs also show how carbon dioxide removal depends on air-to-water ratio, air-water contact time, bubble or droplet diameter (for diffused air and spray decarbonation processes respectively), and the turbulence-dependent film transfer coefficient. Use of these nomographs is illustrated by calculation of the film transfer coefficient from experimental data.

0131 Symmetrical FFT for Migrating Waves

Joe C. Willis, Member, ASCE, (Research Hydr. Engr., U.S. Dept. of Agriculture Sedimentation Lab., Sci. and Education Administration, Oxford, Miss. 38655)

Journal of the Engineering Mechanics Division, Vol. 108, No. EM5, October, 1982, pp. 963-967

A Fast Fourier Transform (FFT) technique is developed to analyze migrating waves for phase relationships as well as spectra. The problem of phase leakage, which arises through the use of finite data records, is more severe than previous analyses have indicated. This severity prevents extraction of meaningful phase relationships by the usual application of asymmetrical, FFT algorithms. The symmetrical method combines the accuracy of the lengthy correlation transform method with the speed of the FFT and gives residual leakage corresponding to that of usual analyses. Adverse effects of the residual leakage can be virtually eliminated by applying a cosine taper to the data records.

0132 Volatilization of Organic Compounds from Streams

Ronald E. Rathbun, Member, ASCE, (Hydro., U. S. Geological Survey, National Space Tech. Lab., NSTL Station, Miss.) and **Doreen Y. Tai**, (Chemist, U. S. Geological Survey, National Space Tech. Lab., NSTL Station, Miss.)

Journal of the Environmental Engineering Division, Vol. 108, No. EE5, October, 1982, pp. 973-989

Mass-transfer coefficients for the volatilization of ethylene and propane were correlated with the hydraulic and geometric properties of seven streams, and predictive equations were developed. The equations were evaluated using a normalized root-mean-square error as the criterion of comparison. The two best equations were a two-variable equation containing the energy dissipated per unit mass per unit time and the average depth of flow and a three-variable equation containing the average velocity, the average depth of flow, and the slope of the stream. Procedures for adjusting the ethylene and propane coefficients for other organic compounds were evaluated. These procedures are based on molecular diffusivity, molecular diameter, or molecular weight. Because of limited data, none of these procedures have been extensively verified. Therefore, until additional data become available, it is suggested that the mass-transfer coefficient be assumed to be inversely proportional to the square root of the molecular weight.

0133 Considerations in Modeling Activated Sludge Process

Gerald E. Speitel, Jr., Assoc. Member, ASCE, (Grad. Student, Dept. of Environmental Sci. and Engrg., Univ. of North Carolina, Chapel Hill, N.C. 27514) and **Michael M. Hughes**, Assoc. Member, ASCE, (Project Engr., Malcolm Pirnie, Inc., 2 Corporate Park Drive, White Plains, N. Y. 10602)

Journal of the Environmental Engineering Division, Vol. 108, No. EE5, October, 1982, pp. 990-1002

Mathematical modeling of the activated sludge process is examined both from a unit process viewpoint and as an integral component of plant-wide mass balance models. Two modeling approaches are reviewed. The first is a standard approach now commonly used in practice. The second is an alternative approach which represents a modification of the first to directly account for influent suspended solids to the activated sludge process. Considerations are presented as to the proper application of both models in the prediction of waste sludge production and mixed liquor concentration. The two approaches also are compared for utility in wastewater treatment plant design. The alternative approach may be more flexible in accounting for sludge handling recycle loads. In many instances it also provides a more conservative estimate of waste sludge production.

0134 Uncertainty Analysis of Dissolved Oxygen Model

Ronald A. Chadderton, Member, ASCE, (Assoc. Prof., Civ. Engrg. Dept., Villanova Univ., Villanova, Pa. 19085), **Arthur C. Miller**, Member, ASCE, (Assoc. Prof., Civ. Engrg. Dept., The Pennsylvania State Univ., University Park, Pa.) and **Archie J. McDonnell**, Member, ASCE, (Prof., Civ. Engrg. Dept., The Pennsylvania State Univ., University Park, Pa. 16802)

Journal of the Environmental Engineering Division, Vol. 108, No. EE5, October, 1982, pp. 1003-1013

First-order uncertainty analysis was applied to 258 stream segments to determine the relative contributions to scatter, or uncertainty, in dissolved oxygen predictions by each of the parameters of the Streeter-Phelps equation. For very sluggish streams with reaeration rates less than about 0.1 per day (base e), the most important parameters for dissolved oxygen predictions were the initial deficit and the deoxygenation rate constant. For streams having reaeration coefficients greater than 0.1 per day the most important parameter corresponded to the predominant aspect of the oxygen balancing mechanism for the stream segment. On the depletion portion of the dissolved oxygen sag curve, where deoxygenation exceeds reaeration, the deoxygenation rate constant contributed most of the uncertainty in D. O. predictions. However, on the recovery portion of the D. O. sag curve, where reaeration exceeds deoxygenation, the reaeration rate constant contributed most of the uncertainty in D. O. predictions. Application of the uncertainty analysis to identify critical water quality parameters for a given stream would lead to more efficient data collection efforts.

0135 Linear Programming in Hazardous Waste Management

J. Jeffrey Peirce, (Asst. Prof., Dept. of Civ. Engrg., Duke Univ., Durham, N.C. 27706) and **Gordon M. Davidson**, (Environmental Analyst, JRB Assoc., McLean, Va.)

Journal of the Environmental Engineering Division, Vol. 108, No. EE5, October, 1982, pp. 1014-1026

Linear programming techniques are applied to investigate the relative costs of regional and state-wide hazardous waste management schemes. The focus is the identification of a cost effective configuration of transportation routes, transfer stations, processing facilities and secure long-term storage impoundments. Wastes generated in North Carolina are studied as a useful example of linear programming applications in general and options available within a given state in particular. The value of the techniques are highlighted, as are their limitations. The usefulness in developing relative costs of alternatives is stressed, particular in the ability of the techniques to conduct sensitivity analyses in a topic area where data may not be generally available. Suggestions are made for overcoming data shortcomings. In the case study, the options are seen to revolve around the state of North Carolina's expressed desire to locate one large centralized storage landfill. From a pure cost standpoint, other management facilities like transfer stations and incinerators appear to be precluded even with optimal routing to and from the facilities. From other viewpoints, including risk aversion to spills while the waste is in transit, the inclusion of these facilities in the state's program can be supported.

0136 Real Time Simulation of Conductivity in River-Lake Systems

Christopher G. Uchrin, Member, ASCE, (Asst. Prof., Dept. of Environmental Science, Cook Coll., Rutgers Univ., New Brunswick, N.J. 08903) and **Walter J. Weber, Jr.**, Member, ASCE, (Prof. of Environmental and Water Resources Engrg. and Chmn., Univ. Program in Water Resources, Univ. of Michigan, Ann Arbor, Mich. 48109)

Journal of the Environmental Engineering Division, Vol. 108, No. EE5, October, 1982, pp. 1027-1044

Conductivity is modeled in real time in a natural water system using the finite section method. Dispersion coefficients are estimated using methods available in the literature. Calibration is performed together with sensitivity analyses. Small scale effects are defined by a dye study. The resultant model is eventually used in the construction of system models describing the transport and fate of various pollutants in the Ford Lake - Melleville Lake (Mich.) System.

0137 Control of Water Pollution from Toxics: Southeast U. S.

David H. Howells, Fellow, ASCE, (Prof. Emeritus, North Carolina State Univ. and the Univ. of North Carolina at Chapel Hill, Chapel Hill, N.C.) and **Paul Zielinski**, Member, ASCE, (Prof. and Dir., Water Resources Research Inst., Clemson Univ., Clemson, S.C.)

Journal of the Environmental Engineering Division, Vol. 108, No. EE5, October, 1982, pp. 1045-1061

A detailed review of the programs of eight southeastern states for the control of water pollution from toxic wastes is presented. The review examines water quality standards, effluent limitations, National Pollution Discharge Elimination System (MPDES) permits for direct and indirect discharges, pretreatment programs, waste residuals management, monitoring, emergency response, and analytic capability. Existing programs fall far short of providing control of 129 priority pollutants. The problem is particularly acute for chemicals with long-term health effects. The state-of-the-art relative to causal relationships and degree of risk primitive is covered. It is concluded that the transition from the earlier era of pollution control to encompass toxic wastes will be very difficult, particularly with cutbacks in federal cost sharing and research.

0138 Computer Analysis of Settling Test Data

Paul M. Berthouex, Member, ASCE, (Prof. of Civ. & Environmental Engrg., Univ. of Wisconsin-Madison, Madison, Wisc.) and **David K. Stevens**, (Research Asst., Civ. & Environmental Engrg., Univ. of Wisconsin-Madison, Madison, Wisc.)

Journal of the Environmental Engineering Division, Vol. 108, No. EE5, October, 1982, pp. 1065-1069

A method of fitting a simple approximating function to quiescent settling test data is given that can serve for either discrete or flocculant settling. The method can be used with the data as either concentrations or as fractional removals. The method is illustrated using concentration data and the results are compared with values calculated by the traditional method.

0139 Current Municipal Sludge Utilization and Disposal

J. Jeffrey Peirce, (Asst, Prof., Civ. and Environmental Engrg. Dept., Duke Univ., Durham, N.C. 27706) and **Stephen Bailey**, (Assoc. Engr., Energy and Environment Div., Booz, Allen, and Hamilton, Inc., Bethesda, Md. 20014)

Journal of the Environmental Engineering Division, Vol. 108, No. EE5, October, 1982, pp. 1070-1073

Current sludge disposal practices are defined by a recently completed survey of 1,008 randomly selected municipal wastewater treatment facilities stratified by each of the 50 states. The size distribution of the facilities is analyzed, and the number of disposal technologies currently utilized at each facility is documented. Prior to this study, municipal sludge utilization and disposal practices are not well inventoried. The U. S. Environmental Protection Agency compiles data in biennial surveys, but the focus is projection of wastewater treatment needs by individual states. Consulting engineering firms keep internal audits on disposal practices, but the data is regional at best and limited to clients previously served by the individual firms. The study documented in this paper presents comprehensive data on current sludge disposal practices.

0140 Nitrification Rates in Activated Sludge

Joseph H. Sherrard, Member, ASCE, (Prof. of Civ. Engrg., Virginia Polytechnic Inst. and State Univ., Blacksburg, Va. 24061), **Deborah K. Manning**, Assoc. Member, ASCE, (Engr., James M. Montgomery Engrs., Inc., Reston, Va. 22091) and **Debra A. Smith**, Assoc. Member, ASCE, (Asst. Engr., Boyle Engrg. Co., Falls Church, Va. 22043)

Journal of the Environmental Engineering Division, Vol. 108, No. EE5, October, 1982, pp. 1074-1078

By using biokinetic and stoichiometric equations it was possible to theoretically develop insight into factors that affect nitrification rate in activated sludge systems. From this study it was shown that nitrifier fraction is not a constant and is dependent on mean cell residence time and the influent COD:TKN ratio, and the percentage of nitrifiers present in an activated sludge biomass does not have an effect on the rate of nitrification.

0141 Stabilization of Solid Waste in Landfills

Todd J. Bookter, Assoc. Member, ASCE, (Project Engr., Foth & Van Dyke and Assoc., Inc., Green Bay, Wisc. 54303) and **Robert K. Ham**, (Prof., Dept. of Civ. and Environmental Engrg., Univ. of Wisconsin, Madison, Wisc.)

Journal of the Environmental Engineering Division, Vol. 108, No. EE6, December, 1982, pp. 1089-1100

A method for the analysis of refuse stability has been developed and tested by analyzing refuse obtained directly from test lysimeters and actual landfills across the United States. Refuse samples from test lysimeters are analyzed to develop the procedures and to test the concepts involved. Results obtained from actual landfill situations are compared to those obtained from the test lysimeters of known refuse age in order to determine the degree of refuse stability in each landfill in relation to depth and location. Each landfill was analyzed as to how environmental conditions such as pH, moisture content, and remaining organic content are affected by location and depth. Also, the results obtained from landfills of different geographical locations, different climatic situations, and different configurations are compared. Because the ultimate use of a sanitary landfill site is of prime concern, the rate of refuse stabilization is important to know from the standpoints of future leachate production, methane gas generation, and differential settlement.

0142 Anaerobic Digestion of Waste Activated Sludge

James M. Gossett, Member, ASCE, (Asst. Prof., Environmental Engrg. Dept., Hollister Hall, Cornell Univ., Ithaca, N.Y. 14853) and **Roy L. Belser**, (Sr. Engr., Environmental Activities, General Motors Assembly Div., General Motors Corp., North Tarrytown, N.Y.)

Journal of the Environmental Engineering Division, Vol. 108, No. EE6, December, 1982, pp. 1101-1120

The purpose of this laboratory study was to evaluate, and model the effect of solids retention time in the activated sludge process (θ_c^{AS}) on the performance of anaerobic digesters fed the resulting waste sludges. Activated sludge systems were operated over a range of θ_c^{AS} values, with the waste sludges used as feed sources for batch and semi-continuous anaerobic digestion studies. Previous models for aerobic biodegradability of activated sludges were found to accurately predict ultimate anaerobic digestibility. However, conventional kinetic models of digestion over-estimated actual performance with this substrate. An activated-sludge anaerobic digestion model is proposed which includes a first-order rate mechanism for preliminary conversion of viable activated sludge biomass into available anaerobic substrate. Decreasing θ_c^{AS} results in increased sludge solids for digestion. However, the solids are of sufficiently greater digestibility that both the quantity of digested solids for ultimate disposal, as well as the cost of digestion, are virtually independent of θ_c^{AS}.

0143 Clarification Failure Control Strategies

Thomas M. Keinath, Member, ASCE, (Prof. and Head, Dept. of Environmental Systems Engrg., Clemson Univ., Clemson, S.C. 29631) and **Vincent D. Laquidara**, (Environmental Engr., Georgia Dept. of Natural Resources, Atlanta, Ga. 30334)

Journal of the Environmental Engineering Division, Vol. 108, No. EE6, December, 1982, pp. 1121-1127

Two solids inventory control strategies for prevention of clarification failure in the activated sludge process are considered. Recycle rate control is shown to be effective for a certain class of operational situations. This control algorithm, however, is not applicable to situations in which an activated sludge system is experiencing severe hydraulic surges. In such situations clarification failure can be effectively controlled only through the application of the step-feed strategy in which the location of the feed to the aeration basin is shifted downstream in a multipass basin.

0144 $CaCO_3$ Neutralization of Acidified Surface Waters

Charles T. Driscoll, (Asst. Prof., Dept. of Civ. Engrg., Syracuse Univ., Syracuse, N.Y. 13210), **Jeffrey R. White**, (Grad. Research Asst., Dept. of Civil Engrg., Syracuse Univ., Syracuse, N.Y. 13210), **Gary C. Schafran**, (Grad. Research Asst., Dept. of Civ. Engrg., Syracuse Univ., Syracuse, N.Y. 13210) and **John D. Rendall**, (Grad. Research Asst., Dept. of Civil Engrg., Syracuse University, Syracuse, N.Y. 13210)

Journal of the Environmental Engineering Division, Vol. 108, No. EE6, December, 1982, pp. 1128-1145

Calcium carbonate was utilized to neutralize an acidic lake and stream in the Adirondack Region of New York state. Theoretical calculations based on aqueous base neutralizing capacity underestimated the $CaCO_3$ dose required to achieve lake neutralization (pH 6.5) for a year. It would appear that lake sediments have a profound influence on base dose requirements. Calcium carbonate neutralization altered water column pH, calcium, acid neutralizing capacity, dissolved inorganic carbon, dissolved organic carbon, free fluoride, aluminum, manganese, zinc, and copper levels. Chemical equilibrium models appear to be a useful tool in predicting changes in aqueous aluminum chemistry. While calcium carbonate appears to be an effective means of accomplishing lake neutralization, there are potentially deleterious consequences associated with its use.

0145 Decomposition of Solid Waste in Test Lysimeters

Robert K. Ham, (Prof., Dept. of Civil and Environmental Engrg., Univ. of Wisconsin, Madison, Wisc. 53706) and **Todd J. Bookter**, Member, ASCE, (Engr., Foth and Van Dyke and Assoc., Green Bay, Wisc.)

Journal of the Environmental Engineering Division, Vol. 108, No. EE6, December, 1982, pp. 1147-1170

Eight test lysimeters containing 100-200 tons of refuse each were monitored for seven years for leachate and runoff rates, leachate and gas composition, refuse temperature, and surface settlement. The lysimeters were designed to indicate the effects of refuse shredding soil cover, and lift thickness on the monitored parameters. More runoff but less evapotranspiration was obtained with soil cover, with the result that leachate rates were approximately the same with or without soil cover. Shredded refuse produced higher peak concentrations of matter in leachate than unprocessed refuse when landfilled under similar conditions. The absence of soil cover resulted in leachate strengths declining rapidly after peak concentrations were produced, with the result that shredded refuse without cover produced less COD than either shredded or unprocessed refuse which was covered. Doubling the depth approximately doubled the leachate COD levels, but greatly increased the stabilization time required to low COD levels to be attained. Placing a second lift of refuse over a relatively stabilized first lift resulted in 75 to 99% reduction in COD produced by the second lift.

0146 Compression Effects in Batch Settling Tests

D. Christopher Dixon, (Sr. Lect., School of Chemical Engrg. and Industrial Chemistry, Univ. of New South Wales, Kensington, N.S.W. 2033, Australia)

Journal of the Environmental Engineering Division, Vol. 108, No. EE6, December, 1982, pp. 1171-1191

The effect of compression resistance on terminal settling velocity results obtained from batch tests has been studied by means of dimensional and numerical analyses, and comparison with experiment. It is concluded that the Talmage and Fitch (single test) method is unreliable when compression effects are significant, because it only gives a satisfactory approximation when the terminal settling velocity - dilution relation is convex downwards, but whether this applies or not to the suspension under test cannot be determined from the tests. To determine whether compression effects are significant, repeat tests using different initial suspension heights are necessary, at least at the smallest dilution for which data are required. When significant effects are present, repeat tests (at each dilution) allow the corresponding terminal settling velocity and compressive stress values to be obtained from the intercept and slope of a plot of initial settling velocity versus reciprocal initial height (the Michaels and Bolger extension of the Coe and Clevenger testing procedure).

0147 Pollutant Loading Model for Highway Runoff

Tai Wik Chui, Assoc. Member, ASCE, (Engr., Systems Architects Engrs., Inc., P. S., Seattle Wash.), **Brian W. Mar**, Member, ASCE, (Prof., Dept. of Civ. Engrg., Univ. of Washington Seattle, Wash.) and **Richard R. Horner**, Member, ASCE, (Research Asst. Prof., Dept. of Civ Engrg., Univ. of Washington, Seattle, Wash.)

Journal of the Environmental Engineering Division, Vol. 108, No. EE6, December, 1982, pp. 1193-1210

A stormwater runoff pollutant loading model has been developed on the basis of results from composite sampling of approximately 500 storms at nine locations in Washington State. One component of the model is an expression for total suspended solids (TSS) load in relation to traffic, runoff coefficient, and surrounding land use contributions. Other contaminants are estimated from TSS load using ratios derived from the data. These ratios may be taken as constants at any Washington State location for some pollutants or as linear functions of traffic for other pollutants. The model as presently constructed is capable of predicting loadings over a time span encompassing a number of storms (monthly or annually), but has little precision when applied to individual storms.

0148 Extraction of Priority Pollutants from Solids

D. Elizabeth Harrold, (Grad. Research Asst., Iowa State Univ., Ames, Iowa 50011) and **James C. Young**, Member, ASCE, (Prof. of Civ. Engrg., Univ. of Arkansas, Fayetteville, Arkansas)

Journal of the Environmental Engineering Division, Vol. 108, No. EE6, December, 1982, pp. 1211-1227

Tests have been conducted to determine the efficiency of and factors affecting extraction of semivolatile organic priority pollutants from solid samples using modified Soxhlet procedures. Test variables included time of solvent contact, particle size of solid matrix, and type of solvent. The results indicated that extraction efficiences varied substantially for different priority pollutants. The most important factor in obtaining maximum recovery seemed to be the contact time during extraction. Particle size had a minor bearing on extraction efficiency. Solvent losses contributed significant errors when using extraction times longer than 24 hours. When proper consideration is given to the extraction efficiencies and to other sources of error, the Soxhlet procedure is useful for providing estimates of the quantities of semivolatile priority pollutants in solids.

0149 Sediment Pollution Control in North Carolina

David H. Howells, Fellow, ASCE, (Prof. Emeritus, North Carolina State Univ., Raleigh, N.C., and University of North Carolina, Chapel Hill, N.C.) and **Harlan K. Britt**, (Chf. Engr., Land Quality Section, Div. of Land Resources, North Carolina Dept. of Natural Resources and Community Development, Raleigh, N.C.)

Journal of the Environmental Engineering Division, Vol. 108, No. EE6, December, 1982, pp. 1228-1247

A review of state and local sediment control programs in North Carolina relative to urban and highway construction and sanitary landfill operations is presented. Law and institutional arrangements, operations and management, technology, economics, enforcement, education, and public acceptance are included. Emphasis is placed on erosion and sediment control as a part of project planning and design. Conclusions and recommendations for strengthened program are given.

0150 Predicting Future Water Demand

Robert M. Clark, Member, ASCE, (Engrg. Systems Analyst, Drinking Water Research Div., Municipal Environmental Research Lab., U.S. Environmental Protection Agency, Cincinnati, Ohio 45268), **James A. Goodrich**, (Research Geographer, Drinking Water Research Div., Municipal Environmental Research Lab., U.S. Environmental Protection Agency, Cincinnati, Ohio 45268) and **James I. Gillean**, (Pres., Act Systems Inc., Winter Park, Fla. 32789)

Journal of the Environmental Engineering Division, Vol. 108, No. EE6, December, 1982, pp. 1248-1264

Decentralization within metropolitan areas has been a major aspect of population movement in the United States over the past two decades. This trend has great significance for all urban service activities. In particular, it affects water supply planning in urban areas. Both numbers of people and their spatial location affect forecasting of water supply demands which, in turn, affect fixed plant investments (source works, treatment works, transmission lines, distribution systems, and so forth) which are frequently built many years in advance of the resulting services. This paper describes, via a case study, the use of population and per capita demands as a basis for forecasting water requirements. Water use is studied according to residential and nonresidential demand. These data, combined with extrapolated demand coefficients, can be used as the basis for demand forecasting that not only provides estimates of quantity of demand but location of demand as well.

0151 Coastal Pollution from Septic Tank Drainfields

Alfred M. Duda, (Supervisor, Nonpoint Source Management Unit., Water Quality Management Branch, North Carolina Div. of Environmental Mgmt., Raleigh, N.C.) and **Kenneth D. Cromartie**, (Soil Sci., Water Quality Management Branch, North Carolina Div. of Environmental Mgmt., Raleigh, N.C.)

Journal of the Environmental Engineering Division, Vol. 108, No. EE6, December, 1982, pp. 1265-1279

Wet weather and dry weather sampling is utilized to monitor densities of coliform bacteria in waters draining residential areas of coastal North Carolina. The bacterial levels are compared to different densities of unsewered residences in each watershed and the limitations of the developed soils for assimilating septic tank effluent. An analysis of the data implicates septic tank drainfields installed in unsuitable soils as a major source of contamination of these shellfish waters. In order to reduce the threat to public health and the multimillion dollar economic loss to the fishing industry, several options are presented for rehabilitating concentrations of failing septic tank systems and for modifying overdesigned drainage systems that carry the contamination directly to shellfish waters. In addition, several common sense management practices that minimize the delivery of bacterial contamination to estuarine waters are suggested for use in siting future coastal residential development.

0152 Activated Carbon for Treatment of Cadmium Wastewater

C. P. Huang, Member, ASCE, (Prof. of Civ. Engrg., Univ. of Delaware, Newark, Del. 19711) and **P. K. Wirth**, (Grad. Student, Environmental Engrg. Program, Dept. of Civ. Engrg., Univ. of Delaware, Newark, Del. 19711)

Journal of the Environmental Engineering Division, Vol. 108, No. EE6, December, 1982, pp. 1280-1299

The use of activated carbon for heavy metal removal was exemplified by studying the adsorption characteristics of Cd(II) from Cd(II)-BF_4 plating solution on both granular and powdered activated carbons. The results indicated that at all pH values, powdered activated carbons such as Nuchar SA and Nuchar SN, exhibited greater Cd(II) removal capacity than the granular ones such as Darco HD 3000 and Filtrasorb 400. The powdered activated carbon was bound to spheres (0.6 mm-1 mm) using a suspension-polymerization technique. The adsorption capacity of the aggregated carbons was 85% - 90% of the unbound ones. The rate of Cd(II) removal at pH 7 is proportional to the cadmium concentration, the carbon dose, and the available surface sites. Used activated carbons were regenerated chemically with strong acids such as HCl, $HClO_4$, and H_2SO_4. Two treatment schemes are proposed with economical analysis for comparison with other common treatment processes.

0153 Coal Conversion Solid Waste Disposal

Chester W. Francis, (Group Leader, Nonnuclear Solid Waste Technology, Environmental Sciences Div., Oak Ridge National Lab., Oak Ridge, TN 37830), **William J. Boegly, Jr.**, (Project Leader, Stored Solids Study, Environmental Sciences Div., Oak Ridge National Lab., Oak Ridge, TN 37830), **Ralph R. Turner**, (Research Assoc., Environmental Sciences Div., Oak Ridge National Lab., Oak Ridge, TN 37830) and **Edward C. Davis**, (Research Assoc., Environmental Sciences Div., Oak Ridge National Lab., Oak Ridge, TN 37830)

Journal of the Environmental Engineering Division, Vol. 108, No. EE6, December, 1982, pp. 1301-1311

The major solid waste produced at coal conversion facilities will be gasification slag or ash. To evaluate the impact of this waste on the environment, the Oak Ridge National Laboratory conducted extensive characterization and leaching studies on ash/slags that had been generated in bench-scale operations, pilot plants, and/or process development units for eight different gasification processes. These studies, designed to assess the consequences of disposal in landfills, showed that none of the leachates from these eight wastes exceeded the U.S. Environmental Protection Agency's toxicity limits. Thus, these ash/slags would be classified as nonhazardous wastes. Quantities of polynuclear aromatic hydrocarbons (PAH) that were observed in the aqueous leachates of selected ash/slags were less than 1 μg/g and appear to be of no significant environmental concern. Column elution studies revealed effluents with pH values less than three and sulfate concentrations greater than 10,000 mg/L for wastes containing sulfur concentrations from 0.3-4%. As a result, the major environmental impact associated with disposal of these solid wastes appears to be the dissolution of sulfate and the potential acidification of ground water.

0154 Determination of Particulate-Nutrient Availability

David A. Hegemann, (Project Engr., L. Robert Kimball and Assocs., Ebensburg, Pa. 15931), **John D. Keenan**, (Assoc. Prof., Dept. of Civ. and Urban Engrg., Univ. of Pennsylvania, Philadelphia, Pa.) and **Arthur H. Johnson**, (Assoc. Prof., Dept. of Landscape Architecture and Regional Planning of Pennsylvania, Philadelphia, Pa.)

Journal of the Environmental Engineering Division, Vol. 108, No. EE6, December, 1982, pp. 1312-1325

A procedure for direct contact algal assay is described which provides quantitative data on phosphorus transfer from soil to algae in serial 14-day incubations. The algae/soil/liquid phases are in direct contact. The procedure requires no special equipment other than that specified for algal assays in the *Selenastrum capricornutum* Printz Algal Assay Bottle Test. The method is a multi-interval assay and, as a result, provides a higher and more realistic value for maximum algal available phosphorus than obtainable using single interval assays.

0155 Effect of Polyelectolytes on Biofilm Growth

Enrique J. La Motta, Member, ASCE, (Chf., Unidad de Ingenieria Ambiental, Facultad de Ingenieria Civil, Escuela Politecnica National, Apartado 2759, Quito, Ecuador), **Robert F. Hickey**, (Grad. Student, Dept. of Civ. Engrg., Univ. of Massachusetts, Amherst, Mass.) and **John F. Buydos**, (Alumni, Dept. of Civ. Engrg., Univ. of Massachusetts, Amherst, Mass.)

Journal of the Environmental Engineering Division, Vol. 108, No. EE6, December, 1982, pp. 1326-1341

In certain biofilm processes, such as nitrification and anaerobic carbon removal, biofilm attachment and growth may take several weeks or even months. This research focuses on determining the effect of synthetic and naturally produced polymers, used as surface conditioners, on the rate of biofilm attachment and growth. Partial success was obtained, and the only practical alternative seems to be to precondition the surface of the media by growing an easy-to-develop biological film before the final biofilm is induced to grow. This paper also reviews the literature which describes the factors affecting the attached growth of bacteria.

0156 Integrated Control of Combined Sewer Regulators

Hoang Nam Nguyen Van, (Montreal Urban Community, Montreal, Quebec, Canada), **Mohamad Osseyrane**, (Montreal Urban Community, Montreal, Quebec, Canada) and **Murray B. McPherson**, Fellow, ASCE, (Consultant, Urban Water Management, Marblehead, Mass.)

Journal of the Environmental Engineering Division, Vol. 108, No. EE6, December, 1982, pp. 1342-1360

Integrated operation was simulated of 10 dynamic combined sewer regulators on a Montreal interceptor. Generally, when no in-line or other in-system storage can be called upon, advantage over local automatic-dynamic regulator control cannot be obtained by integrated operation of regulators. This happens unless expected flowrates to the interceptors can be estimated in advance of their actual occurrence; or unless an operational bias is introduced. These either minimize overflows from only some of the regulators on an interceptor, or favors the timing of overflows from all of the outlets, such as during the initial storm period. Computations for an actual system covering a summer season indicate even when a time-varying pollutant parameter is accounted for, the reduction in mass of a pollutant entering the receiving water under integrated operation of regulators may be only marginal. The potential for pollutant reductions at designated outfalls, however, is promising.

0157 Regeneration of Five Activated Carbons with Methanol

Michael A. Rollor, (Doctoral Student, School of Civ. Engrg., Georgia Inst. of Tech., Atlanta, Ga. 30332), **Makram T. Suidan**, (Assoc. Prof. of Environmental Engrg., Dept. of Civ. Engrg., Univ. of Illinois, Urbana, Ill. 61801), **Wendall H. Cross**, (Research Sci., School of Civ. Engrg., Georgia Inst. of Tech., Atlanta, Ga. 30332) and **Stephan A. Vargo**, (Environmental Engr., Charles Velzy and Assoc., Armonk, N.Y. 10504)

Journal of the Environmental Engineering Division, Vol. 108, No. EE6, December, 1982, pp. 1361-1377

The regeneration characteristics of five commercially available granular activated carbons with methanol were evaluated. Three of the adsorbents were bituminous base carbons, while the remaining two were manufactured from lignite coal and wood. The adsorbates were p-chlorophenol, and two-ring, four-ring and eight-ring azo dyes. The rate of regeneration of the five adsorbents appeared to be sensitive to the molecular weight of the adsorbate; decreasing with increasing molecular weight. During the methanol regeneration of the p-chlorophenol and two-ring dye, an upper limit for capacity recovery was observed for all adsorbents, suggesting that a maximum recoverable capacity has been obtained. This phenomenon was not observed for the remaining two adsorbates due to exceedingly slow extraction rates. The extraction efficiences obtained for p-chlorophenol and the two-ring dye were found to correlate with the adsorption energies of these adsorbates on activated carbon. The extraction efficiency was observed to decreased with increasing adsorption energy.

0158 The Engineer as Moral Agent

Carl Nelson, Member, ASCE, (Dir., Field Services, ASCE, New York, NY 10017) and **Susan Rae Peterson**, (Asst. Prof. of Philosophy, C.W. Post Center, Long Island Univ., Greenvale, NY)

Issues in Engineering Journal of Professional Activities, Vol. 108, No. EI1, January, 1982, pp. 1-5

The ways in which engineers fail to act as moral agents are examined. Discussions and studies show that the self-image frequently found among engineers is that of a rational calculator. Although engineers are rational agents, they are also moral agents, entrusted with public welfare and confidence. The relationship between-solving and conscientious professional behaviour is examined, and the tendency for engineers to make all problems merely technical problems is condemned.

0159 Conflicts of Roles in Engineering Ethics

Carl Nelson, Member, ASCE, (Dir., Field Services, ASCE, New York, NY 10017) and **Susan Rae Peterson**, (Asst. Prof. of Philosophy, C.W. Post Center, Long Island Univ., Greenvale, NY)

Issues in Engineering Journal of Professional Activities, Vol. 108, No. EI1, January, 1982, pp. 7-11

An examination of how ethical dilemmas are created is presented and applied to engineering practice. Cases which are most typical of such conflicts are whistleblowing cases in which the engineer is torn between making a judgement based on his professional standards of excellence and making a judgement which alines itself with management policy. An ethical analysis from a philosophical point of view provides an analysis of the problem of conflicting roles.

0160 If You're an Engineer, You're Probably a Utilitarian

Carl Nelson, Member, ASCE, (Dir., Field Services, ASCE, New York, NY 10017) and **Susan Rae Peterson**, (Assoc. Prof. of Philosophy, C.W. Post Center, Long Island Univ., Greenvale, NY)

Issues in Engineering Journal of Professional Activities, Vol. 108, No. EI1, January, 1982, pp. 13-18

The nature of utilitarianism is described and discussed in regard to engineers' preference of utilitarianism over other ethical theories. The advantages and disadvantages of this attitude are discussed. Theoretical and practical problems in utilitarianism are considered. Some recommendations are made to rectify these problems, which are inherent in utilitarian theory.

0161 A Moral Appraisal of Cost-Benefit Analysis

Carl Nelson, Member, ASCE, (Dir., Field Services, ASCE, New York, NY 10017) and **Susan Rae Peterson**, (Assoc. Prof. of Philosophy, C.W. Post Center, Long Island Univ., Greenvale, NY)

Issues in Engineering Journal of Professional Activities, Vol. 108, No. EI1, January, 1982, pp. 21-25

A cost-benefit analysis is one in which the engineer weighs the possible positive consequences, or benefits, of a given action against the negative consequences, or costs, of that same action. An ethical analysis of the cost-benefit analysis is discussed from a philosophical point of view. Advantages and disadvantages are considered from a moral point of view.

0162 Better Government—Can the Young Engineer Help Make it Happen?

Daniel Yake, Assoc. Member, ASCE, (Assoc., Budinger and Assoc., N. 920 Lake, Spokane, Wash. 99206)

Issues in Engineering Journal of Professional Activities, Vol. 108, No. EI1, January, 1982, pp. 27-31

With increasing repetition, decisions concerning complex engineering issues are being made along purely political lines. To counteract this trend, the engineer must actively become involved in the election of candidates who understand or are sympathetic to the technological consequences of existing and proposed legislation. Engineers must sacrifice a portion of their career to engage in pre-election campaigning, contribute funds, and even run for office to insure that judgments are based on rational, not strictly political, analysis.

0163 Regulation of Engineers in California

Isidore Michael Schulman, (Adjunct Asst. Prof. of Law, Univ. of Southern California, Los Angeles, Calif. 90007)

Issues in Engineering Journal of Professional Activities, Vol. 108, No. EI1, January, 1982, pp. 33-48

State-administered certification programs for the engineering profession are unjustified and should be eliminated. The reasons for this conclusion have nothing to do with the peculiarities of any particular engineering discipline, but are due to the inherent weaknesses of governmental-sponsored certification programs. State-run certification programs are not enforceable, may not reduce information costs to consumers, put barriers on an engineer's ability to practice, and may be duplicative of private certification programs. The second conclusion of this article calls for the registration of all exempt engineers who are in responsible charge in licensed disciplines. Finally, this article proposes that criteria be extablished and followed when determining whether a discipline should be brought under a licensing program. Each state should develop criteria whenever it seeks to impose regulations upon a given discipline, for without such standards it is impossible to establish a rational system of engineering regulations.

0164 Trends in Engineer Registration, Sunset Review: How to survive?

Donald A. Buzzell, Member, ASCE, (Managing Dir., Education and Professional Affairs, ASCE, New York, N.Y. 10017)

Issues in Engineering Journal of Professional Activities, Vol. 108, No. EI1, January, 1982, pp. 51-63

Thirty-one states have legislated periodic review of licensing and regulatory functions, including professional registration or licensure of engineers, land surveyors, and related professionals, to determine if these functions should be continued on the basis of serving the public interest. To date, 23 state boards of registration have been subjected to such review under Sunset Legislation. The reports and recommendations submitted by a number of the reviewing bodies, together with the response from these boards and supporting statements, were reviewed and are summarized. All registration boards reviewed so far have been continued, but numerous changes in composition, procedure and policy have been implemented as a result of the process. These changes are reported, and recommendations related to preparing for and responding to Sunset Law reviews are offered.

0165 Task Committee on Engineering Education: Report to the Board of Direction

Task Committee on Engineering Education

Issues in Engineering Journal of Professional Activities, Vol. 108, No. EI1, January, 1982, pp. 67-92

Engineering education, not civil engineering education alone, is in a crisis which could seriously damage the engineering profession and the national interest. Too few graduates are earning advanced degrees, so the flow of new doctoral degree holders to engineering school faculties is inadequate. Support levels for graduate students in engineering must be increased, and faculty compensation must be improved to make a career in engineering education as attractive as one in practice. Facilities and equipment are also a problem, as technological obsolescence overtakes available equipment and increased class sizes demand more laboratory facilities. External support, from industry and government sources, will be needed to adequately upgrade engineering school laboratories. Another element of the crisis is the strain put on the educational system by the magnitude of current engineering enrollments, now at an all time high. Resources available to engineering schools are too limited to adequately address these problems. ASCE is developing an action plan to address the crisis in engineering education. A three way partnership between industry, government and universities will be needed to effect long-term resolutions.
Discussion: **John M. Hayes**, Honorary Member, ASCE, (Prof. Emeritus and Consulting Structural Engr., 312 Highland Drive, West Lafayette, Ind. 47906) EI Oct. '82, pp. 333-334.

0167 Minority and Female Students in Engineering

Andrew D. Jones, Member, ASCE, (Prof. and Chairman, Civ. Engrg. Dept., Univ. of Texas at El Paso, El Paso, Tex. 79968) and **Braja M. Das**, Member, ASCE, (Assoc. Prof., Civ. Engrg. Dept., Univ. of Texas at El Paso, El Paso, Tex. 79968)

Issues in Engineering Journal of Professional Activities, Vol. 108, No. EI2, April, 1982, pp. 105-110

It appears that the national demand for engineers will increase in future years. To meet this demand, besides traditional prospective engineering students, minorities and women have to be encouraged to accept engineering as a profession. The engineering summer institute is one possible way to expose the high school students to the broad area of engineering. Such a program at The University of Texas at El Paso is discussed here, including its potential for providing future engineering students.

0168 The American Engineer Overseas

Paul M. Reed, Member, ASCE, (Project Mgr., Dames & Moore, Jakarta, Indonesia)

Issues in Engineering Journal of Professional Activities, Vol. 108, No. EI2, April, 1982, pp. 111-114

More and more American engineers will be working overseas, many for the first time. Frequently materials, laboratories and hardware equipment are not available and the engineer must first ascertain what is available, then develop design and specifications on this basis. Too often the engineer working on a foreign assignment disregards local practices which inevitably is an expensive mistake. The author cites mistakes made by engineers and suggests ways to avoid costly pitfalls.

0169 Overseas Marketing Strategy

William L. Kelley, Member, ASCE, (Vice Pres., International Engineering Co., Inc., San Francisco, Calif. 94105) and **Richard E. Kohne**, Fellow, ASCE, (Pres., International Engineering Co., Inc., 180 Howard Street, San Francisco, Calif. 94105)

Issues in Engineering Journal of Professional Activities, Vol. 108, No. EI2, April, 1982, pp. 115-121

The personnel element in achieving and maintaining overseas marketing success is examined, and suggested strategies for selection, preparation, utilization and support of personnel for overseas assignments are presented. Attention is focused on the engineering project manager, and ideal professional and personal characteristics are cited, including technical capability, stability, leadership and communicative ability. A personnel selection process is suggested, in which a competence profile is developed and a remedial program is proposed. A typical overseas compensation package including monetary and material benefits is examined. Home office support of overseas personnel is emphasized. The importance of local representatives for furnishing market intelligence and advice in local customs, legal and tax implications is cited. Integration of local country engineering partners into the total team effort overseas is recognized as a critical element of success.

0170 Project Evaluation in International Development

Branden B. Johnson, (Asst. Prof., Dept. of Geography, Univ. of Pittsburgh, Pittsburgh, Pa.)

Issues in Engineering Journal of Professional Activities, Vol. 108, No. EI2, April, 1982, pp. 122-132

Project evaluation methods used by the World Bank and the United Nations Industrial Development Organization are compared with regard to measures of value, use of shadow prices, discount rates and opportunity costs of investment, and other issues of project evaluation. While there are some similarities, it appears that the choice of method will make a difference in which projects are found acceptable. The UNIDO approach seems to be more flexible but requires considerably more information.

0171 Professional Responsibility of Engineers in Public Works Management

William H. Wisely, Honorary Member, ASCE, (ASCE Exec. Dir. Emeritus and Visiting Scholar, Univ.of Virginia, Charlottesville, Va.)

Issues in Engineering Journal of Professional Activities, Vol. 108, No. EI2, April, 1982, pp. 133-139

The concept "public interest" is defined and examined in regard to professional responsibility and civil engineers. Various issues affecting public works and public interest are discussed. These issues can be financial, political, or ideological in nature. Unionism, ecology, and effects on local communities must also be taken into consideration. The impacts of federal aid and government regulation are emphasized. It is concluded that the principal problems of the public works engineer today stem from government. Recommendations to reconcile these conflicts are listed.

0172 Problems in Administering Overseas Projects

Melvin L. Purtell, (Design Project Mgr., E.I. duPont de Nemours Co., Inc., Engineering Dept., Louviers Building, Wilmington, Del. 19898)

Issues in Engineering Journal of Professional Activities, Vol. 108, No. EI2, April, 1982, pp. 140-144

Management of an overseas project can be successful if potential problems are recognized and considered early in a proposed new venture. Such problems include organizational structure, foreign engineering requirements, world-wide procurement, personnel relocations, host country regulations, construction labor, cost control, and estimating techniques. The paper briefly covers the organizational of the author's company and mentions numerous questions and considerations for major project functions that need resolution early in venture strategy planning. Generally, planning for overseas projects are more complicated, require more up-front time, and need more management attention than comparable domestic projects.

0173 Professional Ethics and the Transportation Engineer

Edmund J. Cantilli, , (editor), (Prof. of Engineering & Transportation Plannning, Polytechnic Institute of New York, 333 Jay Street, Brooklyn, New York, 11201)

Issues in Engineering Journal of Professional Activities, Vol. 108, No. EI2, April, 1982, pp. 145-161

Professional ethics in engineering, their relationship to transportation engineering or transportation planning, and the issues of transportation environmental impact and trans-protation safety are presented and discussed in a panel discussion held at a joint meeting of the Institute of Transportation Engineers, New York Metropolitan section, and the Polytechnic Institute of New York. The danger of trade offs between cost and effectiveness are emphasized. The role of the engineer in controlling and guiding technology is also considered. Other issues raised include the difference between a code of ethics and a legal code, the role of an engineer as an employee on a public works project and the professional engineer's role in the decision making process.

0174 Politicians Are Not Technologists: Technologists Are Not Politicians

Henry Petroski, (Assoc. Prof. of Civ. Engrg., Duke Univ., Durham, N.C. 27706)

Issues in Engineering Journal of Professional Activities, Vol. 108, No. EI2, April, 1982, pp. 162-165

The view that there are too few engineers in the legislative branch of government is confirmed. It is argued that the lack of scientists and engineers in elected public offices creates an imbalance of professional perspectives which will affect decision making on public policy on technological issues ranging from engineering education to advanced weapons systems. It is also shown, however, that the interplay between society and technology is gradually becoming more balanced.

0175 One Uses His Umbrella When It Rains

Kenneth A. Roe, Member, ASCE, (Chmn. and Pres., Burns & Roe, Inc., 550 Kenderkamack Rd., Oradell, N.J. 07649)

Issues in Engineering Journal of Professional Activities, Vol. 108, No. EI2, April, 1982, pp. 166-171

The engineering profession unity movement is reviewed, with a brief explanation of the reasons for forming the American Association of Engineering Societies. A brief historical background is followed by a discussion of the need for unity within the profession, the effects of federal regulations on engineering, how AAES is organized to help engineers work together in areas of common interest, and how concensus in the profession helps both engineers and the public. The specific goals of AAES are outlined and the future development plans of the Association are described which are designed to further the aims and purposes of the Association, eventually to reach the goals described.

0176 Water Resources in Developing Countries

Frank A. Butrico, Fellow, ASCE, (Chf., Div. of Environmental Health Protection, Pan American Health Organization, Washington, D.C. 20037)

Issues in Engineering Journal of Professional Activities, Vol. 108, No. EI3, July, 1982, pp. 175-180

Most of the nations of the world are concerned with providing water to as many people as possible during the Decade of the 1980's, (The International Drinking Water and Sanitation Decade). This ambitious goal and those for economic growth and development will mean increased demands for water which will be difficult to satisfy if governments do not take steps to develop a national policy on water resource management and to incorporate this policy into a national development strategy. This paper reviews the global effort to extend water supply services as well as the possible constraints which may make it difficult to achieve the goals for the Decade. Among these are institutional weaknesses, operation and maintenance problems that affect water quality, the availability of trained personnel, and economic and financial problems. The importance of preserving water quality while proceeding with economic development is also mentioned.

0177 Year-End Report and Valedictory Address

Joseph S. Ward, Fellow, ASCE, (Pres., Converse Ward Davis Dixon, Inc., 221 E. Walnut St., Suite 250, Pasadena, Calif. 91101)

Issues in Engineering Journal of Professional Activities, Vol. 108, No. EI3, July, 1982, pp. 181-189

As out-going President of the American Society of Civil Engineers, Joseph Ward reports on the accomplishments of his administration and enumerates the issues that still need to be addressed by the Society. Among the programs and issues covered are the ASCE Energy Action Program, the role of ASCE and the civil engineering profession in federal and state regulations, the ASCE membership program, engineering education, and active participation in the American Association of Engineering Societies.

0178 Bridge Structural Innovation: Firsthand Report

Jackson Durkee, Fellow, ASCE, (Consulting Struct. Engr., 217 Pine Top Trail, Bethlehem, Pa. 18017)

Issues in Engineering Journal of Professional Activities, Vol. 108, No. EI3, July, 1982, pp. 190-201

Some fundamental aspects of bridge structural innovation and economy are reviewed. Historical examples of bridge innovation are presented and examined broadly. The author cites some of his first-hand experiences with innovative bridgework design and construction, and comments on technical and financial considerations and on the results achieved. The various ways in which owners, designing engineers, and contractors are inclined to view bridge structural innovation are described. Significant factors retarding bridge structural innovation and economy are discussed, and solutions suggested.

0179 Engineer-in-Training Program Alaska District, CoE

Kenneth E. Hitch, Member, ASCE, (Project Engr. and Engineer-in-training Supervisor for the Alaska District, Planning Branch, Alaska Dist., U.S. Army Corps of Engrs., Anchorage, Alaska) and **Dale A. Nelson**, Member, ASCE, (Project Mgr. and Directorate of Facilities Engrg., CH2M Hill, U.S. Army Corps of Engineers, Anchorage, Alaska)

Issues in Engineering Journal of Professional Activities, Vol. 108, No. EI3, July, 1982, pp. 202-208

Rotational Engineer-In-Training Program is a success at the Alaska District, Corps of Engineers. Initiated in 1967, it is better than similar programs conducted by the State of Alaska, Municipality of Anchorage, and local industry and architectural/engineering firms. A high turnover after program completion results from low Federal salaries, lack of professional satisfaction, and discouraged innovation. Data was obtained from questionnaires to engineers who had completed the program, records search and interviews with Alaska District Supervisors as well as representatives of the State, Municipality, and local industrial and engineering firms.

0180 Organizing Engineering Volunteers for Disaster Relief

David H. Philpott, Member, ASCE, (Chartered Engr., 44 Ocean View Road, Bude, Cornwall EX23 8NL, United Kingdom, England)

Issues in Engineering Journal of Professional Activities, Vol. 108, No. EI3, July, 1982, pp. 209-214

National and manmade disasters disrupt people's lives and make them dependant on others. This is particularly severe in the least developed countries often subject to their neighbor's political and racial upheavals. Seldom are such outcomes newsworthy. If assistance is not effective then the problem is compounded through disease and lack of self-sufficiency. There is scope for early engineering participation in this relief work to give safe water, sanitation, and shelter, and so turn refugees to self support. Civil engineers in the United Kingdom have formed a Register of Engineers for Disaster Relief. REDR supplies at short notice (from their normal employment) trained and competent engineers to aid organizations for short term assistance world wide.

0181 Initiating and Maintaining Client Relationships

Wilbur F. Peck, Member, ASCE, (Pres., Technical Management Consultants, 1925 Century Blvd., Suite 4, Atlanta, Ga. 30345)

Issues in Engineering Journal of Professional Activities, Vol. 108, No. EI4, October, 1982, pp. 225-237

This paper describes the developments in our economy in the mid 70's that have forced many businesses to change their views regarding the marketing of professional services. The traditional attitudes of the learned professions that produce barriers to active participation in selling activities are examined. Client maintenance and project development are activities defined and examined. The writer's concepts on marketing responsibilities and planning are outlined in preparation for the more detailed material presenting the types of activities and skills needed to be effective professional salesmen. Misconceptions in professional selling are described along with appropriate activities that should be supportive of a firms' interest in fee growth.

0182 The Ethics of Serving Two Masters: An Engineer's Dilemma

Anthony J. Bellitto, Jr., (Civ. Engrg. Student, Manhattan College, Riverdale, Bronx, N.Y. 10471)

Issues in Engineering Journal of Professional Activities, Vol. 108, No. EI4, October, 1982, pp. 238-242

The question posed is "Does the engineer serve two masters——the employer and the public welfare?" Through analogies of the decision making process to common everday problems and a look at the alternative of a one-sided ideology, it is reasoned that every engineer must indeed serve two masters in his unique role as builder and businessman. To approach the resolution of this dilemma in cases of a conflict of interest, a strategy is suggested which includes an improvement in education, a revision of the Code of Ethics, and a fostering of professionalism through registration and involvement in peer societies. Ethics must necessarily be the concern of all engineers at all levels as an essential component in developing the honesty, dignity, and character expected in a profession relied on so heavily by society.

0183 Contract Arrangements in the Third World

Peter R. Morgan, Member, ASCE, (Sr. Lect. in Civ. Engrg., Univ. of Melbourne, Parkville, 3052, Victoria, Australia) and **M. H. Kamil**, (Civ. Engr., Malaysia)

Issues in Engineering Journal of Professional Activities, Vol. 108, No. EI4, October, 1982, pp. 243-250

Standard contract provisions and communications may not be the most appropriate for all construction work in third world countries. The main criteria for appropriateness reflect, firstly, a favorable degree of local participation and, secondly, the use of methods more suited to local conditions both economically and socially. A review of past and present practices shows a tendency to adhere still to standard western methods. There has been a lack of innovation but the need for more appropriate approaches is being recognized.

0184 Involvement of Engineers in Public Affairs

Milton Mortman, (Chrmn., Civic Affairs Committee, National Capital Section, Greenbelt, Md.)

Issues in Engineering Journal of Professional Activities, Vol. 108, No. EI4, October, 1982, pp. 251-262

A discussion of the opportunities available to engineers in public affairs is presented. Specific suggestions are included for engineers interested in providing volunteer service in their local communities or in running for public office. Also, changes in engineering education to include material on the need for community involvement are recommended.

0185 Defects in Public Building Construction

Daniel L. Schodek, Aff., (Assoc. Prof., Dept. of Architecture, Grad. School of Design, Harvard Univ., Cambridge, Mass.)

Issues in Engineering Journal of Professional Activities, Vol. 108, No. EI4, October, 1982, pp. 263-270

From 1978 - 1980, The Special Commission concerning State and County Buildings in the Commonwealth of Massachusetts made a thorough investigation of maladministration and corruption in the construction of public buildings in the state. In the end, the Commission also suggested legislative reforms for improving public construction. As part of its inquiries, the Commission made a series of extensive studies of physical defects in recently constructed public buildings. One method used to gather information was a mailed survey, which is examined in this paper. The majority of recently constructed public buildings were found to have severe construction defects which impaired building performance or use, were costly to repair, or led to expensive secondary damage, or all three.

0186 Management Communications: Project Manager—Down

J. Robert Moore, Fellow, ASCE, (President, Harland Bartholomew & Assoc., Inc., Memphis, Tenn.)

Issues in Engineering Journal of Professional Activities, Vol. 108, No. EI4, October, 1982, pp. 271-284

Effective communication between the manager of an engineering project and those assigned to the project team is essential to the successful completion of a project. The project manager also must communicate with the general management of his firm and with the client. The manager informs project team personnel of what work is to be done and when and by whom it is to be done. Communication techniques vary with project size, number of personnel, and project organization. Personal, verbal communication is important for all projects; however, written communication, detailed schedules including those using the Critical Path Method, and itemized budgets are especially needed on larger projects. Project management communication tools include the Work Plan, which defines work elements, schedules, and budgets; and Management Control System Reports, which periodically define the status of work. Status reports assist in identifying problem areas. Decisions by the project manager, as well as instructions and guidelines, are communicated from the manager to the project team members.

0187 Employment Conditions from Employee Viewpoint

D. H. Lillard, Member, ASCE, (Partner and Personnel Administrator, Black & Veatch Consulting Engrs.)

Issues in Engineering Journal of Professional Activities, Vol. 108, No. EI4, October, 1982, pp. 285-291

Observations of the engineer's viewpoint regarding employment conditions are presented. Comments are limited to engineer employees because their viewpoint as professional employees are often unique in an organization. General employment relations are reviewed. The viewpoints of the young engineer employee and the experienced engineer employee are compared. It is stated that good employment policies result only when management is able to reconcile its role with the employee's viewpoint.

0188 Professional Schools: The Debate Continues

Robert D. Kersten, Fellow, ASCE, (Prof. and Dean of Engrg., Univ. of Central Florida, Orlando, Fla. 32816)

Issues in Engineering Journal of Professional Activities, Vol. 108, No. EI4, October, 1982, pp. 292-310

The development of a strong unified profession of engineering should start with a careful restructuring of our educational units. Engineering education is in a state of disarray and cannot provide the leadership necessary to unify the profession. All learned professions have some graduate education requirement, and engineering would do well to evaluate the graduate professional school approach of other professions. Abraham Flexner's criteria for a practicing professional are recommended as applicable to the engineering profession.

0189 Preparing U.S. Employees for Overseas Assignments

Edward A. Bryant, Fellow, ASCE, (Vice Pres., Hazen and Sawyer Engrs., 360 Lexington Ave., New York, N.Y. 10017)

Issues in Engineering Journal of Professional Activities, Vol. 108, No. EI4, October, 1982, pp. 311-315

Local bureaucracy, our governments' bureaucracy, and just plain lack of convenience cause overseas employees to be far less productive than in the United States. Yet, they cost about 50% more. If they are unhappy over there, their efficiency drops even more. This paper lists a number of ways to improve the chances of having a relatively satisfied overseas employee. The employee's attitude about going overseas, his preparation for going overseas, the employment agreement, and considerations regarding the spouse are examined.

0190 Communicating Projects Costs to Staff

G. Neil Harper, Assoc. Member, ASCE, (Pres., Harper and Shuman, Inc., Cambridge, Mass. 02138)

Issues in Engineering Journal of Professional Activities, Vol. 108, No. EI4, October, 1982, pp. 316-320

The need for more effective management of engineering, architectural, and planning organizations is becoming increasingly acute as our society becomes more sophisticated and complex. Part of the response to this need has been an increased emphasis in the past few years on communicating project costs to staff. This paper addresses the need and justification for this effort, and discusses organizational patterns and training programs which have proved effective in practice for professional firms.

0191 Iowa's Mandatory Continuing Education Law

Harrison Kane, Member, ASCE, (Prof. of Civ. Engrg., Univ. of Iowa, Iowa City, Iowa 52242)

Issues in Engineering Journal of Professional Activities, Vol. 108, No. EI4, October, 1982, pp. 321-327

Since July 12, 1977 Iowa has had a law requiring continuing education as a condition of license renewal for professional engineers and land surveyors. The requirements and the results of the first year's experience are described. The impacts include a 15% decrease in the number of active registrants and a 93% increase in total attendance in continuation programs.

0192 Scattering of SH Waves by Subsurface Topography

Marijan Dravinski, (Research Asst. Prof., Dept. of Civ. Engrg., Univ. of Southern California, Los Angeles, Calif. 90007)

Journal of the Engineering Mechanics Division, Vol. 108, No. EM1, February, 1982, pp. 1-17

Horizontally polarized shear waves in an elastic alluvial valley of arbitrary shape perfectly bonded to a linearly elastic, homogeneous and isotropic half space are considered. The valley is subjected to a steady state horizontal displacement field. Total displacement field is evaluated along the interface between the half space and alluvial valey. Comparison with known exact solutions for simple geometry provided the following conclusions: 1) the method provides excellent results for wide range of frequencies, 2) for a fixed number of sources, the results are more accurate at lower frequencies, and 3) as the number of sources increases, the accuracy increases.

0193 Scattering of Elastic Waves by an Alluvial Valley

Marijan Dravinski, (Research Asst. Prof., Dept. of Civ. Engrg., Univ. of Southern California, Los Angeles, Calif.)

Journal of the Engineering Mechanics Division, Vol. 108, No. EM1, February, 1982, pp. 19-31

Plane strain model for scattering of steady state plane waves by an elastic alluvial valley of arbitrary shape is presented. The valley is presumed to be perfectly bonded to a linearly homogeneous and isotropic half space. Total displacement and stress fields are evaluated along the interface between the alluvial valey and elastic half space using the source method. The results are compared with known exact solutions for simple geometries to determine the accuracy of the proposed method. It was found that the proposed method provides very good results. As an illustrative example diffraction of a plane P-wave by an elastic semi-elliptical alluvial valley is presented in detail.

0194 Response of Taut Strip Models to Turbulent Wind

Hiroshi Tanaka, (Asst. Prof., Faculty of Science and Engrg., Univ. of Ottawa, Ottawa, Ontario, Canada) and **Alan G. Davenport**, Member, ASCE, (Prof. and Dir., Boundary Layer Wind Tunnel Lab., Univ. of Western Ontario, London, Ontario, Canada)

Journal of the Engineering Mechanics Division, Vol. 108, No. EM1, February, 1982, pp. 33-49

An experimental investigation was conducted to clarify the fundamental behaviour of the proposed taut strip wind tunnel model of long span bridge deck systems. The study shows that this method is relatively simple and convenient and yet effective method of investigating the aerodynamic behaviour of long span bridges with the simulation of natural wind turbulence in laboratory scale. Experimental results are compared with the simple theoretical calculation and some aerodynamic characteristics of bridge response in highly developed turbulent boundary layer flow were observed. Effect of scale of turbulence in particular is found to be significant.

0195 Beams Under Lateral Projectile Impact

Joao G. de Oliveria, (Asst. Prof. of Ocean Engrg., Dept. of Ocean Engrg., Massachusetts Inst. of Tech., Cambridge, Mass., 02139)

Journal of the Engineering Mechanics Division, Vol. 108, No. EM1, February, 1982, pp. 51-71

The influence of transverse shear effects, rotary inertia and the boundary conditions on the dynamic behavior of rigid perfectly plastic beams subjected to projectile impact at midspan is studied. Theoretical solutions for a perfectly clamped and a simply supported beam are presented. Numerical results are given for a rectangular and three wide-flanged I-beams of varying spans. For the particular range of parameters examined it is found that rotary inertia does not affect the response in a significant way. However, shear effects are very important for the particular beams considered, being more pronounced in wide-flanged short I-beams. In particular the maximum lateral displacement can increase substantially due to shear effects, showing that a bending only solution neglecting shear can lead to erroneous results. It is also found that shear effects are slightly more important for short simply supported beams than short clamped beams, but for large spans they are essentially independent of the support conditions.

0196 Torsional Dynamic Response of Solid Media

Robert Henke, Assoc. Member, ASCE, (Research Engr., Exxon Production Research Co., Houston, Tex.) and **E. Benjamin Wylie**, Member, ASCE, (Prof. of Civ. Engrg., Univ. of Michigan, Ann Arbor, Mich.)

Journal of the Engineering Mechanics Division, Vol. 108, No. EM1, February, 1982, pp. 73-94

A new, practical method is proposed for solving the partial differential equations describing the dynamic behavior of either linear elastic or nonlinear inelastic multi-dimensional continua. The ultimate application of the method is the prediction of the dynamic response of foundations to loads having a broad range of intensities. Stresses, particle velocities and partial derivatives are computed numerically for points within the medium giving a detailed description of dynamic response. Since few approximations are introduced, solutions are accurate. The method is applied to the equations describing the axisymmetric torsional behavior of a three-dimensinal medium. Multi-dimensional linear and nonlinear torsional solutions are obtained. Examples in which one-dimensional behavior is approximated are discussed. Linear solutions agree with theory. Nonlinear solutions, for which theoretical solutions are not found, satisfy energy balance requirements. The effects of shearing nonlinear inelasticity on dynamic response are found to be significant.

0197 Stresses on Rapidly Sheared Fluid-Solid Mixtures

Norbert L. Ackermann, Member, ASCE, (Prof., Dept. of Civ. and Environmental Engrg., Clarkson College of Tech., Potsdam, N.Y. 13676) and **Hayley Shen**, (Asst. Prof., Dept. of Civ. and Environmental Engrg., Clarkson College of Tech., Pottsdam, N.Y. 13676)

Journal of the Engineering Mechanics Division, Vol. 108, No. EM1, February, 1982, pp. 95-113

Constitutive equations were derived describing the stress in a rapidly sheared mixture of granular cohesionless solids and a Newtonian fluid. These equations are considered to describe the flow of mixtures such as coal and oil in a pipeline or sediment suspended in a river. The stress is found to depend upon the following properties of the solid: diameter and density of the granular material, coefficient of kinetic friction, coefficient of restitution, and volumetric concentration. The density and viscosity of the interstitial fluid was also found to affect the stress within the granular mixture. The theoretically predicted shear stresses were found to be in good agreement with measured stresses determined from experimental studies where rapid shear flows were produced in mixtures formed from polystyrene spheres in air, polystyrene spheres in water, glass spheres in air and polystyrene spheres in air.

0198 Finite Layer Analysis of Nonhomogeneous Soils

R. Kerry Rowe, (Asst. Prof., Faculty of Engrg., Sci., Univ. of Western Ontario, London, Canada N6A 589) and **John R. Bocker**, (Reader, Univ. of Sydney, Sydney, Australia)

Journal of the Engineering Mechanics Division, Vol. 108, No. EM1, February, 1982, pp. 115-132

A finite layer approach capable of analyzing both layered and continuously varying elastic soil profiles is presented. The technique is shown to be accurate and efficient. It is then used to analyze the behaviour of footings on a soil with a homogeneous crust underlain by a deposit whose modulus increases linearly with depth.

0199 Dynamic Response of Plate on Elastic Half-Space

William L. Whittaker, Assoc. Member, ASCE, (Asst. Prof., Dept. of Civ. Engrg., Carnegie-Mellon Univ., Pittsburgh, Pa.) and **Paul Christiano**, Member, ASCE, (Prof., Dept. of Civ. Engrg., Carnegie-Mellon Univ., Pittsburgh, Pa.)

Journal of the Engineering Mechanics Division, Vol. 108, No. EM1, February, 1982, pp. 133-154

Analytical results are presented for the dynamic response of a plate bearing on an elastic half-space and subjected to harmonic forces. The present work represents a departure from existing analyses in that herein both the flexibility and three-dimensionality of the plate are taken into account. Displacements and contact stresses are presented for square plates having a practical range of flexural stiffness. The harmonic analysis is conducted within the framework of a global stiffness solution, in which the plate and subgrade impedance matrices are formulated independently in accordance with a prescribed discretization pattern. Then compatibility of displacements and equilibrium of forces are enforced at the plate-subgrade interface. Solutions are presented for massless square plates subjected to harmonic point, uniform pressure, and moment loadings. The effect of the plate's mass on its response is studied by starting with a perfectly flexible massive plate and subsequently increasing the stiffness toward the limit of a completely rigid plate.

0200 Generalized Wave Equation in Finite Domains

H. Don Fisher, (Consulting Engr., Mechanical Engineering Department, Combustion Engineering, Inc., Windsor, Conn. 06095)

Journal of the Engineering Mechanics Division, Vol. 108, No. EM1, February, 1982, pp. 155-163

A generalized equation governing one dimensional wave propagation in finite domains is solved by employing the finite Sturm-Liouville transform. The formulation includes the effects of time dependent boundary conditions, non-zero initial conditions and the presence of body or surface forces or both. The modal solutions for the dynamic response are valid for the entire range of the temporal variable. The utility of the analysis is demonstrated by citing more than twenty references in which the derived time dependent response can readily be obtained from the generalized solutions presented here.

0201 Plates of Varying Thickness with Four Simply Supported Edges

Fan Jia-rang, (Assoc. Prof. Engrg. Mech., Hefei Polytechnic Univ., Hefei, Anhui, People's Republic of China)

Journal of the Engineering Mechanics Division, Vol. 108, No. EM1, February, 1982, pp. 167-173

The methods of double finite Fourier transformation and two Trancendental Functions, Six and Cix, are adopted. A general expression of solution is given for a thin plate of variable thickness subject to arbitrary loads. It is suggested that the scope of solvable problems of this nature is expanded using the method described.

0202 Creep of Pipes Under Axial Force and Twisting Moment

John H. Lau, (Sr. Engr., Special Structures Group, Bechtel Power Corp., San Francisco, Calif. 94119) and **Teresa T. Lau**, (Prof., Fairleigh Dickinson Univ., Teaneck, N.J.)

Journal of the Engineering Mechanics Division, Vol. 108, No. EM1, February, 1982, pp. 174-179

The load-deformational behavior of twisting of pipes in the presence of axial force is examined. Based on the small deformation theory, charts and curves, which relate the twisting moment, axial force, and twist rate per unit length, are presented for the design of pipes with creep. The accuracy of the results are demonstrated by comparing the maximum stress and twist rate due to pure tension of fairly heavy pipes, with that of more exact thick-walled cylinder solutions.

0203 Large Deflections of Beams with Combined Loads

John H. Lau, (Sr. Engr., Special Structures Group, Bechtel Power Corp., San Francisco, Calif. 94119)

Journal of the Engineering Mechanics Division, Vol. 108, No. EM1, February, 1982, pp. 180-185

Closed-form solutions are presented for the problem of large deflection of a cantilever beam with a couple and a concentrated load acting simultaneously at the free end. Dimensionless charts are provided which relate the variables maximum vertical deflections, maximum horizontal deflections, couple and force.

0204 Torsion of Hollow Circular Cylinders with Creep

John H. Lau, (Sr. Engr., Special Structures Group, Bechtel Power Corp., P.O. Box 3965, San Francisco, Calif. 94119)

Journal of the Engineering Mechanics Division, Vol. 108, No. EM1, February, 1982, pp. 185-190

An exact analysis is presented for the load-deformational behavior of twisting of hollow circular cylinders made of materials that creep. The hyperbolic sine law is used. Curves which relate the variables, twisting moment, twist rate per unit length, and maximum shear stress are provided. The results presented can also be applied to nonstationary creep under certain conditions.

0205 Creep of Pipes Under Axial Force and Bending Moment

John H. Lau, (Sr. Engr., Special Services Group, Bechtel Power Corp., San Francisco, Calif. 94119) and **Teresa T. Lau**, (Professional Staff, Fairleigh Dickinson Univ., Teaneck, N.J.)

Journal of the Engineering Mechanics Division, Vol. 108, No. EM1, February, 1982, pp. 190-195

The load-deformational behaviour of a thin-walled circular cylinder in the presence of axial force is studied. Based on the small deformation theory, an exact analysis is presented for the bending of the cylinder. Results are presented in the form of the charts with two design parameters. The use of these charts is shown by examples for determining the stress and strain distributions, maximum stress and strain, resultant force and moment, and pipe-wall thickness.

0206 Plane Curved Beam Equations

Debasish K. Roy, Assoc. Member, ASCE, (Asst. Prof. of Civ. Engrg., 210A Anderson Hall, Univ. of Kentucky, Lexington, KY 40506)

Journal of the Engineering Mechanics Division, Vol. 108, No. EM1, February, 1982, pp. 196-201

It is demonstrated that plane curved beam (rectangular cross-section) equilibrium equations can be obtained by reducing the established equations for thin shells. The necessary equations are presented for the special case of curved beams of rectangular cross section having the reference surface located eccentrically to the centroidal surface.

0207 Beams, Frames and Shells on Elastic Foundations

Josef Vykutil, (Research Engr., VITKOVICE—Inst. of Applied Mech. BRNO, Czechoslovakia)

Journal of the Engineering Mechanics Division, Vol. 108, No. EM1, February, 1982, pp. 202-205

Explicit stiffness matrix, present matrix and nodal forces are used as a basis for computer analysis of beams, rings, simple frames, and branched shells of revolution on a three-parameter foundation model. This technique demands neither auxiliary handling nor does it result in a decrease in accuracy when applied to the beam or shell with a general variation of the elastic modulus, thickness, or foundation factors along the beam or shell meridian.

0208 Optimum Membrane Trough

Chang-Yi Wang, (Prof. of Applied Math., Michigan State Univ., E. Lansing, Mich. 48824)

Journal of the Engineering Mechanics Division, Vol. 108, No. EM1, February, 1982, pp. 206-211

A fluid is contained in an inextensible membrane trough. Due to hydrostatics the shape of the trough is governed by a set of nonlinear differential equations which is solved by perturbations and numerical integration.

0209 Simplified Bending Theory for Wire Rope

George A. Costello, Member, ASCE, (Prof., Dept. of Theoretical and Applied Mechanics, Univ. of Illinois at Urbana-Champaign, Ill. 61801) and **Gary J. Butson**, (Grad. Student, Dept. of Theoretical and Applied Mechanics, Univ. of Illinois at Urbana-Champaign, Urbana, Ill. 61801)

Journal of the Engineering Mechanics Division, Vol. 108, No. EM2, April, 1982, pp. 219-227

A theory has been developed to predict the static response of wire rope subjected to tension, torsion and bending. This type of loading occurs frequently in ropes wrapped around sheaves. Since the rope is generally restrained against rotation, a twisting moment is developed in the rope in addition to the tension and bending. Equilibrium requires that a uniformly distributed couple per unit length be applied to the rope. Expressions are presented for the significant stresses in the rope. The results can be extended so as to include more complex cross sections other than the single lay rope considered herein.

0210 Analytic Solution for Continuous Sandwich Plates

Kinh H. Ha, Member, ASCE, (Assoc. Prof., Centre for Building Studies, Concordia University, Montreal, Quebec, Canada), **Raafat Hussein**, Assoc. Member, ASCE, (Grad. Student, Centre for Building Studies, Concordia University, Montreal, Quebec, Canada) and **Paul Fazio**, Member, ASCE, (Prof. and Dir., Centre for Building Studies, Concordia University, Montreal, Quebec, Canada)

Journal of the Engineering Mechanics Division, Vol. 108, No. EM2, April, 1982, pp. 228-241

An analytical solution is presented for simply supported continuous sandwich plate over two spans and subjected to transverse loads. The process involves utilizing the force method in conjunction with the principle of superposition. Solutions for simply supported single sandwich plates subjected to transverse loads and edge moments are found first, and then combined to ensure deformation compatibility along the middle support. The effects of transverse shear deformations on the response of continuous plates are found to be substantial, unlike the case of simply supported single plate. Simple expressions are derived for uniform distributed and concentrated loads.

0211 Vertical Seismic Load Effect on Hysteric Columns

Teng-Yuan Shih, (Consultant, Nuclear Technology, Inc., San Jose, Calif.) and **Y. K. Lin**, (Prof. of Aeronautical and Astronautical Engrg., Univ. of Illinois at Urbana-Champaign, Urbana, Ill.)

Journal of the Engineering Mechanics Division, Vol. 108, No. EM2, April, 1982, pp. 242-254

An analytical method is developed to calculate the statistical properties of the response of hysteretic columns under earthquake excitations, with special emphasis on the effect of the vertical ground acceleration. Both the vertical and horizontal ground accelerations are modeled as amplitude-modulated Gaussian random processes, and the restoring force in a column is assumed to follow a functional relationship proposed by Hata and Shibata. Numerical results are presented for two columns, having two contrasting values of the post-yielding to pre-yielding stiffness ratio of 0.5 and 0.1, respectively. These results indicate that the vertical ground acceleration has an aggravating effect similar to that of the gravitational force in causing greater peak mean-square structural response than one induced by the horizontal ground acceleration alone. The time variation of the mean-square response is found to be extremely sensitive to structural parameters. For softer structures with a small post-yielding to pre-yielding stiffness ratio, large permanent deformations may remain after termination of the earthquake excitations.

0212 Exact Involute Bodies of Revolution

Nicholas J. Pagano, (Materials Research Engr., AFWAL/MLBM, Dept. of the Air Force, Wright-Patterson Air Force Base, Ohio 45433)

Journal of the Engineering Mechanics Division, Vol. 108, No. EM2, April, 1982, pp. 255-276

Involute construction is an approach used in contemporary rocket nozzle technology for the fabrication of exit cones and other bodies of revolution and consists of laminating identical composite sheets of uniform thickness along curved surfaces, called involute surfaces. In this work, the exact solution is derived for these surfaces, which guarantees that the volume of the body being generated is precisely filled, with no gaps or overlaps. The equations of the developed surface, or ply pattern, a formulation of the kinematics of the fabrication debulking process, and an experimental validation of the theoretical work are also presented.

0213 Initial Conditions and Korteweg-de Vries Solitons

Patrick D. Weidman, (Assoc. Prof. of Mechanical Engrg., Univ. of Colorado, Boulder, Colo.) and **Larry G. Redekopp**, (Prof. of Aerospace Engrg., Univ. of S. Calif., Los Angeles, Calif.)

Journal of the Engineering Mechanics Division, Vol. 108, No. EM2, April, 1982, pp. 277-289

A study is made of the asymptotic waveforms described by the Korteweg-de Vries equation ($u_t + 6uu_x = u_{xxx} = 0$) evolving from initial data composed of discrete disturbances of elevation and depression. The analytical approach is based on the inverse scattering transform which relates the initial condition u(x,0) to Schrodinger's equation. Solution of direct scattering problem for an initial disturbance composed of three contiguous rectangles is obtained. The poles of the reflection coefficient determine the number of solitons in the asymptotic state, and specific results are obtained for the following types of initial data: two contiguous rectangles; three contiguous rectangles which are symmetric in x; and two distinct separated rectangles. The competing effect of disturbances of elevation (which give rise to solitons) and depression (which give rise to dispersive waves) is studied in detail. The results can be summarized by stating that the number of evolved solitons depends on the strength of each rectangular disturbance, the relative amplitudes of the rectangular disturbances, and the relative proximity of the disturbances.

0214 Simplified Earthquake Analysis of Suspension Bridge Towers

Ahmed M. Abdel-Ghaffar, Assoc. Member, ASCE, (Asst. Prof., Civ. Engrg. Dept., Princeton Univ., Princeton, N.J.) and **Joel D. Rood**, (Schlumberger, Texas)

Journal of the Engineering Mechanics Division, Vol. 108, No. EM2, April, 1982, pp. 291-308

Free vibration and earthquake response analyses of a simplified continuous analytical model representing suspension bridge towers (fixed at the base) are presented. In addition, various models for seismic analysis of the towers are proposed. These models take into account, among other things, the flexibility of the surrounding soils and the multiple-seismic inputs. The earthquake response of some existing suspension bridges' towers, including the Golden Gate Bridge and the Vincent-Thomas Bridge is analyzed, and stresses (both vibrational and quasi-static) induced by different earthquake ground motions in these two bridges are examined. Both time history response and the classical response spectra are considered.

0215 Probability of Survival in a Blast Environment

Anatole Longinow, (Asst. Prof., Dept. of Civ. Engrg., Valparaiso Univ., Valparaiso, Ind. 46583), **Kuang-Han Chu**, (Prof., Dept. of Civ. Engrg., Illinois Inst. of Tech., Chicago, Ill.) and **Nick T. Thomopoulos**, (Assoc. Prof., Harold Leonard School of Management and Finance, Illinois Inst. of Tech., Chicago, Ill.)

Journal of the Engineering Mechanics Division, Vol. 108, No. EM2, April, 1982, pp. 309-330

The probability of survival of people located in personnel shelters when subjected to the blast effects produced by the detonation of a 1-MT weapon near the ground surface is studied. Personnel shelters analyzed are buried or with the roof slab at grade. The roof slap is the weakest structural component. Its partial or total collapse results in casualties. When the shelter is thus breached, casualties can also be produced by the blast loading when overpressures are sufficiently high. The analysis formulated predicts the probability of roof slab survival when subjected to an airblast loading. Probability of people survival against slab collapse is estimated based on the fraction of shelter area unaffected by collapse. Probability of survival against blast pressure entering the shelter area is determined using available casualty data. Results are combined to yield the probability of survival against both effects. The analysis considers the variability of selected geometric parameters of the roof slab, material properties and airblast parameters.

0216 Vertical Seismic Load Effect on Building Response

Y. K. Lin, (Prof. of Aeronautical and Astronautical Engrg., Univ. of Illinois at Urban-Champaign, Urbana, Ill.) and **Teng-Yuan Shih**, (Consultant, Nuclear Technology Inc., San Jose, Calif.)

Journal of the Engineering Mechanics Division, Vol. 108, No. EM2, April, 1982, pp. 331-343

An analytical procedure is presented for the calculation of the statistical properties of the response of a linear elastic tall building under earthquake excitations. The emphasis is placed on the effect of the vertical ground motion. The restoring force in each story of the structural model is assumed to arise from the bending deformation of the columns whose rigidities are subject to a general reduction due to gravitational forces and to a random variation due to vertical ground accelerations. To illustrate the application of the procedure, numerical results are presented for a 6-story building. Within the limit of linear elastic deformation, the vertical ground acceleration is shown to be capable of causing an increase in certain response variables in the order of 10% for this moderately tall building. The percentage is expected to be larger for a taller building and much larger when the deformations exceed the elastic limit.

0217 Frictional Slip in Elasticity: An Example

John Dundurs, Member, ASCE, (Prof. of Civ. Engrg., Northwestern Univ., Evanston, Ill.) and **Maria Comninou**, Member, ASCE, (Assoc. Prof. of Civ. Engrg., Univ. of Michigan, Ann Arbor, Mich.)

Journal of the Engineering Mechanics Division, Vol. 108, No. EM2, April, 1982, pp. 334-353

In order to study various aspects of propagating slip zones with friction, the unbounded solid with a semi-infinite cut are considered. The solid is first compressed in a direction normal to the cut and subsequently subjected to a concentrated force, which is also normal to the cut and applied at its tip. The concentrated force, which varies with time, induces slip zones. The problem can be solved in closed form, and consequently it is possible to follow not only loading but also unloading, reloading and periodically fluctuating loads.

0218 Sandwich Bean-Columns with Interlayer Slips

Paul Fazio, Member, ASCE, (Prof. and Dir., Centre for Building Studies, Concordia Univ., Montreal, Quebec, Canada), **Raafat Hussein**, Assoc. Member, ASCE, (Grad. Student, Centre for Building Studies, Concordia Univ., Montreal, Quebec, Canada) and **Kinh H. Ha**, Member, ASCE, (Assoc. Prof., Centre for Building Studies, Concordia Univ., Montreal, Quebec, Canada)

Journal of the Engineering Mechanics Division, Vol. 108, No. EM2, April, 1982, pp. 354-366

Elasticity solution for sandwich beam-columns bonded with non-rigid adhesives is presented. The solution satisfies the equilibrium equations of the face and core elements, the compatibility equations of stresses and strains at the interfaces, and the relevant boundary conditions. To investigate the effects of adhesive stiffnesses on the responses of sandwich beam-columns, parametric studies are conducted. The numerical results show that the adhesive stiffness, up to a certain level, has a strong effect on the beam deflection and the column buckling load. Beyond this level, the usual assumption of perfect bonding used in classical theories is quite acceptable. An answer to what constitutes "perfect" bonding is found in terms of the ratio of the core stiffness to the adhesive stiffness. To facilitate the appication of the present solution, simple expressions are presented for the maximum normal force in facings and interlayer shear flux of sandwich beams subjected to various loading types.

0219 Response of MDOF Systems to Vector Random Excitation

Amitabha DebChaudhury, (Asst. Prof., Dept. of Materials Engrg., Univ. of Illinois at Chicago Circle, Chicago, Ill. 60680) and **Dario A. Gasparini**, (Asst. Prof., Dept. of Civ. Engrg., Case Western Reserve Univ., Cleveland, Ohio 44106)

Journal of the Engineering Mechanics Division, Vol. 108, No. EM2, April, 1982, pp. 367-385

A model time history random vibration analysis is presented. The random excitation is nonwhite, nonstationary, vector-valued and correlated. Excitation components may have different frequency contents and variations in intensity with time. The correlation between any two components can be a function of time. The formulation is appropriate for any linear MDOF system which has been decoupled into modes. Analytical expressions are derived for the evolutionary mean and covariance matrices of a modal system augmented by filters and subjected to vector-valued excitation having piece-wise linear strength functions. Then, using modal superposition, evolutionary mean and covariance matrices of any response vector are computed. A three degree-of-freedom system, subjected to vector valued earthquake excitation, is analyzed. The formulation captures the effects of nonstationarity and correlation between earthquake components.

0220 Nonuniform Laminar Free-Surface Flow

Ioannis K. Tsanis, (Grad. Student, Dept. of Civ. Engrg., Univ. of Toronto, Toronto, Ontario M5S 1A4, Canada) and **Hans J. Leutheusser**, Member, ASCE, (Prof., Dept. of Mechanical Engrg., Univ. of Toronto, Toronto, Ontario M5S 1A4, Canada)

Journal of the Engineering Mechanics Division, Vol. 108, No. EM2, April, 1982, pp. 386-398

Industrial flow processes involving molten substances or other highly viscous liquids frequently make use of conveyance by open channels. A one-dimensional analysis is undertaken of laminar flow in horizontal open channels of rectangular cross-section. A closed-form analytical solution is obtained for the non-dimensional surface profile of type H_2 and H_3. The non-dimensional profiles are independent of the Reynolds number of flow but are a distinct function of the aspect ratio of the flow but are a distinct function of the aspect ratio of the flow cross-section. The analytically predicted profiles are found to be in good agreement with the results of experimental observations, except in the immediate vicinity of the critical depth where small discrepancies occur. In distinct contrast to turbulent flow, the location of the critical depth upstream from the brink of a vertical channel drop is found to increase with the Reynolds number of flow.

0221 Schwarz-Christoffel Theory of Flow Past an Opening

Cheng-lung Chen, Member, ASCE, (Hydro., U.S. Geological Survey, Gulf Coast Hydroscience Center, National Space Technology Lab., NSTL Station, Miss. 39529) and **Ming-Yang Su**, (Oceanographer, Naval Ocean Research and Development Activity, Oceanography Div., NSTL Station, Miss. 39529)

Journal of the Engineering Mechanics Division, Vol. 108, No. EM2, April, 1982, pp. 399-418

The Schwarz-Christoffel transformation for a steady two-dimensional plane flow past an eccentric, normal constriction of zero thickness is formulated. The transformation formula contains two unknown complex constants and four unknown parameters (real numbers). It is found that there are five identities for the parameters and two additional equations for two additional physical parameters representing the geometry of the constriction in the channel. Except one of the identities, they are all expressed in the form of elliptic integrals of the first and third kinds; four of them are sufficient for unique solutions of the four parameters. The five identities that do not contain the physical parameters cannot be used in the method of solution alone, but the remaining two equations relating the two physical parameters to the four unknown parameters of the transformation can be reduced to a set of three simple trigonometric relationships, from which the analytical solutions of the parameters are obtained. Once the parameters are determined, the corresponding flow pattern can be readily computed from the transformation.

0222 Suspension Bridge Response to Multiple-Support Excitations

Ahmed M. Abdel-Ghaffer, Assoc. Member, ASCE, (Asst. Prof., Civ. Engrg. Dept., Princeton Univ., Princeton, N.J.) and **Lawrence I. Rubin**, (Grad. Student, Civ. Engrg. Dept., Princeton Univ., Princeton, N.J.)

Journal of the Engineering Mechanics Division, Vol. 108, No. EM2, April, 1982, pp. 419-435

The earthquake-induced vertical response of suspension bridges when subjected to, correlated or uncorrelated, multiple-support excitations is analyzed by means of random vibration theory. Appropriate ground motions describing the basic inputs to such structures are defined through finite Fourier transforms of existing recorded ground displacements at locations within distances similar to the spans of such structures. Spectral correlation analysis of the multiple-support excitations are estimated via these Fourier transforms. It is found that studies of earthquake-induced vertical motion of suspension bridges need to take into account the influences of ground excitations in the longitudinal directions of such bridges. An example is presented in which the response of an existing bridge (in a seismically active area) to earthquake ground motions with characteristics significantly different at the supports is examined. It is observed that the participation of higher modes in the total response is essential to reliably assess the seismic behavior of such structures and that the response values associated with correlated multiple-support excitations are significantly different from those obtained through the uncorrelated case.

0223 Boundary Element Analysis of Fluid Domain

Youssef G. Hanna, (Grad. Student, Dept. of Civ. Engrg., Carleton Univ., Ottawa, Ontario, Canada) and **Jagmohan L. Humar**, (Assoc. Prof., Dept. of Civ. Engrg., Carleton Univ., Ottawa, Ontario, Canada)

Journal of the Engineering Mechanics Division, Vol. 108, No. EM2, April, 1982, pp. 436-450

The boundary element method of analysis is applied to the calculation of hydrodynamic pressures in a reservoir impounded by a gravity dam and subjected to a harmonic ground motion., It is shown that the differential equation governing the small amplitude vibrations of the reservoir reduces to the familiar Helmholtz equation. This equation is converted to an integral equation which involves integrations only on the boundary of the reservoir. The integrals are evaluated numerically by using a series of straight line elements to represent the boundary. The resulting simultaneous equations are solved by standard techniques to obtain the pressures and pressure derivations on the boundary. Analytical results are presented to show that the proposed method gives results that are in good agreement with the known classical solutions. The method is then applied to the solution of vibration problems of reservoirs with irregular boundaries for which classical solutions do not exist. The method is also successful in dealing with cases of infinite reservoirs and in modelling the energy loss in the waves moving to the infinity.

0224 Mixed Formulation of Finite Strip Method

Y. K. Cheung, (Prof. and Head, Dept. of Civ. Engrg., Univ. of Hong Kong, Hong Kong) and **L.C. Tham**, (Post-doctoral Fellow, Faculty of Engrg. Sci., Univ. of Western Ontario, Canada)

Journal of the Engineering Mechanics Division, Vol. 108, No. EM2, April, 1982, pp. 452-456

A least square formation of finite strip method is presented. The method uses a mixed formulation of both stresses and displacements, and minimizes the sum of the square of the residual. The results are compared with the standard finite strip method by displacement model. Two examples illustrate that the method can deal with a wide range of two-dimensional elasto-static problems, and can be easily extended to three dimensional cases. The proposed method is also applicable to other types of boundary value problems.

0225 Matrix Eigenvalue Problem of Multiple-Shaker Testing

Clarence W. de Silva, (Engr., Dept. of Mech. Engrg., Carnegie-Mellon Univ., Pittsburgh, PA.)

Journal of the Engineering Mechanics Division, Vol. 108, No. EM2, April, 1982, pp. 457-461

It is shown that the problem of dynamic testing using multiple shaker excitations can be represented by a matrix-eigenvalue formulation. A problem is formulated and an optimal solution is described. The normalized eigenvector corresponding to the maximum eigenvalue provides the direction of excitation for an equivalent single shaker. The corresponding optimal intensity of excitation is expressed in terms of the eigenvalues.

0226 Cyclic Bending of Plates under Transverse Loading

Yuhshi Fukumoto, Member, ASCE, (Prof., Dept. of Civ. Engrg., Nagoya Univ., Nagoya, Japan) and **Haruyuki Kusama**, (Grad. Student, Dept. of Civ. Engrg., Nagoya Univ., Nagoya, Japan)

Journal of the Engineering Mechanics Division, Vol. 108, No. EM3, June, 1982, pp. 477-492

The cyclic bending behavior of simply supported steel square plates under transverse loads is studied. Three cyclic conditions, the fixed deflection amplitude, and the fixed load amplitude are considered. A method of elastic and perfectly plastic large deflection small strain analysis is explained and the hysteretic characteristics of plates are obtained for three width-thickness ratios ($b/t=40$, 60 and 80) under two loading conditions, uniformly distributed load and center patch load. The influence of membrane action and the spread of the plastic zone on the shape of the hysteretic loops, the variation of the force and bending moment intensities, and the deterioration of plates are also theoretically clarified.

0227 Progressive Collapse of Rigid-Plastic Circular Foundations

George Gazetas, Assoc. Member, ASCE, (Assoc. Prof. of Civ. Engrg., Rensselaer Polytechnic Inst., Troy, N.Y. 12181)

Journal of the Engineering Mechanics Division, Vol. 108, No. EM3, June, 1982, pp. 493-508

The paper presents an analytical study of the behavior of a rigid-perfectly plastic circular foundation plate indenting an elastic two-parameter soil layer (Vlasov soil) under the action of a statically increasing applied load. Closed-form solutions are obtained for critical Loads, maximum surface settlements, foundation deflections and soil reactions in terms of the load intensity, after the plate has been transformed into a mechanism. Two distinct phases of metaplastic behavior are identified depending on whether or not full contact is maintained between soil and foundation. The results are compared with those obtained for a Winkler space and an elastic continuum halfspace.

0228 Crack Growth in Fiber-Reinforced Materials

James A. Mandel, Member, ASCE, (Prof., Dept. of Civ. Engrg., Syracuse Univ., Syracuse, N.Y. 13210) and **Steven C. Pack**, (Grad. Research Asst., Syracuse Univ., Syracuse, N.Y. 13210)

Journal of the Engineering Mechanics Division, Vol. 108, No. EM3, June, 1982, pp. 509-526

Engineering models for initiation of crack growth, crack arrest, and the effective fracture toughness of fiber reinforced materials are developed. Emphasis is placed on brittle matrix material, reinforced with a small percentage by volumm of short, uniformly distributed fibers of a material with a higher Young's modulus. The analysis is first reduced to two dimensions by considering fiber-reinforced plates and assuming the fiber areas to be distributed uniformly through the thickness of the plate. A parametric micro-mechanical finite element study is conducted using a mesh which includes quarter-point triangular crack tip elements, quadratic isoparametric quadrilateral elements, bar elements, and bond interface elements. An experimental program is then conducted. Edge notched speciments of methyl methacrylate with a row of closely spaced fibers near the crack tip and control specimens without fibers are cast and tested. The close agreement between the results of the finite element analysis and the experimental study show that the finite element method, if properly applied, can be a useful tool in micromechanical studies of fiber reinforced materials.

0229 Random Shrinkage Stresses in Aging Viscoelastic Vessel

Tatsuya Tsubaki, (Asst. Prof. of Civ. Engrg., Yokohama National Univ., 156 Yokiwadai, Hodogaya-Ku, Yokohama, Japan 240) and **Zdenek P. Bazant**, Fellow, ASCE, (Prof. of Civ. Engrg., Northwestern Univ., Evanston, Ill. 60201)

Journal of the Engineering Mechanics Division, Vol. 108, No. EM3, June, 1982, pp. 527-545

Shrinkage stresses caused by random variation of environmental humidity in a long cylindrical concrete wall sealed at the internal surface are analyzed, taking creep and aging of concrete into account. The random time-variation of humidity and stress is analyzed by the spectral method using the elastic-viscoelastic analogy. Spatial variations are described by Bessel and Kelvin functions. Numerical examples typical of a 1 m thick reactor containment and a thin shell of 10 cm thickness are given. The environmental humidity varies with a dominant period of one year. The results confirm that, due to low moisture diffusivity, as well as aging, the problem must be analyzed as nonstationary. The mean and the standard deviation are obtained as functions of time as well as location. Random stress fluctuations are found to be significant in all cases and the influence of aging to be strong. In the thicker containment only about 30 percent of the wall thickness ever feels the random variation, while in the thin shell the entire thickness is affected.

0230 Body Wave Excitations of Embedded Hemisphere

Vincent W. Lee, (Research Asst. Prof., Dept. of Civ. Engrg., Univ. of Southern California, University Park, Los Angeles, Calif. 90007) and **Mihailo D. Trifunac**, Aff., (Prof., Dept. of Civ. Engrg., Univ. of Southern California, University Park, Los Angeles, Calif. 90007)

Journal of the Engineering Mechanics Division, Vol. 108, No. EM3, June, 1982, pp. 546-563

Solutions of three-dimensional vibrations of an embedded hemispherical foundation are presented for excitation by incident plane P, SV, and SH waves with arbitrary angles of incidence. The effects of embedment are illustrated by comparing the motion of the embedded hemispherical foundation with those of a rigid disc bonded to a half space of equal radius and excited by the same waves. The impedence matrix and the foundation motion are analyzed and compared with similar and reLated approximate solutions. In the context of Fourier synthesis, the results can be viewed as representing the transfer functions for the forces and motions of the foundation. For arbitrary transient excitation which can be represented by plane body waves, the results can be used to calculate the transient motion of the foundation.

0231 Vertical Structure of Internal Modes in Stratified Flows

Angelos N. Findikakis, Assoc. Member, ASCE, (Sr. Engr., Hydr./Hydrology Grp., Bechtel Civ. and Minerals Inc., San Francisco, Calif.) and **Robert L. Street**, Member, ASCE, (Prof., Dept. of Civ. Engrg., Stanford Univ., Stanford, Calif. 94305)

Journal of the Engineering Mechanics Division, Vol. 108, No. EM4, August, 1982, pp. 583-595

An analytic solution of the linearized two-dimensional equations for stably stratified flows driven by a surface shear stress which is periodic in the horizontal direction, and in time, is developed. The solution is based on the assumptions of a fluid with infinite Prandtl number and a linear thermal stratification which is constant in each horizontal plane. The analytic solution is used to examine the joint effect of the viscosity value, the frequency of the driving force and the intensity of the stratification on the vertical structure of the internal modes. It is shown that as the viscosity value increases, the formation of higher-order modes is prohibited. The validity of the assumptions of the analytic solution is tested by comparing it with numerical solutions of the same problem.

0232 Elastic and Particulate Media

James M. Hill, (Sr. Lect., Dept. of Mathematics, Univ. of Wollongong, New South Wales, Australia) and **Milton E. Harr**, Fellow, ASCE, (Prof. of Geotechnical Engrg., Purdue Univ., Lafayette, Ind.)

Journal of the Engineering Mechanics Division, Vol. 108, No. EM4, August, 1982, pp. 596-604

Constitutive relations for soil are proposed which given rise to a diffusion-like equation for the normal stress component. This partial differential equation includes both particulate media theory and the classical linear theory of elasticity, in the sense that the solutions of concentrated load problems for half-spaces, coincide with those due to Flamant and Boussinesq for special values of the contained parameters. This means that all stress boundary value problems for infinite half-spaces in the classical linear theory of elasticity arise as special cases of the theory described here. On the other hand the particulate media theory arises from probabilistic considerations and the results given present for the first time a formal connection with classical elasticity theory.

0233 Dynamic Analysis of Underwater Cables

Alain Lo, (Research Asst., Dept. of Civ. Engrg., Oregon State Univ., Corvallis, Oreg. 97331) and **John W. Leonard**, Assoc. Member, ASCE, (Prof., Dept. of Civ. Engrg., Oregon State Univ., Corvallis, Oreg. 97331)

Journal of the Engineering Mechanics Division, Vol. 108, No. EM4, August, 1982, pp. 605-621

A general approach is presented for the finite element analysis of cable systems under hydrodynamic influences. The drag and inertia effects of fluid are included using a Morrison type equation. Curved cable elements are used to improve the geometric approximation of cable segements so that less elements are required than if straight elements are used. The formulation can also handle geometric and material nonlinearities and the effects of any discrete elements can be lumped at nodal points. The principle of virtual work is used to formulate the governing equations and an incremental solution scheme is adopted to simulate the nonlinear effects. The development of the model is presented in a general setting so that further extension or inclusion of other hydrodynamic models is possible. Numerical examples demonstrate the validity and capability of the mathematical formulation. Excellent comparison to experimental results are obtained using a relatively coarse grid of curved cable elements.

0234 Stationary and Transient Response Statistics

Peter H. Madsen, (Ph.D. Student, Riso National Lab., 4000 Roskilde, Denmark) and **Steen Krenk**, (Research Engr., Riso National Lab., 4000 Roskilde, Denmark)

Journal of the Engineering Mechanics Division, Vol. 108, No. EM4, August, 1982, pp. 622-635

The covariance functions for the transient response of a linear MDOF-system due to stationary time limited exitation with an arbitrary frequency content are related directly to the covariance functions of the stationary response. For rational spectral density functions closed form expressions for the covariance functions are obtained. In case of white noise excitation the expressions simplify considerably, and an approximative method for including the main effects of an actual spectral density function is derived based on a white noise analogy. The theory is applied to a multistorey frame structure, and it is demonstrated that it offers an efficient way to calculate the time-dependent second-order moments. The approximation appears to yield fully adequate results for many practical applications involving lightly damped structures.

0235 Large Deflection of Curved Beam with Shear Deformation

Dr. Izhak Sheinman, (Sr. Lect., Faculty of Civ. Engrg., Technion-Israel Inst. of Tech., Haifa, Israel)

Journal of the Engineering Mechanics Division, Vol. 108, No. EM4, August, 1982, pp. 636-647

A general analytical and numerical procedure, based on large deflections, small strains and moderately small rotations, is developed for an arbitrary plane curved beam made of linear elastic material and subjected to arbitrary conservative static loading. The equilibrium equations admit shear deformation, geometric imperfections, and the parameters of the linear elastic foundation. The numerical solution is obtained by a modification of Newton's method and by finite differences. Three numerical examples are presented, and the influence of shear stiffness is considered.

0236 Bridge Impact due to Wheel and Track Irregularities

Apiwan Wiriyachai, (Grad. Student, Civ. Engrg. Dept., Illinois Inst. of Tech., Chicago, Ill.), **Kuang-Han Chu**, Fellow, ASCE, (Prof. of Civ. Engrg., Illinois Inst. of Tech., Chicago, Ill.) and **Vijay K. Garg**, Fellow, ASCE, (Mgr., Dynamics Research Div., Tech. Center, Assoc. of American Railroads, Chicago, Ill.)

Journal of the Engineering Mechanics Division, Vol. 108, No. EM4, August, 1982, pp. 648-666

The impact factors in bridge members due to flat wheels and track irregularities were investigated. The maximum wheel flat height was taken as 0.2 in. (5 mm), maximum pier settlement as –0.25 in. (–6.3 mm) and maximum camber error as –0.5 in. (–12.5 mm). The track roughness spectra for class 6 track was used in the analysis, and only selected members of the 175 ft. pinned connections truss bridge were studied. The train of three 70 ton freight cars with spring constants of 11.2 kip/in. (1.96 kn/mm) per wheel were used. The train was moving at various speeds and with initial displacements of 0.25 in. (6.3 mm) and roll of 0.02 rad. Damping of the bridge and vehicle were neglected. The impact factors for the train moving at 50 mph (22.35 m./sec.) seem to agree fairly well with the current American Railway Engineering Association specification. Other conclusions are summarized.

0237 Cylindrical Shells of Bimodulus Composite Material

Charles W. Bert, (Benjamin H. Perkinson Prof. of Engrg., Univ. of Oklahoma, Norman, Okla.) and **V. Sudharkar Reddy**, (Grad. Research Asst., Mech. Engrg., Univ. of Oklahoma, Norman, Okla.)

Journal of the Engineering Mechanics Division, Vol. 108, No. EM5, October, 1982, pp. 675-688

Certain fiber-reinforced materials, especially those with slightly curved fibers in very soft matrices, exhibit considerably smaller stiffnesses when loaded in compression than when loaded in tension. Examples are tire cord-rubber, wire-reinforced solid propellants, and certain soft biological tissues. For purposes of analysis and design, such materials can be modeled as a bimodulus material, i.e., one having one set of stiffnesses when the fiber direction strain is tensile and another set when this strain is compressive. Using the fiber-governed bimodulus-material model introduced several years ago and verified for cord-rubber composites, the previous work is extended on the deflection of single-layer and cross-ply laminated rectangular plates to circular cylindrical shells of the same construction. A closed-form solution is presented for the case of a simply supported cylindrical panel under sinusoidally distributed loading. Numerical results are presented to show the effect of shell curvature on the neutral-surface position and deflection.

0238 Potential Flow over Weirs of Moderate Curvature

Peter A. Davidson, (Research Engr., T.I. Research Lab., Hinxton, England) and **George D. Matthew**, (Reader in Engrg., Univ. of Aberdeen, Marischal Coll., Aberdeen, Scotland)

Journal of the Engineering Mechanics Division, Vol. 108, No. EM5, October, 1982, pp. 689-707

A review is given of some salient studies of critical flow over round-crested weirs, in particular Lenau's extension of Matthew's results. The basic equations of two-dimensional, incompressible, potential flow are recapitulated and applied to general free-surface, gravity flows. Approximate solutions in the form of series expansions about the free surface for ψ, ln v, and Θ in terms of x and y are derived. An assumed free surface profile of simple exponential form is investigated and is shown to represent a weir type of flow provided certain relationships exist between the free surface equation parameters. These are related to the ratio of total head above crest to radius of curvature of that crest (H/R) and the coefficient of discharge and the weir surface profile are deduced as series functions of that ratio. Compatibility with the Lenau theory is shown. This method brings out both the attractions and difficulties inherent in an inverse method remaining within a cartesian framework.

0239 System Reliability Bounding by Conditioning

Ove Ditlevsen, (Prof. of Applied Mathematics in Civ. Engrg., Danmarks Ingeniorakademi, Building 373, DK 2800 Lyngby, Denmark)

Journal of the Engineering Mechanics Division, Vol. 108, No. EM5, October, 1982, pp. 708-710

Series systems that may fail in one of several unimodes and which have normally distributed unimode safety margins may be analysed with respect to reliability in terms of bounds reported in literature. For the majority of the correlation coefficients between the safety margins smaller than about 0.5 these bounds are very narrow Provided the total reliability is high and even if there is not just some very few dominating unimodes. It is shown how these bounds may be applied in a method of conditional sounding to give close bounds also in a case where most or all of the correlation coefficients are high. In a specific example the method is compared to another known method based on product form correlation coefficients.

0240 Active Control of Large Structures

Mohamed Abdel-Rohman, Assoc. Member, ASCE, (Asst. Prof., Dept. of Civ. Engrg., Coll. of Engrg. and Petroleum, Kuwait Univ., P.O. Box 5969, Kuwait)

Journal of the Engineering Mechanics Division, Vol. 108, No. EM5, October, 1982, pp. 719-730

In this paper, it is shown that the pole assignment method through which one can control high degrees of freedom structures can be used to determine an optimal control law. The finding of this paper is that among all possible gain matrices which can be found by the pole assignment method there is only one gain matrix which is optimal. This result is proven when the poles of an optimal control system are assigned as the desired poles and the optimal gain matrix is found among the all possible gain matrices. This finding may solve the difficulties met for controlling large structures in an optimal way. An application to optimal control of two-story building frame is given.

0241 Mechanics of Three-Dimensional Soil-Structure Interaction

C. S. Desai, (Prof., Dept. of Civ. Engrg. and Engrg. Mech., Univ. of Arizona, Tucson, Ariz. 85721), **H. V. Phan**, (Assoc. Prof. of Agr. Engrg., Virginia Tech., Blacksburg, Va.) and **J. V. Perumpral**, (Bechtel Power Corp., San Francisco, Calif.)

Journal of the Engineering Mechanics Division, Vol. 108, No. EM5, October, 1982, pp. 731-747

A finite element procedure for the general problem of three-dimensional soil-structure interaction involving nonlinearities caused by material behavior, geometrical changes, and interface behavior is presented. The formulation is based on the updated Lagrangian or approximate Eulerian approach with appropriate provision for constitutive laws. Consideration is given to the mathematical and experimental aspects related to the important topic of development and adoption of a constitutive law for the geologic medium. development and use of an interface element to describe the three-dimensional interface behavior are described. The influence of cracking and tensile stress conditions in the soil and interfaces is allowed for by using the stress transfer approach. The procedure is verified by comparing predictions with observation of a structure (tool) pushed in the soil in a special soil-bin test facility.

0242 Constitutive Relationships for Fluid-Solid Mixtures

Hayley Shen, (Asst. Prof., Dept. of Civ. and Environmental Engrg., Clarkson Coll. of Tech., Potsdam, N.Y. 13676) and **Norbert L. Ackermann**, Member, ASCE, (Prof., Dept. of Civ. and Environmental Engrg., Clarkson Coll. of Tech., Potsdam, N.Y. 13676)

Journal of the Engineering Mechanics Division, Vol. 108, No. EM5, October, 1982, pp. 748-763

Constitutive equations are derived for the rapid shear flow of a mixture formed from granular solids in a fluid. The stress state within the mixture is considered to be created by the momentum exchange between colliding solids. The constitutive equations describe the shear and normal stresses as a function of the velocity gradient, friction and restitution coefficients of the solids, the fluid drag coefficient for the particle shape and the density of the solid and fluid constituents. The theoretical results agree well with the results obtained from numerous laboratory experiments.

0243 Strain-Rate Effect in Rapid Triaxial Loading of Concrete

Zdenek P. Bazant, Fellow, ASCE, (Prof. of Civ. Engrg. and Dir., Center for Concrete and Geomaterials, Northwestern Univ., Evanston, Ill. 60201) and **Byung H. Oh**, (Grad. Research Asst., Northwestern Univ., Evanston, Ill. 60201)

Journal of the Engineering Mechanics Division, Vol. 108, No. EM5, October, 1982, pp. 764-782

Presented is a generalization of time-independent nonlinear triaxial constitutive relations for concrete to model short-time viscoelastic effects. The effect of strain-rate magnitude upon the initial elastic modulus, the peak stress, and the sharpness of the stress peak is taken into account. To obtain a model that also exhibits short-time rapid creep and stress relaxation, the incrementally linear constitutive law fully characterized by tangential moduli is replaced by a nonlinear first-order differential equation in stress, strain, stress rate and strain rate. This requires introduction of the (apparent) instantaneous modulus, the relaxation time and retardation time. The irreversibility at unloading, caused by the short-time viscoelastic mffects, is also analyzed. The model allows more realistic dynamic finite element analysis of concrete structures.

0244 Dynamic Characteristics of Liquid Storage Tanks

Medhat A. Haroun, Assoc. Member, ASCE, (Asst. Prof., Civ. Engrg. Dept., Univ. of California, Irvine, Calif. 92717) and **George W. Housner**, Member, ASCE, (C. F. Braun Prof. of Engrg., California Inst. of Tech., Pasadena, Calif. 91125)

Journal of the Engineering Mechanics Division, Vol. 108, No. EM5, October, 1982, pp. 783-800

A reliable and computationally effective method for calculating the dynamic characteristics of ground-supported cylindrical tanks is developed. The method offers a rigorous treatment of the interaction problem and provides a starting point for the consideration of the complicating factors which influence the dynamic behavior of tanks. The liquid region is treated analytically and only the shell is modeled by finite elements. In this approach, the number of unknowns is substantially less than in those analyses where both tank wall and liguid are subdivided into finite elements. The reliability of the analysis is illustrated by computing modes and natural frequencies of full-scale tanks and comparing them with the results of vibration tests. The effectiveness of the method has led to a development of an improved seismic design procedure for tanks.

0245 Complications in Free Vibration Analysis of Tanks

Medhat A. Haroun, Assoc. Member, ASCE, (Asst. Prof., Civ. Engrg. Dept., Univ. of California, Irvine, Calif. 92717) and **George W. Housner**, Member, ASCE, (C. F. Braun Prof. of Engrg., California Inst. of Tech., Pasadena, Calif. 91125)

Journal of the Engineering Mechanics Division, Vol. 108, No. EM5, October, 1982, pp. 801-818

A simple and computationally effective method for computing the dynamic characteristics of ground-supported, cylindrical liquid storage tanks has been developed. A generalization of this method is presented herein to include some complicating factors which affect these characteristics. It has been shown that the initial hoop stress due to the hydrostatic pressure and the in-plane rigidity of the roof system may affect considerably the cosnΘ-type modes of the tank wall. In addition, the flexibility of the foundation soil can reduce measurably the fundamental natural frequency of the cosΘ-type modes due to the rocking motion. It is also concluded that the coupling between liquid sloshing modes and shell vibrational modes is weak; and consequently, the "convective" pressure can be evaluated with reasonable accuracy by considering the tank wall to be rigid, and the "impulsive" pressure can be determined by analyzing the liquid-shell system and neglecting the sloshing motion. The validity of the method of analysis has been confirmed by both scale model testing and field measurements of the vibrational characteristics of full-scale tanks.

0246 TheoreticaL Model for Collapse of Shear Webs

Cedric Marsh, (Prof., Centre for Building Studies, Concordia Univ., Montreal, Canada)

Journal of the Engineering Mechanics Division, Vol. 108, No. EM5, October, 1982, pp. 819-832

The ultimate load capacity of shear webs in a plate girder with stiffeners is modelled as the sum of the strengths of the web and the flange. The web is considered to be held in the buckled configuration by a shear stress that varies along the boundaries, while the bending strength of the flanges make an independent contribution. This model of the collapse mechanisms gives a direct prediction of the shear force at collapse without the need for empirically established relationships.

0247 Control of Tall Building Under Earthquake Excitation

J. N. Yang, Member, ASCE, (Prof., Dept. of Civ., Mech., and Environmental Engrg., The George Washington Univ., Washington, D.C. 20052)

Journal of the Engineering Mechanics Division, Vol. 108, No. EM5, October, 1982, pp. 833-849

A method of analysis for tall buildings implemented by active control systems and excited by earthquake ground motions is proposed. The effectiveness of the active mass damper and the active tendon system is investigated. Demonstrations show that both control systems can be used to substantially reduce the response of tall buildings under strong earthquake excitations, if the control parameters are designed appropriately. The control law suggested herein is referred to as the semi-active control in the sense that on-line computations to regulate active control forces are not needed. Thus, the control systems investigated are more reliable, although the control law is not optimal. Numerical examples for two tall buildings are worked out to demonstrate the feasibility of active control systems.

0248 Three-Parameter Failure Criterion for Concrete

Poul V. Lade, Member, ASCE, (Prof., Mechanics and Structures Dept., School of Engrg. and Applied Sci., Univ. of Calif., Los Angeles, Calif.)

Journal of the Engineering Mechanics Division, Vol. 108, No. EM5, October, 1982, pp. 850-863

A general, three-dimensional failure criterion for plain concrete and mortar is proposed. This criterion is formulated in terms of the first and the third stress invariants of the stress tensor, and it involves only three independent material parameters. Although these parameters interact with one another, each parameter corresponds to each of three failure characteristics of concrete behavior. These material parameters may be determined from simple tests such as uniaxial compression and triaxial compression or biaxial tests. For the purpose of including reasonable values of tensile strengths in the failure criterion, it is advisable to include the uniaxial tensile strength in the parameter determination. A simple expression for evaluation of the uniaxial tensile strength on the basis of the uniaxial compressive strength is given. Twenty-two sets of good quality data for concrete and mortar have been included in this study, and comparisons between the proposed failure criterion and the experimental data are made in biaxial, triaxial, octahedral, and τ - σ planes. The ability of this criterion to capture the characteristics of failure in concrete and mortar appears to be quite good with accuracies generally within the natural scatter of the test data.

0249 Numerics for Common First-Passage Problem

Polychronis-Thomas D. Spanos, Member, ASCE, (Assoc. Prof. of Engrg. Mechanics, Univ. of Texas, Austin, Tex. 78712)

Journal of the Engineering Mechanics Division, Vol. 108, No. EM5, October, 1982, pp. 864-882

Numerical analysis aspects of the Kolmogorov backward partial differential equation associated with the first-passage problem of the response amplitude of a lightly damped linear structure are considered. It is assumed that the structure is excited by a stationary broad-band random process. A formula is presented for the analytical estimation of the eigenvalues of the boundary value problem constructed by a separation of variables procedure on the Kolmogorov equation. The analytical estimates are used as initial values in an iterative scheme which determines the eigenvalues numerically, for several values of the circular barrier of the first-passage problem. An efficient algorithm for the numerical computation of the corresponding eigenfunctions is presented. This algorithm is used, as well, to compute the constant coefficients of a solution of the Kolmogorov equation in the form of an eigenvalues-eigenfunctions series expansion. The numerical data obtained are examined in context with the physics of the problem.

0250 Cavitation at Rigid Boundary by Tensile Pulses

Thambirajah Balendra, Member, ASCE, (Lect., Dept. of Civ. Engrg., Univ. of Singapore, Singapore)

Journal of the Engineering Mechanics Division, Vol. 108, No. EM5, October, 1982, pp. 883-893

Closed form solutions are obtained for the size and location of the cavitation bubbles caused by travelling tensile pulses in liquid otherwise at rest and reflecting at a rigid surface. It is found that the bubbles collapse at the trailing edges of the pulses and the interface pressure rises discontinuously at the points where the bubbles disappear.

0251 Simplified Dynamic Analysis of Buildings with Basements

Thambirajah Balendra, Member, ASCE, (Lect., Dept. of Civ. Engrg., National Univ. of Singapore, Kent Ridge, Singapore), **Yong-Phai Tan**, (Grad. Student, Dept. of Civ. Engrg., National Univ. of Singapore, Kent Ridge, Singapore) and **Seng-Lip Lee**, Fellow, ASCE, (Prof. and Head, Dept. of Civ. Engrg., National Univ. of Singapore, Kent Ridge, Singapore)

Journal of the Engineering Mechanics Division, Vol. 108, No. EM5, October, 1982, pp. 895-914

The effects of basements on the dynamic soil-structure interaction of buildings are investigated and simplified models in which each interaction force is represented by a set of spring and dashpot in parallel are proposed. The frequency independent stiffness and damping coefficients to be used in the simplified models are obtained from transient analyses by comparing the average work done and the average rate of energy dissipated by the interaction forces in a more accurate lumped parameter model and the simplified model. To take into account the inertial resistance of the soil against the basement walls, virtual masses are added to the basement floors to obtain more accurate shear forces acting on the flexible basement columns. Using the proposed simplified models, the influence of the number of basements on the structure-foundation system is investigated.

0252 Plastic Models for Structural Analysis: Facts and Fallacies

Ronald L. Bannister, (Mgr., Westinghouse Electric Corp., STGD, Lester, Pa.), **Ish K. Aneja**, Member, ASCE, (Sr. Engr., Westinghouse Electric Corp., STGD, Lester, Pa.) and **Kazuhiro Shiraki**, (Mgr., Mitsubishi Heavy Industries, Ltd., Tokyo, Japan)

Journal of the Engineering Mechanics Division, Vol. 108, No. EM5, October, 1982, pp. 915-926

Laboratory and field measurements have shown that scale plastic models can be used to measure static deflections and stress levels natural frequencies, response levels, mode shapes and the effect of terminating impedance of complex structures before they are built. Tests on a realistic model enable a designer to achieve an optimum design before construction of the full-size structure. Model accuracy requires a dedication to understanding and controlling the variables which can alter the results. Poor model workmanship, misunderstood material properties, improper instrumentation, or inadequate experimental techniques may invalidate a complete testing program, not just alter the expected model accuracy.

0253 Simple Buckling Theory for Pressurized Cylinders

Ronaldo C. Batista, (Adjunct Prof. of Civ. Engrg., COPPE-Engenharia Civil Universidade Federal do Rio de Janeiro, Rio de Janeiro, Brazil) and **James G. A. Croll**, (Lect. in Civ. Engrg., Dept. of Civ. Engrg., Univ. College London, London, England)

Journal of the Engineering Mechanics Division, Vol. 108, No. EM5, October, 1982, pp. 927-944

Energy distribution within the critical modes of a classical energy eigenvalue analysis for circular cylindrical seashells under external pressure shows that a significant contribution to the stability of the shell is derived from membrane stiffness. It is argued from physical reasoning that the unstable post-critical behavior is the result of the loss of this membrane stiffness. To account for the combined effect of mode coupling and imperfections in eroding this membrane stiffness, a simplified theoretical analysis is described in which appropriate terms in the membrane potential energy are neglected. It is shown, by comparison with the writers' experiments, and with a collection of data found in the literature, that this reduced stiffness method provides lower bounds which, for shells with moderate imperfections, are reliable lower bounds of the experimental buckling pressures in the modes observed to trigger buckling. This lower boundedness combined with the conceptual and analytic simplicity of the reduced stiffness method make it a particularly attractive basis for the design.

0254 Neutral Holes: Theory and Design

Rowland Richards, Jr., Member, ASCE, (Prof., Dept. of Civ. Engrg., State Univ. of New York at Buffalo, Buffalo, N.Y. 14260) and **Gordon S. Bjorkman, Jr.**, Assoc. Member, ASCE, (Sr. Consultant, CYGNA Energy Services, Boston, Mass.)

Journal of the Engineering Mechanics Division, Vol. 108, No. EM5, October, 1982, pp. 945-960

As previously demonstrated it is possible to leave the geometry of an opening unspecified and invert the standard problem in elasticity to solve for shapes which achieve an optimum design condition on the final stress state. This fundamental strategy is applied to the problem of reinforced holes to determine optimum liner shape and stiffness properties. The design specification used is that the neutral liner completely eliminates any perturbation in the field giving no stress concentration whatsoever. The general theory is developed and the resulting expressions for equilibrium and compatibility are, in fact, not limited to the neutral condition but are valid for any thin reinforcement and, therefore, are fundamental to the general interaction problem. Closed form solutions are obtained for both the circular and the membrane shapes for a variety of free fields. New neutral liner designs are demonstrated for two basic situations important in practice: a) Circular holes in deviatoric and biaxial fields (flexural liner), and b) the membrane deloid shape and liner for the gradient field with an isotropic component. The second case is particularly useful in that it allows a design within a bending field and closely resembles the geostatic field condition for shallow pipes and tunnels.

0255 Postbuckling of Rings under Two Uniform Loads

Robert Schmidt, Member, ASCE, (Prof. of Engrg. Mechanics, Dept. of Mech. Engrg., Univ. of Detroit, Detroit, Mich.)

Journal of the Engineering Mechanics Division, Vol. 108, No. EM5, October, 1982, pp. 967-971

The interaction effect of the normal and centrally directed pressures on the initial postbuckling behavior of a thin elastic circular ring buckling freely in its own plane is investigated by means of the perturbation method applied to the exact nonlinear differential equations of the engineering theory of curved beams. A dimensionless value of the postbuckling deflection is employed as the perturbation parameter. The analysis has yielded a simple formula representing the relationship between small increases in the buckling loads. This formula characterizes the initial postbuckling behavior of the ring subjected to the two interacting, uniformly distributed loads. Heretofore, the limited technical literature on the postbuckling characteristics of rings has dealt only with single distributed external pressure loads. In practice, the two types of static loads considered can be realized by compressing the ring by means of closely spaced, perfectly flexible spokes and simultaneously applying a gas pressure.

0256 Optimal Grillage Layout for Given Natural Frequency

Niels Olhoff, (Assoc. Prof., Dept. of Solid Mech., Technical Univ. of Denmark, 2800 Lyngby, Denmark) and **George I. N. Rozvany**, Fellow, ASCE, (Reader, Dept. of Civ. Engrg., Monash Univ., Clayton, Victoria, Australia)

Journal of the Engineering Mechanics Division, Vol. 108, No. EM5, October, 1982, pp. 971-975

The problem of minimizing the volume of structural material of an elastic grillage subject to a prescribed value or the fundamental frequency of free transverse vibrations, is considered. The domain and the boundary conditions are given, and the grillage is equipped with a specified, distributed non-structural mass, which is large in comparison with the structural mass. It is the objective to determine the optimal layout of the grillage, and to optimize the beams, which are of given depth and varying width. It is shown that the optimal grillage layout is the same as that found earlier for three different types of design problems, namely plastic design for given collapse load, elastic design for given permissible stress, and elastic design for given compliance. An an example, a simply supported, square grillage with uniformly distributed mass loading is optimized, and it is shown that the amount of structural material of the optimal grillage is significantly smaller than that to be used for a uniform plate with the same fundamental natural frequency of transverse vibrations.

0257 Intersecting Submerged Jet Velocity Profiles

W. Hall C. Maxwell, Fellow, ASCE, (Prof., Dept. of Civ. Engrg., Univ. of Illinois, Urbana, Ill. 61801) and **Arni Snorrason**, (Research Asst., Illinois State Water Survey, Urbana, Ill.)

Journal of the Engineering Mechanics Division, Vol. 108, No. EM5, October, 1982, pp. 975-980

To gather experimental evidence on the details of the velocity field near the intersection of two submerged turbulent jets, this study measured time average velocity magnitudes and directions of two perpendicular interesting axisymmetric submerged turbulent incompressible air jets of approximately equal strength. Because of the need to detect reverse flows, a three-dimensional pitot-type probe was used. This could sense yaw and pitch angles as well as velocity magnitudes. The data show that the reverse flow spreads much more rapidly perpendicular to the nozzle plane than in the nozzle plane, whereas the forward flow is fairly symmetric. Similarity profiles were found in both the forward and reverse flows. In the forward flow the distribution was essentially Gaussian. This was also true in the backward flow in the direction normal to the plane of the nozzles. In the plane of the nozzles the backward flow profiles were close to self-elliptical or semi-circular, depending on the scales for plotting.

0258 Ring Loaded Laminated Cylindrical Shells

Daryl L. Logan, Member, ASCE, (Assoc. Prof., Civ.-Mech. Div., Rose-Hulman Inst. of Tech., Terre Haute, Ind. 47803) and **G. E. O. Widera**, Member, ASCE, (Prof., Dept. of Materials Engrg., Univ. of Illinois at Chicago Circle, Chicago, Ill. 60680)

Journal of the Engineering Mechanics Division, Vol. 108, No. EM5, October, 1982, pp. 980-986

The stress analysis of a laminated anisotropic cylindrical shell subjected to a circumferentially varying line load is considered. The solution state is obtained through a super-position of two specialized theories derived by the method of the asymptotic integration along with Reissner's variational principle for an anisotropic elastic solid. Each theory is seen to yield a fourth order differential equation in terms of the radial displacement. A proper superposition of the respective solutions allows for the satisfaction of all boundary conditions. Numerical results are presented for a glass-epoxy material system. It is also shown that the results presented here reduce to those of a previous method of analysis for an isotropic cylinder.

0259 Internal Hydraulic Jump in Stratified Counterflow

Joseph H. W. Lee, Assoc. Member, ASCE, (Lect., Dept. of Civ. Engrg., Univ. of Hong Kong, Hong Kong, China)

Journal of the Engineering Mechanics Division, Vol. 108, No. EM5, October, 1982, pp. 986-990

Based on a Boussinesq approximation of the governing momentum equations for an internal hydraulic jump in a two-layer counterflow, two important properties of an equal counterflow are demonstrated analytically. Corresponding to a given supercritical upstream state, there can be either two possible downstream conjugate states within the physically realizable range, or none at all. Further, the energy loss criterion across the jump is always satisfied. Consequently, the uniqueness of the downstream state cannot be determined by energy considerations, as done previously for the internal jump in a coflowing stratified flow. Physical arguments, coupled with some experimental observations indicate strongly that the conjugate state closer to the upstream state is the one that will be physically realized. The concept presented herein has been successfully applied to the study of stability and mixing of buoyant discharges in shallow water.

0260 Approximate Solution of First-Passage Problem

Mircea Grigoriu, Member, ASCE, (Assoc. Prof. of Civ. Engrg., Cornell Univ., Ithaca, N.Y.)

Journal of the Engineering Mechanics Division, Vol. 108, No. EM5, October, 1982, pp. 991-999

The first-passage time probability is approximated for Markovian processes of diffusion type from the mean crossing rate of random sequences or moving averages obtained from these processes by sampling or averaging. Simple rules are developed for selecting sampling intervals and averaging periods. The approach is applied with satisfactory results to the analysis of the reliability of a brittle prismatic bar, with random strength, loaded axially.

0261 Separation Angle Effect on Vortex Shedding Frequency

Vittore Cossalter, (Prof. of Engrg., Inst. of Applied Mech., Padua Univ., 35100 Padua, Italy)

Journal of the Engineering Mechanics Division, Vol. 108, No. EM5, October, 1982, pp. 999-1003

Vortex shedding frequencies were measured for a number of two-dimensional bluff bodies whose boundary layers separation angles ranged from 0°-90° to the oncoming flow. Strouhal numbers are presented as a function as Reynolds number in the range of about 350-7500. The results are summarized in a figure from which one can obtain the Strouhal number of vortex shedding for a given value of the angle of the boundary layer separation.

0262 Buckling of Corrugated Open Cylindrical Shells

Kuang-Han Chu, Fellow, ASCE, (Prof. of Civ. Engrg., Illinois Inst. of Tech., Chicago, Ill.), **B. P. Jain**, Assoc. Member, ASCE, (Sr. Struct. Engrg. Specialist, Sargent & Lundy, Chicago, Ill.) and **T. L. Wang**, (Grad. Student, Dept. of Civ. Engrg., Illinois Inst. of Tech., Chicago, Ill.)

Journal of the Engineering Mechanics Division, Vol. 108, No. EM5, October, 1982, pp. 1004-1008

The purpose of this study is to determine the critical buckling stress of open cylindrical shells with sinusoidal corrugations with free longitudinal edges subjected to uniform compression in the direction of the corrugations. Practical considerations lead to the decision of limiting the length (L) to radius (R) ratio to 1 to 10, radius (r) to thickness (t) ratio of 2,500 to 15,000, and thickness (t) to amplitude of corrugation (f) ratio of 0.04 to 0.36. The last ratio has a dominant effect on the buckling stress in the range of $L^2/Rt < 10^5$. The buckling stress is found to be less sensitive to the magnitude of the half subtended angle (30° to 90°) and insensitive to other shell parameters such as curved to chord length ratio, the reduction factor of shear rigidity and the Poisson's ratio. The results are given in figures in log-log plotting with buckling coefficient as ordinate and geometry property parameter as abscissa. This may be of use to the designers.

0263 Solution for Clamped Infinite Strip Loaded by a Couple

Robert H. Bryant, Member, ASCE, (Assoc. Prof. of Materials Engrg., Univ. of Illinois at Chicago Circle, Chicago, Ill.)

Journal of the Engineering Mechanics Division, Vol. 108, No. EM5, October, 1982, pp. 1009-1013

The problem of a clamped infinite strip subjected to an in-plane couple is presented. The solution is obtained by Fourier Transform techniques. An example of a linearly varying distributed couple is solved and a graph of the clamping moment at the fixed edge versus the axial coordinate is presented.

0264 Sampling with Decomposed Variance

Gregory B. Baecher, C.M., (Assoc. Prof. of Civ. Engrg., Massachusetts Inst. of Tech., Cambridge, Mass.)

Journal of the Engineering Mechanics Division, Vol. 108, No. EM5, October, 1982, pp. 1013-1017

In many problems of statistical sampling in civil engineering a prior distribution reflecting some global population is available and a small sample of observations is made on a particular subset of that global population. From this combined sample one estimates the behavior of the particular subset. Despite the familiarity of the problem, it presents statistical difficulties because the subsct specific data are masked by the more numerous global data. These difficulties may be overcome by decomposing the data variance into and among subset parts and within a subset part. Common Bayesian sampling techniques are modified to achieve this goal.

0265 Dynamic Fracture of Beam under Transverse Load

Hojjat Adeli, Assoc. Member, ASCE, (Assoc. Research Prof., Dept. of Civ. Engrg., The Univ. of Utah, 3012 Merrill Engrg. Bldg., Salt Lake City, Utah 84112)

Journal of the Engineering Mechanics Division, Vol. 108, No. EM6, December, 1982, pp. 1025-1035

The dynamic fracture response of a long beam of brittle elastic material subjected to a transverse time-dependent concentrated load is investigated. If the magnitude of this load is increased to a critical value, a crack will propagate from the tensile side of that cross section. An analysis is presented which relates the transverse force, the bending moment, the axial force, and the crack length at the fracturing section. Consequently, the crack length and other parameters of interest can be determined for any given force. The analysis is applied to several specific examples and typical results are presented. If the value of the elastic modulus is appropriately modified, the results also apply for plane strain fracture of a plate subjected to a line load.

0266 Dynamic Stability Boundaries for Shallow Arches

William E. Gregory, Jr., (Engineer, CASDE Corporation, Alexandria, Va. 22303) and **Raymond H. Plaut**, Member, ASCE, (Prof., Dept. of Civil Engrg., Virginia Polytechnic Inst. and State Univ., Blackburg, Va. 24061)

Journal of the Engineering Mechanics Division, Vol. 108, No. EM6, December, 1982, pp. 1036-1050

The interactive effects of multiple dynamic loads on the snap-through instability of shallow elastic arches are investigated. Impulse loads and step loads of infinite duration are considered. The loads are applied either as concentrated loads at the quarter points of the arch, or as concentrated loads at the third points together with a uniformly distributed load. The arch is sinusoidal and has pinned ends. Critical load combinations, based on the Budiansky-Roth criterion, are computed numerically and plotted as interaction curves (i.e., stability boundaries). Curves of constant maximum response are also presented. The results for step loads are compared to the static critical loads, while the critical impulse loads are compared to some lower bounds. In addition, a single concentrated load at an arbitrary position of the arch is treated, and the locus of critical loads is determined for various arch rises. Both for step and impulse loads, the positions corresponding to minimum critical load are usually near the third points of the arch.

0267 Suspension Bridge Response to Moving Loads

Toshiro Hayashikawa, (Research Assoc., Dept. of Civ. Engrg., Hokkaido Univ., Nishi 8 Kita 13 Kita-Ku, Sapporo, 060, Japan) and **Noboru Watanabe**, (Prof., Dept. of Civ. Engrg., Hokkaido Univ., Nishi 8 Kita 13 Kita-Ku, Sapporo, 060, Japan)

Journal of the Engineering Mechanics Division, Vol. 108, No. EM6, December, 1982, pp. 1051-1066

An analytical method for determining natural frequencies and mode shapes of multispan suspension bridges is developed by using a solution for the fourth order linearized differential equations. This method takes into account the support conditions of stiffening girders and the effect of flexural stiffness of towers and can be calculated to higher eigenvalues accurately. The dynamic response of multispan suspension bridges traversed by moving loads with constant velocity is studied. The analysis is conducted by the method of modal analysis. A numerical example is presented to demonstrate the applicability of the analysis and to investigate the dynamic characteristics of suspension bridges.

0268 Torsional Dynamic Response of Solid Systems

Robert Henke, Assoc. Member, ASCE, (Sr. Research Engr., Exxon Production Research Co., Houston, Tex. 77001), **E. Benjamin Wylie**, Member, ASCE, (Prof. of Civil Engrg., Univ. of Michigan, Ann Arbor, Mich. 48109) and **Frank E. Richart, Jr.**, Fellow, ASCE, (Prof. of Civil Engrg., Univ. of Michigan, Ann Arbor, Mich. 48109)

Journal of the Engineering Mechanics Division, Vol. 108, No. EM6, December, 1982, pp. 1067-1085

A practical numerical method which solves the multidimensional axisymmetric torsional wave equation for linear and nonlinear inelastic solids was developed by Henke. The application of this procedure is the prediction of the dynamic response of soil-foundation systems subjected to torsional loads having a broad range of intensities. Here the method is further validated. Two solutions giving the steady state response of a linear half space to different loadings are found to agree with analytical solutions. A third study considers the transient response of a rigid disk on a nonlinear as well as linear half space. Slip is permitted at the interface between the disk and half space. Since analytical solutions were not found, examples are validated using an energy balance. In all studies physical and mathematical aspects of the solutions are discussed. It is concluded that to predict accurately the dynamic response of a soil-foundation system, nonlinear inelasticity must be taken into account if large shearing strains are developed and that the developed numerical method is practical and accurate.

0269 Fluctuating Wind Loads on Buildings

Ahsan Kareem, Member, ASCE, (Asst. Prof., Dept. of Civ. Engrg., Univ. of Houston, Houston, Tex.)

Journal of the Engineering Mechanics Division, Vol. 108, No. EM6, December, 1982, pp. 1086-1102

A method for evaluating the fluctuating wind forces and their distribution on buildings has been developed and applied to the understanding and quantification of crosswind loads on a square cross-section building. Wind loads are determined by integrating simultaneously monitored pressure fluctuations on a building model surface in a boundary layer wind tunnel. An example is presented to illustrate that a suitable combination of a limited number of pressure transducers can be used to monitor pressure fluctuations for the estimation of multi-level wind loads on building models. This methodology also provides information on the local spatio-temporal variation of pressure fluctuations which can provide useful input to the cladding design. Power spectral density and chordwise variation of co-spectra between locations at one level on the model surface are presented. The cross power spectral density matrices of the crosswind forcing function are developed for urban and suburban flow conditions. An increase in the incident turbulence has dampening effects on the crosswind forcing function, that is, a slight spectral peak. Methods of random vibration analysis can be used to compute the probabilistic response of a building modeled as a single or a multi-degree-of-freedom system.

0270 Vibration of Flexible Plate on Viscoelastic Medium

Michio Iguchi, (Assoc. Prof., Dept. of Architectural Engrg., Faculty of Sci. and Engrg., Science Univ. of Tokyo, Noda City, Japan) and **J. Enrique Luco**, (Assoc. Prof., Dept. of Applied Mechanics and Engrg. Sci., Univ. of California, San Diego, La Jolla, Calif. 92093)

Journal of the Engineering Mechanics Division, Vol. 108, No. EM6, December, 1982, pp. 1103-1120

The dynamic response of a massless flexible circular plate with a rigid core supported on a layered viscoelastic half-space and subjected to harmonic vertical and rocking excitation is studied. The mixed boundary-value problem for the case of relaxed contact conditions between the plate and the half-space is reduced to Fredholm integral equations of the second kind which are solved numerically. The effects of flexibility of the plate on the force-displacement relationship, on the motion of different points on the plate, and on the distribution of contact stresses beneath the plate are studied numerically. In general, it has been found that all of these aspects of the response of the plate are highly dependent on the flexibility of the plate relative to that of the supporting half-space.

0271 Noise in Earthquake Accelerograms

Vincent W. Lee, (Research Asst. Prof., Dept. of Civ. Engrg., Univ. of Southern California, Los Angeles, Calif. 90007), **Mahailo D. Trifunac**, (Prof., Dept. of Civ. Engrg., Univ. of Southern California, Los Angeles, Calif. 90007) and **Ali Amini**, (Grad. Student, Dept. of Civ. Engrg., Univ. of Southern California, Los Angeles, Calif. 90007)

Journal of the Engineering Mechanics Division, Vol. 108, No. EM6, December, 1982, pp. 1121-1129

The amplitudes of processing and digitization noise in strong motion earthquake accelerograms are reviewed for hand and automatic digitization. By finding the period bands for which the signal-to-noise ratio in digitized accelerograms is greater than one, dependence of cut-off periods for accelerogram data processing is presented for amplitude scaling in terms of: (1) Earthquake magnitude and epicentral distance; and (2) modified mercalli intensity at the recording site.

0272 Deformations and Stresses in Finite Porous Medium

M. Yavuz Corapcioglu, Member, ASCE, (Asst. Prof. of Civ. Engrg., Dept. of Civ. Engrg., Univ. of Delaware, Newark, Del. 19711), **Turgut Tokdemir**, (Asst. Prof. of Engrg. Sci., Middle East Technical Univ., Ankara, Turkey) and **A. Namik Ural**, (Grad. Student, Middle East Technical Univ., Ankara, Turkey)

Journal of the Engineering Mechanics Division, Vol. 108, No. EM6, December, 1982, pp. 1130-1151

Deformations and stresses in a fluid saturated porous medium of finite width are analyzed. The porous medium is assumed to be a binary mixture consisting of a linear elastic fluid and a linear elastic solid. The equilibrium equations for the solid constituent and all constitutive equations are expressed in terms of the solid displacements in a two-dimensional space. In addition to these equations the boundary conditions for a surface line load are obtained by separating the problem into symmetric and antisymmetric parts. Governing equilibrium equations are solved using standard and extended finite sine and cosine transformations for symmetric and antisymmetric parts, respectively. The final expressions are obtained by superposing the symmetric and antisymmetric solutions resulting from inverse standard and extended finite sine and cosine transformations. The stresses for the solid and fluid constituents and diffusion forces are obtained in terms of solid displacements by means of constitutive equations. The problem is of practical importance especially in the case of elevated roads which are supported by retaining walls from both sides.

0273 Fracture Toughness of Concrete: K_{Ic} Revisited

V. E. Saouma, Assoc. Member, ASCE, (Asst. Prof., Dept. of Civ. Engrg., Univ. of Pittsburgh, Pittsburgh, Pa. 15261), **A. R. Ingraffea**, Assoc. Member, ASCE, (Asst. Prof. and Manager of Experimental Research, Cornell Univ., Ithaca, N.Y. 14853) and **D. M. Catalano**, (Grad. Student, Cornell Univ., Ithaca, N.Y. 14853)

Journal of the Engineering Mechanics Division, Vol. 108, No. EM6, December, 1982, pp. 1152-1166

It is now widely accepted that many of the experimental investigations into the applicability of linear elastic fracture mechanics to cement paste, mortar, and concrete were performed before specimen size limitations were understood. Another less recognized problem was the lack of a simple but accurate stress-intensity factor analysis capability necessary to calibrate non-standard test configurations. In this paper, the results of a significant experimental investigation of a decade ago, one which concluded that linear elastic fracture mechanics concepts are not directly applicable to paste, mortar, and concrete, and reevaluated on the basis of a correct stress-intensity factor calibration not then available. The simplicity and accuracy of currently available analysis techniques for problems in linear elastic fracture mechanics are detailed. It is shown that, when evaluated with the corrected analysis, the previously reported experimental results indicate exactly the opposite of what had previously been concluded.

0274 Optimal Critical-Mode Control of Building under Seismic Load

Jann N. Yang, Member, ASCE, (Prof., Dept. of Civ., Mech., and Environmental Engrg., The George Washington Univ., Washington, D.C. 20052) and **J. Lin Mingchien**, (Grad. Student, Dept. of Civ., Mech., and Environmental Engrg., The George Washington Univ., Washington, D.C. 20052)

Journal of the Engineering Mechanics Division, Vol. 108, No. EM6, December, 1982, pp. 1167-1185

An analysis of optimal open-loop critical-mode control for tall buildings excited by earthquakes is presented. It is shown that both the active mass damper and the active tendon system are effective in reducing the building response. Furthermore, the optimal critical-mode control is likely to be as effective as the optimal global control, since the building response under earthquake excitations is usually dominated by a few lowest modes.

0275 Progressive Damage of a Spalling Rod

D. Krajcinovic, (Prof., Univ. of Illinois, Chicago Circle, Chicago, Ill. 60680), **M. G. Srinivasan**, (Mech. Engr., Argonne National Lab., Bldg. 335, Argonne, Ill. 60439), **G. U. Fonseka**, (Mech. Engr., NUTECH, San Jose, CA) and **R. A. Valentin**, (Sr. Mech. Engr., Argonne National Lab., Argonne, Ill. 60439)

Journal of the Engineering Mechanics Division, Vol. 108, No. EM6, December, 1982, pp. 1186-1197

A phenomenological theory describing the process of gradual damage of brittle rods leading to their ultimate rupture is developed and applied to the problem of spatial and temporal evolution of damage in a free-free rod subjected to a pressure pulse at one of its ends. As a result of nucleation, growth and coalescence of voids, de-cohesions and dislocations, irreversible structural changes occur in a polycrystalline material subjected to certain types of loading. A special internal variable is introduced to characterize the gradual loss of mechanical strength or stiffness. A system of quasi-linear partial differential equations describing the problem is solved numerically using the method of characteristics. Numerical results of damage evolution, energy dissipation, etc. are presented and discussed.

0276 Bayesian Analysis of Regional Wind Energy Potential

Harish G. Rao, Member, ASCE, (Asst. Prof., Old Dominion Univ., Norfolk, Va. 23508) and **Ross B. Corotis**, Member, ASCE, (Prof., Dept. of Civ. Engrg./Materials Sci. and Engrg., The Johns Hopkins Univ., Baltimore, Md. 21218)

Journal of the Engineering Mechanics Division, Vol. 108, No. EM6, December, 1982, pp. 1198-1214

The Bayesian method of statistical analysis has been applied to the problem of site assessment of the wind speed distribution. A Rayleigh distribution is used for wind speeds of interest for energy conversion. The mean wind speed is selected as the random parameter and a prior distribution selected to enable computation of the Bayesian distribution of wind speed reflecting both existing regional data and sampling observations. The convergence of the Bayesian distribution to the Rayleigh with increasing data is illustrated in terms of a sufficient statistic of sampling. Utility functions are computed for the power generated by a wind turbine. These are then combined with the Bayesian wind speed distribution to demonstrate two types of decision analysis: (1) Selection of the optimal site for a given wind turbine; and (2) selection of the optimal turbine for a given site.

0277 Suspension Bridge Vibration: Continuum Formulation

Ahmed M. Abdel-Ghaffar, Assoc. Member, ASCE, (Asst. Prof., Civ. Engrg. Dept., Princeton Univ., Princeton, N.J.)

Journal of the Engineering Mechanics Division, Vol. 108, No. EM6, December, 1982, pp. 1215-1232

The methodology of the free vibration analysis of suspension bridges with horizontal decks is refined and extended, utilizing a continuum approach to include both the coupled vertical-torsional vibration and the effect of cross-sectional distortion. Variational principles are used to obtain the coupled equations of motion in their most general, and nonlinear form. The general equations are linearized, eliminating the coupling effect, and solved for a specific bridge; the resulting natural frequencies and modes of vertical and torsional vibrations are compared with those calculated using the finite element approach and with those estimated from low-amplitude full-scale ambient vibration tests which were conducted on the bridge. In addition, the general, nonlinear equations of motion governing the lateral vibration of suspension bridges are presented and the linearized forms are obtained; a numerical example and a comparison among the analytical, numerical and full-scale test results are also presented.

0278 Plastic Design under Multiple Loads

Donald E. Grierson, Member, ASCE, (Prof., Solid Mechanics Div., Dept. of Civ. Engrg., Univ. of Waterloo, Waterloo, Ontario, Canada) and **Ahmed A. Aly**, (Research Asst., Solid Mechanics Div., Dept. of Civ. Engrg., Univ. of Waterloo, Waterloo, Ontario, Canada)

Journal of the Engineering Mechanics Division, Vol. 108, No. EM6, December, 1982, pp. 1233-1251

A mathematical model and a solution algorithm are developed for the rigid-plastic optimal design of skeletal structures under multiple loading conditions. Iterative linear programming techniques are applied to account for nonlinear variations in element plastic capacities under combined stress states as the structural design is modified to meet specified strength requirements. The design concepts are illustrated for two framework examples.

0279 Elastic Creep Bending Analysis of Plates

Hany Hamza, (Ice and Offshore Engrg. Consultant, Calgary, Alberta, Canada) and **Derek B. Muggeridge**, (Assoc. Prof., Faculty of Engrg. and Applied Sci., Memorial Univ. of Newfoundland, St. John's, Newfoundland, Canada, A1B 3X5)

Journal of the Engineering Mechanics Division, Vol. 108, No. EM6, December, 1982, pp. 1252-1261

The finite element method is used to solve the nonlinear problem of creep bending of a plate. The initial strain approach has been adapted to solve the resulting nonlinear simultaneous equations. The algorithm leads to a time incremental set of equations. An isoparametric thick plate element was used. Good agreement was obtained between the present work and some of the available literature for plates with and without elastic foundations. A load bearing test, on a fresh water ice cover, was carried out and good agreement again was noted between the field data and the numerical model.

0280 Stochastic SH Waves along a Frictional Interface

Richard K. Miller, Member, ASCE, (Assoc. Prof., Dept. of Civ. Engrg., Univ. of Southern California, Los Angeles, Calif. 90007)

Journal of the Engineering Mechanics Division, Vol. 108, No. EM6, December, 1982, pp. 1262-1276

The reflection, refraction, and adsorption of a planar SH wave with stochastic time dependence at a frictionally bonded interface between elastic solids is considered. The incident stress wave is assumed to be a sample function from a zero-mean Gaussian random process with known power spectral density. Exact and approximate solutions are presented for the probability density functions of the transmitted and reflected stress waves and for the rate of slip at the interface. Results are also presented for the stationary and nonstationary mean-square values of these quantities, the mean energy flux partitioning between the various waves, and the mean rate of energy dissipation at the interface. It is shown that for small levels of incident stress, the interface behaves as though perfectly welded, but, for high levels, it behaves as though perfectly lubricated.

0281 Motion Characteristics of Floating Structures

Akira Nishitani, (Post Doctoral Research Assoc., Dept. of Civ. Engrg. and Engrg. Mechanics, Columbia Univ., New York, N.Y. 10027) and **Masanobu Shinozuka**, Member, ASCE, (Renwick Prof. of Civ. Engrg., Dept. of Civ. Engrg. and Engrg. Mechanics, Columbia Univ., New York, N.Y. 10027)

Journal of the Engineering Mechanics Division, Vol. 108, No. EM6, December, 1982, pp. 1277-1296

An engineering approach to the time domain analysis which accounts for the effects of damping is developed for the purpose of evaluating the rigid body response motion of a two-dimensional floating body confined in a fixed basin. The frequency dependent added mass coefficients are first obtained under the assumption of an ideal fluid. Then the equations of motion in the frequency domain are constructed by artificially introducing the damping coefficients. Frequency response functions then follow logically from these equations of motion and are inverse-Fourier transformed to produce the impulse response functions. Essentially, three forms of damping coefficients are considered and experiments are then performed to determine how well these forms reproduce the experimental results. The optimum values of the damping ratios are also estimated from the experiments and their use for the analysis of large scale floating bodies is suggested. Finally, the causality conditions are verified numerically and it is found that they are satisfied at acceptable levels of engineering accuracy.

0282 Nonlinear Transient Phenomena in Elastic-Plastic Solids

Jean H. Prevost, Member, ASCE, (Assoc. Prof., Dept. of Civ. Engrg., Princeton Univ., Princeton, N.J. 08544)

Journal of the Engineering Mechanics Division, Vol. 108, No. EM6, December, 1982, pp. 1297-1311

A finite element model is proposed to analyze nonlinear transient phenomena in elastic-plastic solids due to both material and geometrical nonlinearities. Time integration of the semi discrete finite element equations is performed by using an implicit-explicit predictor-multicorrector algorithm developed by Hughes and co-workers. It is shown that the algorithm can be simply and concisely implemented. The technique allows for a convenient selection of implicit and explicit elements, and for a convenient implicit-explicit split of the various operators appearing in the equations of motion. The procedure proves to be extremely effective in dealing with incompressible plastic flow problems. Numerical results which demonstrate the versatility and accuracy of the method are presented.

0283 Earthquake Motions Using a New Data Processing Scheme

S. Shyam Sunder, Assoc. Member, ASCE, (Asst. Prof. of Civ. Engrg., Massachusetts Inst. of Tech., Rm. 1-274, Cambridge, Mass. 02139) and **Betsy Schumacker**, (Sr. Lect., Dept. of Civ. Engrg., Massachusetts Inst. of Tech., Cambridge, Mass. 02139)

Journal of the Engineering Mechanics Division, Vol. 108, No. EM6, December, 1982, pp. 1313-1329

The significant improvements in the reliability of ground motion and response spectrum estimates that arise from a new procedure for processing strong-motion earthquake signals using state-of-the-art filter design and implementation techniques are illustrated through a series of case studies. A comparison of results from an existing model, and the proposed model show significant differences in the ground motion and response spectrum characteristics for the same set of filter limits. Drifts in integrated velocity and displacement characteristics and theoretically incorrect asymptotic behavior of response spectrum curves arising out of the existing processing scheme have been eliminated. In addition to the importance of appropriately selecting a low-frequency limit for band-pass filtering the signals, this work demonstrates the sensitivity of the acceleration trace to the particular choice of a high-frequency limit.

0284 Dynamic Rail Overturning: Modelling and Application

Morteza A. M. Torkamani, Member, ASCE, (Asst. Prof. of Civ. Engrg., Univ. of Pittsburgh, Pittsburgh, Pa.), **Majeed H. Bhatti**, Assoc. Member, ASCE, (Grad. Student of Civ. Engrg., Illinois Inst. of Tech., Chicago, Ill.) and **Allan M. Zarembski**, (Mgr.-Track Research, Assoc. of American Railroads, Chicago, Ill.)

Journal of the Engineering Mechanics Division, Vol. 108, No. EM6, December, 1982, pp. 1330-1350

An analytical model for examining the dynamic equilibrium is utilized to determine the lateral deflection and rotation of the rail subjected to time dependent lateral and vertical forces, and constant axial force. Galerkin's method was applied to remove the partial dependents of the differential equations of motion. The method resulted in a system of coupled ordinary differential equations in the time domain. Normal mode method of dynamic analysis was utilized to decouple the system of differential equations and convert them to a set of second order differential equations with constant coefficients. This model together with other defined parameters of the rail and fasteners is utilized to investigate several case studies of dynamic behavior of the rail. Comparison between these dynamic cases and static tests has been done. It is recommended that multiple axial load should be considered in the future dynamic studies.

0285 Flexural-Torsional Stability of Curved Beams

Chai Hong Yoo, Member, ASCE, (Assoc. Prof. of Civ. Engrg., Auburn Univ., Auburn, Ala. 36849)

Journal of the Engineering Mechanics Division, Vol. 108, No. EM6, December, 1982, pp. 1351-1369

General fourth order coupled differential equations are presented to determine the elastic buckling loads of in-plane as well as out-of-plane buckling modes of spatial thin-walled curved beams. Included are the warping contribution to the buckling loads, and the effects of antisymmetry of cross-section (found to be significant). The governing differential equations are based on the variational principle. Critical load formulas are given for a few special combinations of loadings and boundary condition. Comparative studies were made and possible sources of discrepancies were traced.

0286 Identification of Linear Structural Dynamic Systems

Masanobu Shinozuka, Member, ASCE, (Renwick Prof. of Civ. Engrg., Columbia Univ., New York, N.Y. 10027), **Chung-Bang Yun**, (Research Engr., Amoco Production Co., Tulsa, Okla.) and **Hiroyuki Imai**, (Research Assoc., Kyoto Univ., Kyoto, Japan)

Journal of the Engineering Mechanics Division, Vol. 108, No. EM6, December, 1982, pp. 1371-1390

This paper studies methods of parameter estimation for linear multi-degree-of-freedom structural dynamic systems, based on observed records of the external forces and the structural responses. The auto-regressive and moving-average (ARMA) model is used for this purpose. It is found that the ARMA model is a convenient model representing linear multi-degree-of-freedom structural dynamic systems and that the model is highly compatible with the instrumental variable method and the maximum likelihood method of identification. In order to check the accuracy of the estimation methods, analytical simulation studies are performed on the basis of simulated data dealing with the aerodynamic coefficient matrices that appear in the equations of motion of a two-dimensional model of a suspension bridge. Then, these methods are applied to the same equations to identify the coefficient matrices using the field measurement data yielding good estimates of the system parameters even under large output noise conditions.

0287 Estimating Underground Power Station Costs

Robert N. Reid, (Civ. Engr., Utah Projects Office, Planning Div., U.S. Bureau of Reclamation, Provo, Utah)

Journal of the Energy Division, Vol. 108, No. EY1, March, 1982, pp. 1-10

Planners involved in evaluating pumped-storage alternatives found the capital cost of the underground powerstation directly related to the installed capacity. Estimates for three underground powerplant sizes are tabulated and discussed. These sizes are 450MW, 800MW, and 2100MW each with 1652 feet (504 m.) of head. Costs are given for various types of powerplant machinery including reversible pump-turbines, generator-motors, transformers, electrical switchgear, and powerplant acccssories. The costs are being used in planning several pumped-storage complex systems in the Rocky Mountain region. Some discussion is included as to why the designers favored an underground site and why these costs would be applicable to planning other pumped-storage systems.

0288 Biomass as Alternative Energy Source

John D. Keenan, Assoc. Member, ASCE, (Assoc. Prof. of Civ. Engrg., Univ. of Pennsylvania, Philadelphia, Pa. 19104)

Journal of the Energy Division, Vol. 108, No. EY1, March, 1982, pp. 11-22

The potential for biomass utilization as a source of fuel, petrochemicals and petroleum-sparing substances is reviewed. The choices for cultivating biomass, utilizing it directly as an energy source, and converting it to fuels are considered against the background of the principal metabolic pathways. Engineering options for the production of biomass derived fuels include the direct use of photosynthetic materials, the photosynthetic production of hydrogen, the fermentation of organic matter, and thermochemical conversions. It is concluded that biomass offers a feasible renewable energy source which can make a significant future contribution to the U.S. energy economy. Process economics, however, prevent the widespread use of such systems. Improvements in bioconversion efficiencies and reductions in harvesting and handling costs are needed to improve the economic situation.

0289 Design of Fan Foundations

Donald B. Walters, Member, ASCE, (Civ. Engr., Burns and Roe, Inc., 5520 Los Santoes Way, Jacksonville, Fla. 32211) and **Jeffrey B. Kirby**, Assoc. Member, ASCE, (Civ. Engr., Burns and Roe, Inc., 800 Kinderkamack Rd., Oradell, N.J. 07649)

Journal of the Energy Division, Vol. 108, No. EY1, March, 1982, pp. 23-36

A design approach for satisfactory and cost effective rotary machine foundations, especially fan foundations, is presented. The often used minimum stiffness criterion for the foundation is replaced by a criterion on bearing deflections. After the properties of the foundation and soil are determined, a dynamic analysis is performed, utilizing a multi-degree of freedom system which accounts for the fan rotor, oil film, bearing housing and supports, foundation and soil. The adequacy of the foundation is determined by comparing the calculated bearing deflections to allowable limits. These calculated and limiting deflections are then used in conjunction with calculated dynamic forces to obtain foundation design loads. Items to be incorporated in project specifications, computer costs, and additional manhours required for this design approach and future design recommendations are given. An illustrative example is provided.

0290 Ash Pond Construction to Meet Performance Requirements

Myron S. Temchin, Member, ASCE, (Consulting Civ. Engr., Ebasco Services Inc., 3731 Brier Park Dr., Houston, Tex.) and **Joseph L. Ehasz**, (Chf. Consulting Civ. Engr., Ebasco Services Inc., Two World Trade Center, New York, N.Y.)

Journal of the Energy Division, Vol. 108, No. EY1, March, 1982, pp. 37-48

The establishment of environmentally acceptable ash disposal reservoirs for a steam electric station is achieved through the design of impervious embankments and foundation cutoffs. Field and laboratory data are evaluated with respect to geotechnical and hydrologic aspects. The systems are designed to meet the water management and water pollution requirements. Rigid field and laboratory quality control activities are developed to control construction and achieve the design criteria. An instrument monitoring and surveillance program is used to evaluate long term performance and regulatory compliance.

0291 Coal-Fired Reference Plants: Project Benefits

Clyde B. Tatum, Member, ASCE, (Construction Superintendent, Ebasco Services, Inc., Elma, Wash.) and **D.J. Bracey**, (Sr. Business Representative, Ebasco Services, Inc., Portland, Oreg.)

Journal of the Energy Division, Vol. 108, No. EY1, March, 1982, pp. 49-57

It is expected that requirements for new electric generating capacity will necessitate shortened engineering and construction schedules for future coal fired projects. The reference plant approach, as described in this paper, is one means of meeting this challenge. Under this method, conceptual design information in the form of preengineered arrangements and system design descriptions is available at project inception. This data supports both earlier licensing submittals and earlier initiation of detailed design. The reference plant approach therefore provides for a more orderly design process and also presents benefits of improved constructibility. This paper describes the engineering/design process under this approach and the major elements of the reference plant package for a 600 MW(e) unit. Project cost and schedule benefits are also presented.

0292 Cooling Tower Practice in Germany: State of the Art

Wolfgang Zerna, (Prrof., Ruhr-Universitat Bochum, Institut fur Konstruktiven Ingenieurbau, D-4630, Bochum 1, Germany), **Wilfried B. Kratzig**, (Prof., Ruhr-Universitat Bochum, Institut fur Konstrukiven Ingenieurbau, D-4630, Bochum 1, Germany) and **Ihsan Mungan**, (Asst. Prof., Wissenschaftlicher Angestellter, Ruhr-Universitat Bochum, Institut fur Konstructiven Ingenieurbau, D-4630, Bochum 1, Germany)

Journal of the Energy Division, Vol. 108, No. EY1, March, 1982, pp. 59-68

A rapid development in design and construction of natural draught cooling towers has taken place in Germany. Research has been concentrated on theory and analysis of shells, on acting forces, especially on wind effects, on buckling behaviour and constructional problems. An approximate earthquake analysis allows a quick estimation of seismic response. Assuming rigid-body base excitation a discretization of the cooling tower into a beam like model is possible. The earthquake analysis itself is carried out by the response-spectrum-method. All design methods develop construction methods minimizing the imperfections and their control and correction during the erection process. Due to restrictions with respect to the height of the cooling towers there exists a trend towards larger diameters. By arranging stiffening rings the buckling resistance and the lowest natural frequency of this new generation of cooling towers can be improved.

0293 Selecting Sites for Coal-Fired Power Plants

Magnus B. Bennedsen, (Sr. Project Engr., Woodward-Clyde Consultants, Three Embarcadero Center, Suite 700, San Francisco, Calif. 94111) and **Craig W. Kirkwood**, (Sr. Project Engr., Woodward-Clyde Consultants, Three Embarcadero Center, Suite 700, San Francisco, Calif. 94111)

Journal of the Energy Division, Vol. 108, No. EY2, June, 1982, pp. 69-78

Methods that have been used to select sites for coal-fired power plants are reviewed. A specific site selection procedure is presented that has been successfully applied in a number of cases. This procedure addresses the challenging features of site selection, including multiple evaluation concerns, significant uncertainties, multiple interest groups, data limitations, and regulatory requirements. Screening criteria and expert judgment are utilized to determine candidate sites within a large study area. These candidate sites are then evaluated using multiobjective decision analysis methods to select preferred and alternate sites for the power plant. With this approach, evaluation measures are determined and a multiattribute utility function assessed over these which reflects tradeoffs among the evaluation measures and attitude toward risktaking. Levels are assessed for each evaluation measure for the candidate sites, including use of probabilities to indicate uncertainties. An illustrative application is presented to show how the site selection procedure is carried out.

0294 A Look at Co-Disposal

Brian P. Butz, Assoc. Member, ASCE, (Asst. Prof., Coll. of Engrg., Tech., Temple Univ., Philadelphia, Pa. 19122) and **Lucien B. Calhoun**, (Vice Pres., Public Financial Mgmt. Inc., Philadelphia, Pa. 19103)

Journal of the Energy Division, Vol. 108, No. EY2, June, 1982, pp. 79-80

The quantity of sewage sludge and solid waste that the minicipalities of the United States must dispose of is increasing while the conventional disposal methods are being restricted or eliminated. A number of Federal laws, enacted over the past few years have directed waste disposal toward the use of new and innovative technologies that incorporate resource recovery and energy recovery systems. One alternative for the disposal of solid waste and sludge is combined disposal, or codisposal. This paper is intended as an introduction to codisposal technology for municipalities considering this option. The two basic thermal codisposal techniques are described and seven codisposal processes are discussed. Options available for financing municipal codisposal plants are outlined and potential environmental impacts of these plants are noted. Applications of codisposal technology in the United States and in Europe are described.

0295 Nuclear Fuel Cycle: Issues and Some Answers for the U.S.A.

William O. Harms, (Dir., Nuclear Reactor Tech. Programs, Oak Ridge National Lab., Oak Ridge, Tenn. 37830)

Journal of the Energy Division, Vol. 108, No. EY2, June, 1982, pp. 91-103

The principal issues regarding nuclear power center on reprocessing and recycling of spent fuel, deployment of breeder reactors, and waste disposal. Analysis of future energy needs using a minimum societal expectation approach indicates that economically recoverable domestic uranium resources probably will have been exhausted early in the next century. Even if speculative resources are assumed to become available, deployment of commercial breeders may be required in the 2010-2020 time frame. Accordingly, a vigorous breeder and fuel recycle research, development and demonstration (RD&D) program is needed at this time. The plutonium needs for current breeder RD&D projects are a matter of some concern. The estimated requirements for the period 1987-1995 range from 12 to 30 tons of plutonium. A facility for producing U-Pu oxide fuel subassemblies for the Clinch River Breeder Reactor Plant must be operable by 1986 (there is no existing industrial capability for fabricating U-Pu fuels), and a pilot-scale breeder fuel reprocessing plant must be ready to accept CRBRP fuel in 1993. For waste disposal, an engineering-scale demonstration for high-level nuclear wastes is urgently needed.

0296 Thermal Plume Area Calculation

George H. Ward, Jr., (Assoc., Espey, Huston and Assoc., Inc., P.O. Box 519, Austin, Tex. 78767)

Journal of the Energy Division, Vol. 108, No. EY2, June, 1982, pp. 104-115

A simple formula for the surface area of a thermal plume is proposed, appropriate for application to a depth-limited surface discharge. The formula is based upon thermodynamic and kinematic considerations neglecting hydrodynamics, and will therefore overpredict by the extent to which hydromechanical dilution affects plume size. It therefore provides a conservative estimate for engineering and environmental purposes. An example application to a power plant in Nueces Bay, Texas, is given.

0297 Bayesian Persistence Analysis for Wind Energy

Harish G. Rao, Assoc. Member, ASCE, (Asst. Prof., Dept. of Civ. Engrg., Old Dominion Univ., Norfolk, Va. 23508) and **Ross B. Corotis**, Member, ASCE, (Prof. and Head of Civ. Engrg., Dept. of Civ. Engrg./Materials Sci. and Engrg., The Johns Hopkins Univ., Baltimore, Md. 21218)

Journal of the Energy Division, Vol. 108, No. EY2, June, 1982, pp. 116-127

The persistence of wind speed is an important consideration in the assessment of a wind enery conversion site. Since data are often limited at such a site, a practical Bayesian formulation is presented whereby regional and site-specific mean wind speed may be used to calibrate a probability model of persistence. Prior estimates for the principal model parameter are obtained from the ratio of wind speed level considered for persistence to site mean wind speeds using weighted regression. The Bayesian approach is in terms of specific persistence events, such as a wind speed run above a given wind speed exceeding a particular duration. Updating may be through an improved estimate of the site mean wind speed or, if available, a site persistence analysis.

0298 Dry Ash Handling System for Power Plants

Tariq Rafay, Member, ASCE, (Project Civ. Engr., INTEG-Intercontinental Engrg. Limited, Vancouver, British Columbia, Canada)

Journal of the Energy Division, Vol. 108, No. EY3, November, 1982, pp. 129-142

The most commonly used wet ash handling system in which slurried ash is pumped to ash lagoons is compared with a relatively new dry ash handling system in which moistened fly ash and dewatered bottom ash are continuously transported to the disposal area and compacted. The proposed system for a typical 2,000 MW power plant features submerged drag bar conveyors under the boiler hopper and a system of fixed and movable conveyors. Two sets of independent conveyors are provided to enable conveyance of fly ash and bottom ash separately when required, and to minimize chances of complete breakdown of the system. These conveyors are fully enclosed to reduce noise and dust. Manned equipment is limited to bulldozers which spread and compact the ash. Problems in conveying, storing, and compacting ash are enumerated. Control of moisture content, use of suitable compactors, and provision of lined drainage courses in the ash pile are emphasized. Conveyor movements are shown with reference to a valley disposal area. As soon as the top of the ash pile is reached, it is covered with topsoil and seeded. Runoff to be treated is restricted to that over the exposed ash area by means of interception ditches.

0299 Building Energy Reductions: Windbreak Optimization

David T. Harrje, (Sr. Research Engr. and Lect., Center for Energy and Environmental Studies, Princeton Univ., Princeton, N.J.), **Charles E. Buckley**, (PhD candidate, Stanford Univ., Stanford, Calif.) and **Gordon M. Heisler**, (Research Forest Meteorologist, U.S. Dept. of Agr. Forest Service, Northeastern Forest Experiment Station, University Park, Pa.)

Journal of the Energy Division, Vol. 108, No. EY3, November, 1982, pp. 143-154

Wind tunnel model studies of houses are used to determine how one can best reduce the surface pressure variations from wind and the associated air infiltration, emphasizing the correct placement of suitably modeled coniferous trees. The trees are shown to have energy-saving potential by converting directed kinetic energy of the approaching wind into random turbulent energy by passing the air through tortuous paths in the tree crowns. This reduces the pressure gradients on the windward walls which are a prime region for air infiltration. The most cost- and energy-effective windbreaks are those that selectively protect against the prevailing winter winds, use as few trees as possible, and are placed at least two tree heights upstream. Underpruning and poorly designed openings through the trees, such as driveway access, were shown to enhance air infiltration effects. Using as few as three trees, 3% fuel savings were estimated. When a planting pattern affording general protection was analyzed, it was estimated to achieve higher than 10% savings in energy for heating.

0300 Reconversion to Coal Firing at Crystal River

Robert P. McBean, Member, ASCE, (Project Struct. Engr., Power Div., Black & Veatch, Consulting Engrs., Kansas City, Mo.) and **Ronald J. Ott**, (Project Mech. Engr., Power Div., Black & Veatch, Consulting Engrs., Kansas City, Mo.)

Journal of the Energy Division, Vol. 108, No. EY3, November, 1982, pp. 155-168

Units 1 and 2 at the Crystal River Plant of Florida Power Corporation were originally designed for coal firing in the 1960's, were converted to heavy fuel oil firing in the early 1970's when oil was readily available and economically priced, and subsequently were reconverted to coal firing by 1980 when coal firing again became more favorable. This paper describes the plant modifications and additional construction required for the reconversions of both units. Specifically described are the boiler modifications, ash handling system modifications, precipitator and associated ductwork and induced draft fan additions, coal handling system modifications, and foundation and superstructure designs.

0301 Modeling Flow Characteristics of Reactor Sumps

James B. Nystrom, (Lead Research Engr. and Instr. of Mech. Engrg., Alden Research Lab., Worcester Polytechnic Inst., Holden, Mass. 01520), **Mahadevan Padmanabhan**, (Lead Research Engr. and Asst. Prof. of Mech. Engrg., Alden Research Lab., Worcester Polytechnic Inst., Holden, Mass. 01520) and **George E. Hecker**, Member, ASCE, (Dir. and Assoc. Prof. of Civ. Engrg., Alden Research Lab., Worcester Polytechnic Inst., Holden, Mass. 01520)

Journal of the Energy Division, Vol. 108, No. EY3, November, 1982, pp. 169-184

After a postulated loss of coolant accident in a nuclear powerplant, heated water from the core is removed through the emergency reactor building floor sump, cooled, and reinjected into the core or the containment building via the spray system. Reduced scale models of reactor sumps are demonstrated to be an efficient means to assure the hydraulic performance of the emergency core cooling system is adequate for long term operation. Similitude criteria for model operation are described, considering the phenomena of interest such as free surface and submerged vortices, swirling flow in the pump suction lines, and energy losses in the flow path. Typical model scale ratios, construction techniques, instrumentation, and operating procedures are described. Test results from several typical recent studies are used to illustrate design features which are desirable and those that are undesirable.

0302 Densification of Paper Mill Sludge Materials

Orlando B. Andersland, Member, ASCE, (Prof. of Civ. Engrg., Michigan State Univ., E. Lansing Mich.), **Abdul-Amir W.N. Al-Khafaji**, Assoc. Member, ASCE, (Asst. Prof. of Civ. Engrg., Univ. of Detroit, Mich.) and **Richard K. Lowe**, (Asst. Project Engr., Soil Testing Services of Wisconsin, Inc., Green Bay, Wisc.)

Journal of the Geotechnical Engineering Division, Vol. 108, No. GT1, January, 1982, pp. 1-15

Dewatered paper industry primary sludges have a consistency similar to very soft remolded clays. Densification involves removal of additional water, usually by placement of a surcharge load after deposition in a landfill. Solids composition includes a noncombustible fraction, primarily kaolin clays, and the organic constituents which include wood fibers and other components. Use of kaolinite-pulp fiber mixtures to simulate these sludges permitted control over organic content, fiber type and size, and mineral type. Decomposition was accelerated by the addition of nutrients and seed microorganisms. Use of the ignition test to determine current organic fractions permitted changes in the degree of decomposition to be monitored concurrent to changes in the model sludge compressibility. The effects of organic content, pressure, and decomposition on sludge densification was shown by compression tests. Experimental data permitted development of a method for computation of void ratio as a function of pressure and organic content. Excellent agreement was observed with limited field data. Using the degree of decomposition, procedures for settlement prediction of decomposing paper mill sludges were formulated. Examples are included.

0303 Design of Filters for Clay Cores of Dams

Peter H. Vaughan, Member, ASCE, (Reader in Soil Mechanics, Imperial College, London, England) and **Hermusia F. Soares**, (Civ. Engr., Rio de Janeiro, Brazil)

Journal of the Geotechnical Engineering Division, Vol. 108, No. GT1, January, 1982, pp. 17-31

The clay core of the Balderhead Dam Suffered cracking and internal erosion in 1967. Segregation of the eroded debris was observed and the filter failed to retain the finer particles that reached it. This observation has lead to the development of a method of filter design for clays in flocculating conditions, based on the retention of clay flocs, which are the smallest particles which can arise during erosion. Such a filter will operate irrespective of cracking and the amount of segregation of the eroded debris. Factors controlling floc size are described, and an experimental relationship between permeability of a filter and the size of particle which will pass through it is presented. It is suggested that the retention characterists of a filter can be described more generally by its permeability than by its grading.

0304 Swell of Wyoming Montmorillonite and Sand Mixtures

Mehmit A. Sherif, Member, ASCE, (Prof. of Civ. Engrg., Univ. of Washington, Seattle, Wash. 98195), **Isao Ishibashi**, Member, ASCE, (Assoc. Prof. of Civ. Engrg., Univ. of Washington, Seattle, Wash. 98195) and **B. W. Medhin**, (Grad. Student, Dept. of Civ. Engrg., Univ. of Washington, Seattle, Wash. 98195)

Journal of the Geotechnical Engineering Division, Vol. 108, No. GT1, January, 1982, pp. 33-45

Eight blends of silica sand and Wyoming montmorillonite clay (volclay) at varying proportions by weight were tested at three different temperatures (24°C, 38°C, and 66°C) to investigate their swell characteristics. The study showed that the mixtures with liquid limits below 35 percent are not susceptible to expansion. An analytical equation and a general procedure are shown for estimating the percent of swell these soil mixtures exhibit as a function of time and their liquid limits. A general procedure and an illustrative example are presented to acquaint the reader with the use of laboratory swell test data to evaluate the swell behavior of in-situ field soils.

0305 Safety Factor of Dams and Retaining Walls

Alexander Danilevsky, (Prof., Inst. of Transportation Engrg., School of Engrg., Univ. of Buenos Aires, Buenos Aires, Argentina)

Journal of the Geotechnical Engineering Division, Vol. 108, No. GT1, January, 1982, pp. 47-61

Whatever the definition of the safety factor against overturning of dams and retaining walls, the state of minimum stability occurs when there is some water present in front of the wall. The correct definition of the safety factor of a gravity wall calls "stabilizing moment" the algebraic sum of moments of all the remaining forces applied to it. In the case of an L-shaped wall, the earth behind it is an intrinsic part of the structure, and its weight also contributes to the "stabilizing moment". The forces intrinsic to the structure and those applied to it may not be indiscriminately shifted. They form groups not to be torn apart.

0306 Sand Liquefaction: Inelastic Effective Stress Model

Christos A. Katsikas, Assoc. Member, ASCE, (Engr., Giffels Associates Inc., Southfield, Mich.) and **E. Benjamin Wylie**, Member, ASCE, (Prof. of Civ. Engrg., Univ. of Michigan, Ann Arbor, Mich. 48109)

Journal of the Geotechnical Engineering Division, Vol. 108, No. GT1, January, 1982, pp. 63-81

An effective stress numerical model for soil liquefaction, which provides interaction between shearing deformation and transient pore-water pressure development, is presented. Some attractive features of this one-dimensional model are the preservation of the inelastic soil character during shearing, the coupling between shear wave propagation and pore-water pressure development and seepage, and the association of excess pore-water pressure development with the inelastic volumetric deformation of the solid matrix. A comparison is made with results from another effective stress model and a second comparison is made with experimental results from a large scale shaking table.

0307 Compression of "Free-Ends" During Triaxial Testing

Robert W. Sarsby, (Principal Lect. in Geotechnical Engrg., Dept. of Civ. Engrg., Bolton Inst. of Technology, Deane Rd., Bolton, England), **Nikolas Kalteziotis**, (Geotechnical Engr., Athens Public Works Dept., Athens, Greece) and **Essam H. Haddad**, (Engrg. Asst., Univ. of Basrah, Basrah, Iraq)

Journal of the Geotechnical Engineering Division, Vol. 108, No. GT1, January, 1982, pp. 83-107

When lubricated platens (free-ends) are used to load particulate media there is an associated error in any measured axial deformation due to compression and distortion of the rubber/grease layer. This extraneous deformation is termed the bedding error and it may account for as much as 80% of the total axial deformation recorded during a constant R stress path. Two independent methods of determining the bedding error have been used. In the first method, additional sets of free-ends are incorporated into samples during their preparation. In the second method, samples with various heights are employed. For a Constant R stress path the bedding error is linearly proportional to log $\sigma_1{}^1$ for the whole of the stress range investigated. This linear relationship is true regardless of sample porosity, particle shape and particle size, however, the magnitude of the bedding error varies non-linearly with particle size and the value for a sandy gravel is approximately 30 times the value for a fine silt.

0308 El Infiernillo and La Villita Dams: Seismic Behavior

Daniel Resendiz, Member, ASCE, (Dir. and Research Prof., Institutio de Ingenieria, UNAM, Mexico 20, D.F.), **Miguel P. Romo**, Assoc. Member, ASCE, (Research Prof., Instituto de Ingenieria, UNAM, Mexico, 20, D.F.) and **Edmundo Moreno**, (Head, Dept. of Experimental Studies, Comision Federal de Electricidad, Mexico, D.F.)

Journal of the Geotechnical Engineering Division, Vol. 108, No. GT1, January, 1982, pp. 109-131

El Infiernillo and La Villita rockfill dams in Southern Mexico underwent an earthquake of magnitude 7.6 (Ricter) on March 14, 1979, with epicenter about 68 mi(110 km) from each dam. As a consequence measurable deformations occurred in both dams, including settlement, slope distortion and cracking. These deformations and the earthquake accelerations on the rock foundations and on the embankments were recorded. The patterns of permanent deformation are analyzed and the relative contributions of compaction and shear distortion are estimated. In both dams the fraction of the observed crest settlement attributable to compaction was larger than that due to shear distortion, and a clear correlation was observed between cracking and settlement patterns. The observed settlements attributable to shear distortion were found to lie between those computed from sliding-block methods (lower estimate) and those from a method of accounting for overall embankment deformations (upper estimate).

0309 CW Microwave Location of Voids Beneath Paved Areas

Robert M. Koerner, Member, ASCE, (Prof. of Civ. Engrg., Drexel Univ., Philadelphia, Pa. 19104), **Arthur E. Lord, Jr.**, (Prof. of Physics, Drexel Univ., Philadelphia, Pa. 19104), **John J. Bowders, Jr.**, (Grad. Research Asst., Drexel Univ., Philadelphia, Pa. 19104) and **William W. Dougherty**, (Stone and Webster, Cherry Hill, N.J.)

Journal of the Geotechnical Engineering Division, Vol. 108, No. GT1, January, 1982, pp. 133-144

The problem of locating voids beneath paved areas is a perplexing and costly one. Usually voids are investigated by test borings, which is a discontinuous hit-or-miss process. The use of a continuous nondestructive testing method is a very attractive alternative. One such method is based upon the use of continuous wave (CW) microwaves which can penetrate the upper paved material and, depending upon the nature of the underlying material (soil, air or water), will give different return reflections. Such a method is described and results based on large scale laboratory tests are presented. It is concluded that the method could be used to indicate differences between the underlying material (soil, air or water), but fell short of assessing the thickness of the air void when it was present.

0310 Application of Screw Plate to Stiff Clays

J. Neil Kay, Member, ASCE, (Senior Lect., Dept. of Civ. Engrg., Univ. of Adelaide, S. Australia) and **Derek L. Avalle**, (Professional Officer, Dept. of Civ. Engrg., Univ. of Adelaide, S. Australia)

Journal of the Geotechnical Engineering Division, Vol. 108, No. GT1, January, 1982, pp. 145-154

The screw plate load test application to stiff clays has been further developed to provide more accurate results with greater efficiency in testing. From the results of this test direct calculation may be made of undrained and drained modulus, undrained shear strength and coefficient of consolidation. Temperature effects have been minimized and electronic load and deflection measurement have been incorporated. An alternative proposal has been made for interpretation of settlement-time results. Results for undrained parameters obtained from field tests in stiff Adelaide clay have been favourably compared with those from self-boring pressuremeter tests.

0311 Laterally Loaded Pile Behavior

Michael Georgiadis, (Geotechnical Engr., D'Appolonia, S.A., Brussels, Belgium) and **Roy Butterfield**, (Prof. of Soil Mechanics, Univ. of Southampton, England)

Journal of the Geotechnical Engineering Division, Vol. 108, No. GT1, January, 1982, pp. 155-165

An analysis of laterally loaded piles is presented in which the pile is treated as an elastic beam supported on non-linear, shear coupled, horizontal springs. The edge forces which act along the pile length are also taken into account and prove to be of significant importance to the pile behaviour. The characteristics of the axial springs and the shear coupling are determined from the interpretation of horizontal plate bearing tests carried out, in pairs, at different depths along a 2 in. (50 mm.) diameter pile which is advanced into the soil. The results obtained from laterally loaded pile tests on six piles of three different diameters and stiffnesses are compared to the analytically predicted pile behaviour.

0312 Investigation of Failures

Gerald A. Leonards, Fellow, ASCE, (Prof., School of Civ. Engrg., Purdue Univ., West Lafayette, Ind. 47907)

Journal of the Geotechnical Engineering Division, Vol. 108, No. GT2, February, 1982, pp. 185-246

The problems associated with determining what actually happended after an unexpected instability has occurred are illustrated by case records. Examples are taken from failures of cut slopes in, and embankments on, soft clays and from the failure of a large pile foundation during proof loading. It is then shown how more information can be gleaned from post-failure investigations than has generally been the case up to now, and the important lessons learned are highlighted. It is proposed that a National Center for Investigating Civil Engineering Failures can be created to build up concentrated expertise and to develop rewarding methodologies for investigation of failures, to make the lessons learned cumulative and accessible to the profession, and to provide focal points for rewarding research to reduce significantly the frequency of unexpected failures.

0313 Undisturbed Sampling of Saturated Sands by Freezing

Sukhmander Singh, Member, ASCE, (Sr. Engr. and Assoc., Dames & Moore, San Francisco, Calif.), **H. Bolton Seed**, Fellow, ASCE, (Prof. of Civ. Engrg., Univ. of California, Berkeley, Calif.) and **Clarence K. Chan**, Member, ASCE, (Research Engr., Dept. of Civ. Engrg., Univ. of California, Berkeley, Calif.)

Journal of the Geotechnical Engineering Division, Vol. 108, No. GT2, February, 1982, pp. 247-264

From considerations of the basic mechanisms of soil behavior during freezing, it is suggested that undisturbed samples can only be obtained by undirectional freezing with no impedance of drainage while the confining pressure is maintained. Experimental data illustrate that this method can lead to laboratory samples which maintain their in situ characteristics. Care is required to ensure the effectiveness of the method and it is noted that variables such as the coolant temperature, the percentage of fines in the sand and the depth of sampling can influence the efficiency and effectiveness of the method. It is shown that the beneficial effects of freezing samples by surrounding the sampling tube with dry ice prior to shipping can only be achieved if the samples are drained prior to freezing. If a sample contains a large percentage of fines, insufficient drainage may cause all-around freezing to have deleterious effects on the quality of the sample.

0314 Considerations in the Undisturbed Sampling of Sands

H. Bolton Seed, Fellow, ASCE, (Prof. of Civ. Engrg., Univ. of California, Berkeley, Calif.), **Sukhmander Singh**, Member, ASCE, (Sr. Engr. and Assoc., Dames & Moore, San Francisco, Calif), **Clarence K. Chan**, Member, ASCE, (Lect. and Research Engr., Dept. of Civ. Engrg., Univ. of California, Berkeley, Calif.) and **T. F. Vilela**, (Visiting Scholar, Dept. of Civ. Engrg., Univ. of California, Berkeley, Calif.)

Journal of the Geotechnical Engineering Division, Vol. 108, No. GT2, February, 1982, pp. 265-283

The effect of sampling disturbance on the cyclic strength characteristics of sands is investigated by sampling from deposits of known dynamic characteristics prepared in the laboratory. Push-tube (Hvorslev fixed piston) sampling caused density changes (densification) of Monterey 0 sand with a relative density about 60 percent. A relatively larger relative density increase (about 10 percent) was noticed in the case of virgin samples as compared with about 6 percent increase for samples having a prior strain history. A combination of advance trimming and block sampling techniques caused little, if any, change in density. Some disturbance, resulting in only a partial loss of the resistance gained due to previous seismic effects, did take place. Based on the experimental results, an estimation of sampling disturbance effects for natural deposits is presented.

0315 Shallow Foundation on Frozen Soils: A Bibliography

Committee on Shallow Foundations of the Geotechnical Engineering Division of ASCE

Journal of the Geotechnical Engineering Division, Vol. 108, No. GT2, February, 1982, pp. 285-291

This bibliography, prepared by the Committee on Shallow Foundations, provides a list of current publications on the analysis and design of shallow foundations on frozen soils. The bibliography is divided into two parts. The first part lists references on foundations designed for frost heave in the regions of temperate climate where only seasonal frost is experienced. The second section lists references on design of foundations on permafrost.

0316 Pumped Bentonitic Mixture Used to Seal Piezometers

Claud A. Fetzer, Fellow, ASCE, (Consulting Geotechnical Engr., 1425 Holman View Dr., Cincinnati, Ohio 45215)

Journal of the Geotechnical Engineering Division, Vol. 108, No. GT2, February, 1982, pp. 295-299

Methods used to install the bentonite seal and sand-cement grout seal for multi-level piezometers are described. After problems developed in trying to use bentonite pellets to install a seal, powdered bentonite was mixed with water for form a paste which was pumped to the bottom of the hole. Additives were used in the sand-cement grout above the bentonite seal to prevent shrinkage of the grout away from the wall of the hole and to retard the initial set until the casing could be pulled.

0317 Utilization of Steel Roads as AE Waveguides

Arthur E. Lord, Jr., (Prof. of Physics, Drexel Univ., Philadelphia, Pa. 19104), **Chales L. Fisk**, (Undergrad. Physics Student, Drexel Univ., Philadelphia, Pa. 19104) and **Robert M. Koerner**, Member, ASCE, (Prof. of Civ. Engrg., Drexel Univ., Philadelphia, Pa. 19104)

Journal of the Geotechnical Engineering Division, Vol. 108, No. GT2, February, 1982, pp. 300-305

The matter of soil covering a steel rod, and of its influence on the wave carrying capability of the rod itself, is addressed. It is found the attenuation values of a soil covered steel rod are intermediate to that of the soil or the steel rod acting alone. It is recommended that steel rod waveguides can be used for monitoring of soil structures.

0318 Membrane Penetration in Triaxial Tests

K. V. Ramana, (Reader, S.V. Univ. Coll. of Engrg., Tirupati, India) and **V. S. Raju**, (Head, Ocean Engrg. Centre, Indian Inst. of Tech., Madras, India)

Journal of the Geotechnical Engineering Division, Vol. 108, No. GT2, February, 1982, pp. 305-310

The possibility of predicting volume changes due to membrane penetration in sands of different grain sizes is explored. It is shown that these changes are influenced by grain size, effective containing pressure, density index, and geometry of sample. The effect of membrane penetration can be minimized by adopting large diameter test samples and with other parameters remaining constant.

0319 Method to Predict Ultimate Compression in Clays

S. Narasimha Rao, (Asst. Prof., Dept. of Civ. Engrg., Indian Inst. of Tech., Madras 600036, India) and **K. Kondandaramaswamy**, (Project Assoc., Dept. of Civ. Engrg., Indian Inst. of Tech., Madras 600036, India)

Journal of the Geotechnical Engineering Division, Vol. 108, No. GT2, February, 1982, pp. 310-314

A simple two constant hyperbolic equation is proposed to characterize the compression tests and to predict the ultimate compression of clays under an imposed external load. Laboratory consolidation test data on four samples support this theory.

0320 Pore Pressure Rise During Cyclic Loading of Sands

Christos P. Tsatsanifos, (Research Student, Dept. of Civ. Engrg., Imperial Coll., London SW7 2BU, England) and **Sarada K. Sarma**, Member, ASCE, (Lect., Dept. of Civ. Engrg., Imperial Coll., London, SW7 2BU, England)

Journal of the Geotechnical Engineering Division, Vol. 108, No. GT2, February, 1982, pp. 315-319

A new model for the quantitative determination of the pore pressure increase during undrained cyclic loading of saturated loose sands is proposed. The model depends on four parameters which are determined from undrained cyclic loading tests. It is shown that the parameters should be determined from tests where the conditions are representative of the actual loading conditions.

0321 Mapping Landslide Hazards in Forest Watersheds

Timothy J. Ward, Member, ASCE, (Assoc. Prof. of Civ. and Geological Engrg., New Mexico State Univ., Las Cruces, N.M.), **Ruh-Ming Li**, Member, ASCE, (Assoc. Prof. of Civ. Engrg., Colorado State Univ., Fort Collins, Colo.) and **Daryl B. Simons**, Fellow, ASCE, (Prof. of Civ. Engrg., Colorado State Univ., Fort Collins, Colo.)

Journal of the Geotechnical Engineering Division, Vol. 108, No. GT2, February, 1982, pp. 319-324

A physically realistic methodology is presented for estimating and mapping landslide hazards for watersheds. This methodology also considers the variability of the controlling factors. A demonstration of this methodology on an area in Southeast Alaska is described.

0322 Full-Scale Tests on Ground Anchors in Fine Sand

Allen G. Davis, (Head, Div. of Soil Mechanics and Highway Pavements, Centre Experimental de Recherches et d'Etudes du Batiment, Paris, CEDEX 15, France) and **Claude Plumelle**, (Research Engr., Centre Experimental de Recherches et d'Etudes du Batiment, Paris, France)

Journal of the Geotechnical Engineering Division, Vol. 108, No. GT3, March, 1982, pp. 335-353

A series of ground anchors of varying lengths were installed in a medium dense sand, with a veiw to study as many as possible of the variables influencing anchor behaviour and load transfer. The anchor tie-bars, bulbs and the surrounding sand were instrumented using strain gauges, load cells and marker bands. Critical anchor loads as defined by the French National Code of Practice increased linearly with anchor bulb length. A model for load transfer from the bulb to the ground is proposed.

0323 Single Pile Response to Cyclic Lateral Load

Harry G. Poulos, Member, ASCE, (Reader in Civ. Engrg., School of Civ. and Mining Engrg., Univ. of Sydney, N.S.W., Australia, 2006)

Journal of the Geotechnical Engineering Division, Vol. 108, No. GT3, March, 1982, pp. 355-375

An analysis is presented for the deflection of a pile in clay subjected to quasi-static cyclic lateral loading. This analysis is an extension of the elastic continuum analysis for static pile response, and allows for the reduction of soil modulus and yield pressure with increasing cyclic strain or deflection. A number of theoretical solutions are presented to illustrate the factors influencing the increase in deflection and bending moment in the pile with increasing cycles and cyclic load level. Some of these solutions are presented in dimensionless form as correction factors which may be applied to the computed static response of a pile. Some comparisons are made between the theoretical behaviour and that measured in field tests, and these demonstrate that the theory reproduces the observed characteristics of behaviour of the pile. They also give an indication of the order of magnitude of some of the key variables in the cyclic response analysis.

0324 Shallow Foundations for Tall Structures in Florida

Dennis J. Leary, Member, ASCE, (Principal, Langan Engrg. Assocs., Clifton, N.J. 07018) and **Bernard F. Langan**, Member, ASCE, (Principal, Langan Engrg. Assocs., Clifton, N.J. 07018)

Journal of the Geotechnical Engineering Division, Vol. 108, No. GT3, March, 1982, pp. 377-393

The use of shallow foundation systems with measurable but tolerable settlement of major building complexes constructed in the Miami, Florida area is demonstrated by three case histories, The Omni International, Turnberry Isle Complex, and the James L. Knight Convention Center. Subsurface exploration programs included conventional borings, static cone penetrometer tests, and large diameter coring of soft limerock zones. The compressibility of the subsurface materials can be confidently estimated through the proper interpretation of the field test results and the results of full scale instrumented preloads at geologically similar sites. High-rise structures can be constructed on shallow foundation systems with acceptable settlement behavior. The structures will have measurable but tolerable settlements. Settlements tend to become intolerable when the structures exceed 10 to 15 stories. However, effective precompression of the underlying soils by preloading can be achieved within a few weeks. This reduces total and differential settlements to tolerable limits and makes shallow foundation systems economically desirable for high-rise structures.

0325 Probabilistic Analysis of Deposit Liquefaction

Michael N. Fardis, Assoc. Member, ASCE, (Asst. Prof. of Civ. Engrg., Massachusetts Inst. of Tech., Cambridge, Mass. 02139) and **Daniele Veneziano**, Assoc. Member, ASCE, (Assoc. Prof. of Civ. Engrg., Massachusetts Inst. of Technology, Cambridge, Mass. 02139)

Journal of the Geotechnical Engineering Division, Vol. 108, No. GT3, March, 1982, pp. 395-417

A probabilistic methodology is presented for liquefaction analysis of horizontally layered sand deposits, subjected to vertically propagating seismic S waves. Three different deposit models are developed. The models are effectively one-dimensional and differ in that they include or neglect pore pressure diffusion and stiffness reduction due to pore pressure buildup. Conclusions are drawn about sensitivity of the results to different assumptions about the mechanical behavior of the deposit, the effect of vertical and horizontal variations of soil properties, and about the importance of statistical variability of the response spectrum at bedrock given peak acceleration. An approximate procedure is proposed for the calculation of the probability of almost complete layer liquefaction, for given input motion and given layer average SPT results. Predictions from the approximate procedure are in good agreement with historical data.

0326 Dynamic Formula for Pressure Injected Footings

Reymond L. Nordlund, Member, ASCE, (Vice. Pres. and Chf. Engr., Franki Foundation Co., Arlington, Va.)

Journal of the Geotechnical Engineering Division, Vol. 108, No. GT3, March, 1982, pp. 419-437

The mathematical derivation of a dynamic pile-driving formula for the installation of Pressure Injected Footings (Franki Piles) is presented. The usefulness of such a formula, despite the inherent limitations of all dynamic formulas, is demonstrated. The necessity of differentiating between the two fundamental methods used to make the shaft of a Pressure Injected Footing is proposed and load-test data then presented which support the proposition. An example of how to use the formula is given. The formula contains a constant which is a function of the type of soil and type of shaft. If the result of a load-test is available, the constant can be evaluated from the test data. For those cases where no load-test result is available, a table is presented giving suggested values of the constant for various types of soil and shaft.

0327 Horizontal Stiffness and Damping of Single Piles

Ricardo Dobry, Member, ASCE, (Prof., Dept. of Civ. Engrg. Rensselaer Polytechnic Inst., Troy, N.Y. 12181), **Ernesto Vicenti**, (Grad. Student, Dept. of Civ. Engrg., Rensselaer Polytechnic Inst., Troy, N.Y. 12181), **Michael J. O'Rourke**, Member, ASCE, (Assoc. Prof., Dept. of Civ. Engrg., Rensselaer Polytechnic Inst., Troy, N.Y. 12181) and **Jose M. Roesset**, Member, ASCE, (Prof., Dept. of Civ. Engrg., Univ. of Texas, Austin, Tex. 78712)

Journal of the Geotechnical Engineering Division, Vol. 108, No. GT3, March, 1982, pp. 439-459

Simple expressions are developed for the equivalent horizontal spring and damping coefficients at the top of an end-bearing pile embedded in a uniform linear soil. Beam on elastic foundation (BEF) and dynamic finite element analyses are used for this purpose. It is demonstrated that, for long end-bearing piles, the top displacement response at high frequencies is independent of the pile length, and the pile behaves identically to a long floating pile in a half-space. A numerical criterion is presented to decide when a pile is long or short. All results are expressed as a function of the pile and soil Young's Moduli, E_p and E_s, and they are valid for a range of Poisson's Ratios of the soil between 0.25 and 0.49. Finally, expressions are also presented for the soil springs and dashpots required for more complicated BEF analyses of piles in multilayer soils. The results presented in the paper show good agreement with previous studies.

0328 Numerical Models for Track Support Structures

C. S. Desai, Member, ASCE, (Prof., Dept. of Civ. Engrg. and Engrg. Mechanics, Univ. of Arizona, Tucson, Ariz.) and **H.J. Siriwardane**, Assoc. Member, ASCE, (Asst. Prof., Dept. of Civ. Engrg., West Virginia Univ., Morgantown, W. Va.)

Journal of the Geotechnical Engineering Division, Vol. 108, No. GT3, March, 1982, pp. 461-480

Three finite element formulations based on one-, two-, and three-dimensional idealization for analysis of railroad track support structures are presented with major emphasis on the three-dimensional procedure. The formulations incorporate nonlinear constitutive behavior by using four different elastic-plastic laws and thin interface elements for simulation of behavior at junctions. The parameters for defining nonlinear constitutive laws are obtained from comprehensive tests by using a truly triaxial device. The predictions are compared with other available models, and with field observations from the Urban Mass Transportation Administration Test Section at Transportation Test Center, Pueblo, Colorado.

0329 Elasto-Plastic Settlement of Rigid Footings

Madhira R. Madhav, (Prof., Dept. of Civ. Engrg. Indian Inst. of Tech., Kanpur-208016, India) and **R.S. Karmarker**, (Grad. Student, Dept. of Civ. Engrg., Indian Inst. of Tech., Kanpur-208016 India)

Journal of the Geotechnical Engineering Division, Vol. 108, No. GT3, March, 1982, pp. 483-488

A simple method of estimating elasto-plastic settlement of circular or annular centrally loaded footings is presented. The method incorporates the elasto-plastic behavior manifested through the contact pressures reaching yield value over certain regions of the footings. The results agree with those of finite element methods and experiments.

0330 Piping Potential of Weak Zones Under Concrete Dams

Gregg A. Scott, Assoc. Member, ASCE, (Geotechnical Engr., U.S. Bureau of Reclamation, Denver, Colo.)

Journal of the Geotechnical Engineering Division, Vol. 108, No. GT3, March, 1982, pp. 488-493

A method is presented for evaluating the piping potential of shear zones crossing the foundation of a concrete dam from upstream to downstream. The steps described for this method involves determination of the exit gradient for the zone in question, determination of the exit gradient for the zone under reservoir conditions, and comparison of the exit gradient under reservoir conditions to the critical exit gradient for determining required foundation treatment.

0331 Interaction of Soil and Penetration Projectile

David Z. Yankelevsky, (Prof., Civ. Engrg. Dept., Technion-Israel Inst. of Tech., Haifa, Israel 32000)

Journal of the Geotechnical Engineering Division, Vol. 108, No. GT3, March, 1982, pp. 493-497

The motion of a terradynamic penetrator is formulated, and some interaction characteristics as the added mass and the interaction pressure distribution are calculated. The magnitude of the added mass is found to be small compared to the penetrator's mass. The expression for the interaction pressure is shown to be composed of a static and a dynamic term. The calculated stresses and masses can be used for the dynamic analysis of the penetrator's structure.

0332 Pulverized Coal Ash as Structural Fill

Gerald A. Leonards, Fellow, ASCE, (Prof., School of Civ. Engrg., Purdue Univ., West Lafayette, Inc.) and **Bruce Bailey**, Member, ASCE, (Sr. Project Engr., ATEC Assoc., Inc., Indianapolis, Ind.)

Journal of the Geotechnical Engineering Division, Vol. 108, No. GT4, April, 1982, pp. 517-531

Untreated pulverized coal ash, consisting of fly ash with varying percentages of bottom ash, was used as a structural fill material to support the foundation of a new precipitator for a power generating station in Indianapolis, Indiana. Geotechnical studies for this fill included evaluation of index and chemical properties, strength, compresibility, compaction and aging characteristics. Field density and plate load tests were conducted during fill construction. Standard and static cone penetration tests were used to evaluate the consistency of the compacted fill and total settlements of the foundation were measured. For this project, compacted coal ash was used successfully as a structural fill material. Although the ash was inherently variable, it could be compacted satisfactorily if the moisture content was maintained below the optimum value and if the percentage of fines was below about 60%. The load-settlement relation for the foundation on compacted ash could not be inferred from standard or static cone penetration tests, largely because these tests do not adequately sense the effects of prestressing due to compaction. Plate load tests proved to be a reliable indicator of the observed settlements.
Errata: GT July '82, pp. 998.

0334 Stresses and Displacements in Cross-Anisotropic Soils

George Gazetas, Assoc. Member, ASCE, (Assoc. Prof. of Civ. Engrg., Rensselaer Polytechnic Inst., Troy, N.Y. 12181)

Journal of the Geotechnical Engineering Division, Vol. 108, No. GT4, April, 1982, pp. 532-553

An analytic investigation is presented on the effect of soil cross-anistropy on surface displacements and stress distributions in a homogeneous thick soil deposit (halfspace) subjected to axisymmetric parabolic vertical surface loading. Rigorous treatment is given to deformational soil anisotropy and particular emphasis is accorded to the sensitivity of calculated stresses and deformations to assumed realistic values of the five independent material constants, under both drained and undrained loading conditions. For a given vertical Young's modulus, it is found that the horizontal Young's modulus, the shear modulus on vertical planes and the Poisson's ratio for horizontal due to vertical strain have a profound effect on all stress and displacement components, especially under undrained conditions. The dependence of normal vertical stresses on material properties and particularly on one of the Poisson's ratios disproves the opposite conclusion of isotropic elasticity.

0335 Seismic Response of End Bearing Piles

Raul Flores-Berrones, Assoc. Member, ASCE, (Prof., School of Engrg., National Univ. of Mexico, Mexico 20, D.F., Mexico) and **Robert V. Whitman**, Fellow, ASCE, (Prof. of Civ. Engrg., Massachusetts Inst. of Tech., Cambridge, Mass.)

Journal of the Geotechnical Engineering Division, Vol. 108, No. GT4, April, 1982, pp. 554-569

A very simple analysis of dynamic soil-pile-supported mass interaction is presented as an aid to understanding pile foundations behaviour during earthquakes. The equilibrium equations of the soil and pile movements during seismic motions are established and the response of the soil-pile system is obtained as a function of the existing frequency; for simplicity the analysis was based on Winkler's model although the same procedure can be applied using a more refined model. The effect of the supported mass is studied, the natural frequency of the system is obtained, and some practical design considerations are presented. A method for considering group effects is also presented.

0336 Vertical Vibration of Pile Groups

Mohamed Sheta, (Research Asst., Faculty of Engrg. Sci., Univ. of Western Ontario, London, Ontario, Canada) and **Milos Novak**, Member, ASCE, (Prof., Faculty of Engrg. Sci., Univ. of Western Ontario, London, Ontario, Canada)

Journal of the Geotechnical Engineering Division, Vol. 108, No. GT4, April, 1982, pp. 570-590

An approximate theory for vertical vibration of pile groups is presented. The theory accounts for dynamic interaction of piles in a group, weakening of the soil around the pile due to high strain, soil layering and arbitrary tip conditions. The theory is used to study the effects of pile interaction on stiffness and damping of pile groups, distribution of internal forces in the piles and response of pile supported foundations to harmonic excitation. The theory indicates that dynamic group effects can considerably differ from static group effects particularly in that the group stiffness and damping can be not only reduced by pile-soil-pile interaction but also increased depending on frequency, pile spacing and the weakening of the soil around the piles. The effect of frequency is much stronger for pile groups than for single piles.

0337 Hydrodynamic Effects in Earthquake Response of Embankment Dams

John F. Hall, (Research Fellow, California Inst. of Tech., Pasadena, Calif.) and **Anil K. Chopra**, Member, ASCE, (Prof. of Civ. Engrg., Univ. of California, Berkeley, Calif.)

Journal of the Geotechnical Engineering Division, Vol. 108, No. GT4, April, 1982, pp. 591-597

Utilizing a recently developed analytical procedure, the significance of hydrodynamic effects in the earthquake response of embankment dams is evaluated. Effects of dam-fluid interaction, water compressibility and fluid-foundation interaction are all considered in this study. Numerical results for the hydrodynamic forces on rigid dams and the horizontal acceleration response of the dam crest due to harmonic ground motion are presented. Hydrodynamic effects are shown to be relatively unimportant in the response of embankment dams to horizontal ground motion, and the response to vertical ground motion to be small compared to that due to horizontal ground motion.

0338 SPT Hammer Energy Measurement

William D. Kovacs, Member, ASCE, (Research Geotechnical Engr., National Bureau of Standards, Geotechnical Engrg. Grp., Washington, D.C. 20234) and **Lawrence A. Salomone**, Member, ASCE, (Research Geotechnical Engr., National Bureau of Standards, Geotechnical Engrg. Grp., Washington, D.C. 20234)

Journal of the Geotechnical Engineering Division, Vol. 108, No. GT4, April, 1982, pp. 599-620

A field measurement system and procedure which measures the energy delivered by a drill rig system was developed and successfully used to study the factors which affect delivered energy. Results are presented which indicate that the energy delivered by certain drill rig systems varies widely in engineering practice. The energy delivered to the drill stem varied with the number of turns of rope around the cathead, the fall height, drill rig type hammer type, and operator characteristics. The type of hammer had a strong influence on the energy transfer mechanism between the anvil and the drill stem. It appears that the safety (sleeve enclosed) hammer is more efficient in transmitting the available kinetic energy through the anvil to the drill stem than the donut hammer.

0339 Embankment Construction: New Melones Dam

Clark E. Stanage, (Geotechnical Engr., Embankment Design Section, U.S. Army Engr. Dist., Sacramento, Calif.)

Journal of the Geotechnical Engineering Division, Vol. 108, No. GT4, April, 1982, pp. 621-636

New Melones Dam, located on the Stanislaus River in central California, is a zoned earth and rockfill structure with a maximum height above foundation at embankment centerline of 637 feet and with a volume of about 15.9 million cubic yards. The dam consists of a rolled rockfill embankment with a central impervious core inclined upstream. This paper describes the site geology, foundation treatment, embankment construction procedures, embankment properties and instrumentation. The influence of site conditions and fill requirements on the degree of foundation excavation, surface treatment and grouting patterns are discused. Sources of construction materials and methods of excavation, stockpiling, materials processing and fill placement procedures are described.

0340 Residual Undrained Drformation from Cyclic Loading

Ching S. Chang, Assoc. Member, ASCE, (Asst. Prof. of Civ. Engrg., Univ. of Massachusetts, Amherst, Mass 01003)

Journal of the Geotechnical Engineering Division, Vol. 108, No. GT4, April, 1982, pp. 637-646

Residual undrained deformation of anisotropically consolidated soil samples under cyclic loading is investigated. The magnitude of residual deformation is influenced by factors such as initial stress condition, excess pore pressure generated by the cyclic loading, and soil properties. A semi-empirical model derived from hyperbolic fitting is developed for predicting residual undrain deformation during cyclic loading. This analytical model consists of five parameters: /c, /ϕ, k, n, and R_f, in which k, n, and R_f can be determined either from static or cyclic tests. Based on the results of this study, the values of parameters determined from the cyclic test show close agreement with the values determined from static tests. The comparison of predicted and measured residual undrained deformation of samples during cyclic tests shows excellent agreement. This semi-empirical model can be used to estimate residual strains after cyclic loading of soil elements subjected to anisotropic initial stress conditions, such as soil elements under a footing or in slopes.

0341 Interface Properties of Sands

Yalcin B. Acar, Member, ASCE, (Asst. Prof., Civ. Engrg. Dept., Louisiana State Univ., Baton Route, La.), **H. Turan Durgunoglu**, Member, ASCE, (Assoc. Prof., Civ. Engrg. Dept., Bagazici Univ., Istanbul, Turkey) and **Mehmet T. Tumay**, Member, ASCE, (Prof., Civ. Engrg. Dept., Louisiana State Univ., Baton Route, La.)

Journal of the Geotechnical Engineering Division, Vol. 108, No. GT4, April, 1982, pp. 648-654

Previous studies are reviewed and experiments are performed on the interface properties of sand. It is shown that shear stiffness and roughness values are of primary importance in the analysis and design of structures in contact with soil using the finite element method. Hyperbolic representation of the stress-displacement relationship closely reflects the actual behaviour. It is verified that the initial tangent stiffness relates to normal stress. Other conclusions are also obtained and demonstrated.

0342 Axisymmetric Parabolic Loading of Anistropic Halfspace

George Gazetas, (Prof., Dept. of Civ. Engrg., Rensselaer Polytechnic Inst., Troy, N.Y., 12181)

Journal of the Geotechnical Engineering Division, Vol. 108, No. GT4, April, 1982, pp. 654-660

The complete formulation and solution of a boundary value problem, referred to as a frictionless loading case, are presented. An analytical solution to the problem of adhesive parabolic loading of a cross-anisotropic halfspace is outlined. The effect of soil anistropy on vertical and radial stresses and displacements under both frictionless and adhesive loading conditions are shown.

0343 Stiffness Coefficients of Elastic Medium

M. V. Nagendra, (Research Scholar, Dept. of Civ. Engrg., Indian Inst. of Sci., Bangalore 560012, India) and **A. Sridharan**, (Assoc. Prof., Dept. of Civ. Engrg., Indian Inst. of Sci., Bangalore, 560012, India)

Journal of the Geotechnical Engineering Division, Vol. 108, No. GT4, April, 1982, pp. 661-668

Using the theory of elasticity, values of stiffness coefficients for vertical and horizontal modes are evaluated for cases for which solutions are not available. Three types of pressure distributions and three types of displacement conditions are considered. It is shown that the computed values can be used to predict the response of footings subjected to vibration, using discrete models comprising of equivalent mass, stiffness, and damping coefficients.

0344 Earth Pressures against Rigid Retaining Walls

Mehmet A. Sherif, Member, ASCE, (Prof. of Civ. Engrg. and Adjunct Prof. of Quaternary Research, Seattle, Wash.), **Isao Ishibashi**, (Assoc. Prof. of Civ. and Environmental Engrg., Cornell Univ., Ithaca, N.Y.) and **Chong Do Lee**, Member, ASCE, (Grad. Student, Dept. of Civ. Engrg., Univ. of Washington, Seattle, Wash.)

Journal of the Geotechnical Engineering Division, Vol. 108, No. GT5, May, 1982, pp. 679-695

This paper outlines the results of unique shaking table experiments to determine neutral and active static and dynamic stresses and the points of application of corresponding total thrust generated by granular soils against rigid retaining walls and compares the experimental data with the well-known Coulomb and Mononobe-Okabe equations. A new criterion is proposed for identification of the wall deformation level at which the active state of stress develops. This criterion states that active stress develops when the angle of friction between the wall and the backfill soil reaches its maximum value. Based on the above criterion, experimental results show that the backfill soil enters the active state at much lower wall displacement levels than generally believed. An equation is proposed for calculating the displacement necessary to initiate active static or dynamic stress. This equation reveals that the displacement necessary to initiate active state increases with the height of the wall and decreases with backfill soil strength.

0345 Resilient Characteristics of Brazilian Soils

Jacques de Medina, Member, ASCE, (Prof. of Civ. Engrg., COPPE-UFRJ, Federal Univ. of Rio de Janeiro, Brazil) and **Ernesto S. Preussler**, (Research Engr., COPPE-UFRJ, Federal Univ. of Rio de Janeiro, Brazil)

Journal of the Geotechnical Engineering Division, Vol. 108, No. GT5, May, 1982, pp. 697-712

Resilient properties of twenty soil samples from subgrades, sub-bases, and mechanical stabilized soil bases of flexible pavements in the South and Southeast of Brazil were evaluated. Soils are mostly residual in origin. Correlations of resilient moduli with confining pressure or deviator stress were established. Correlations of resilient moduli with CBR values were established for finegrained soil samples. A tentative soil classification according to resiliency is developed. Also reference is made to the applications to pavement deflection calculations.

0346 Slope Safety Prediction under Static and Seismic Loads

Dimitri A-Grivas, Assoc. Member, ASCE, (Assoc. Prof., Dept. of Civ. Engrg., Rensselaer Polytechnic Inst., Troy, N.Y.) and **Akira Asaoka**, (Assoc. Prof., Dept. of Civ. Engrg., Nagoya Univ., Nagoya, Japan)

Journal of the Geotechnical Engineering Division, Vol. 108, No. GT5, May, 1982, pp. 713-729

A simplified procedure permitting the determination of the probability of failure of natural or man-built slopes under static and seismic conditions is presented. Limit equilibrium is expressed as a function of the soil strength parameters (random variables), the numerical values of which are obtained from strength tests under drained conditions. Seismic load is introduced in terms of the maximum horizontal acceleration expected to occur at the site of the slope. The procedure is applied in a case study to predict the probability of failure of a slope during an earthquake. The seismic load is determined in two different ways, namely, (a) from the seismic map of the region (deterministic), and (b) from an engineering analysis of the seismic history of the site of the slope (probabilistic). The results are presented and discussed.

0347 Simulation of Earthquake Motions in the Centrifuge

Kandiah Arulanandan, Member, ASCE, (Prof. of Civ. Engrg., Univ. of California, Davis, Calif.), **Jeffrey Canclini**, (Research Engr., National Aeronautics and Space Agency) and **Annalingam Anandarajah**, (Grad. Student, Univ. of California, Davis, Calif.)

Journal of the Geotechnical Engineering Division, Vol. 108, No. GT5, May, 1982, pp. 730-742

The present study is mainly concerned with the design of an earthquake simulator package for the generation of a sinusoidal motion during a centrifuge test. It was then used to excite the dynamic response of a clay embankment in the centrifuge. A comparison of measurements made during the tests with results calculated using a computer program with an equivalent-linear material model (QUAD IV), for the test condition is presented. The results demonstrate the potential beneficial use of dynamic testing using the centrifuge.

0348 Spatial Variability of the Undrained Strength of Clays

Akira Asaoka, (Assoc. Prof., Dept. of Civ. Engrg., Nagoya Univ., Nagoya, Japan) and **Dimitri A-Grivas**, Assoc. Member, ASCE, (Dept. of Civ. Engrg., Rensselaer Polytechnic Inst., Troy, N.Y.)

Journal of the Geotechnical Engineering Division, Vol. 108, No. GT5, May, 1982, pp. 743-756

The objective of this study is to provide a simple yet accurate procedure for the description of the spatial variability of the undrained shear strength S_u of soft clays. This is achieved on the basis of experimental findings and field observations of the depth-dependent nature of S_u. The main features of the method are a linear with depth variation of the mean value and standard deviation of S_u and a smooth exponential function that describes the correlation between values received by S_u at different vertical locations. The developed procedure is illustrated in a case study involving actual strength data. Two sites are selected from the general area investigated in connection with the West Side Highway in New York City. Each site is characterized by the values of S_u obtained along two boreholes using field vane shear tests. The results of the applied analysis verify previous findings on the vertical variability of S_u and demonstrates the applicability of the present procedure in the commonly encountered situations with limited data.

0349 New Models for Rock Joints and Interfaces

Francois E. Heuze, Member, ASCE, (Leader, Rock Mechanics Program, Lawrence Livermore National Lab., Univ. of California, Livermore, Calif. 94550) and **Timothy G. Barbour**, Assoc. Member, ASCE, (Sr. Engr., Science Application Inc., Golden, Colo.)

Journal of the Geotechnical Engineering Division, Vol. 108, No. GT5, May, 1982, pp. 757-776

Two new developments and refinements in the modeling of rock structures are presented: a new model for axisymmetric interfaces, such as found in footing designs, and a new model which describes the dilatant effects of rock joints, such as encountered in underground caverns and reinforced rock slopes. The two new models were incorporated into a finite element program and test-case analyses were performed. The results indicate that the use of the new models can lead to more accurate estimates of footing settlements, to more economical designs of rock reinforcement, and to safer analysis of rock structures. The new dilatant joint model provides a refined estimate of joint opening and closing, and thus can also be applied to the analysis of hard rock hydraulics, in which flow is very sensitive to fracture aperture.

0350 Electrical Resistivity Used to Measure Liquefaction of Sand

Ronald A. Erchul, Member, ASCE, (Assoc. Prof., Dept. of Civ. Engrg., Virginia Military Inst., Lexington, Va. 24450) and **Ronald C. Gularte**, Member, ASCE, (Asst. Prof., Dept. of Naval Systems Engrg., U.S. Naval Academy, Annapolis, Md. 21402)

Journal of the Geotechnical Engineering Division, Vol. 108, No. GT5, May, 1982, pp. 778-782

Laboratory data is presented to show that liquefaction of cohesionless soils can be evaluated by electrical resistivity techniques. In addition this same material can be monitored by electrical resistivity techniques during energy input and subsequent stabilization. The instrumentation used in this study can be adapted to field measurements. The results (resistivity measurements) in this investigation are repeatable, and increase with increasing energy input. Applications of this work include soil improvement techniques such as deep dynamic compaction.

0351 Earthquake Magnitude Scales

Otto W. Nuttli, (Prof. of Geophysics, Dept. of Earth and Atmospheric Sci., St. Louis Univ., St. Louis, Mo. 63103) and **Robert B. Herrman**, (Assoc. Prof. of Geophysics, Dept. of Earth and Atmospheric Sci., St. Louis Univ., St. Louis, Mo. 63103)

Journal of the Geotechnical Engineering Division, Vol. 108, No. GT5, May, 1982, pp. 783-786

The use and meaning of earthquake magnitude scales are examined. Common misunderstandings regarding local magnitude, surface wave magnitude, and body-wave magnitude, are discussed and clarified. It is shown that no single magnitude will adequately describe the size of an earthquake, and that different geographic regions pose different requirements. Inter-relations between seismic moment and the various types of magnitude are examined. Magnitudes are recommended for equations or curves which specify the strong ground-motion acceleration, velocity, displacement, or response spectra as a function of distance from the earthquake source.

0352 Elastic and Plastic Strain Properties of Sand

Chin-Su Ting, (Lect., Dept. of Hydraulic Engrg., Qinghua Univ., Beijing, China)

Journal of the Geotechnical Engineering Division, Vol. 108, No. GT5, May, 1982, pp. 787-793

This paper describes the results of tests using conventional triaxial apparatus with loading and unloading, on samples of medium sand with two different unit weights. The method is used to study the elastic and plastic volumetric and axial strains under the conditions of hydrostatic and deviator stress. It is shown that under hydrostatic compression, the elastic volumetric strain of medium sand is always greater than its plastic volumetric strain. The ratio of elastic volumetric strain to plastic volumetric strain may decrease as the hydrostatic pressure increases. Other conclusions are also described.

0353 Probabilistic Model of Progressive Failure of Slopes

Robin N. Chowdhury, Member, ASCE, (Reader, Dept. of Civ. Engrg., Univ. of Wollongong, New South Wales 2500, Australia) and **Dimitri A-Grivas**, Member, ASCE, (Assoc. Prof., Rensselaer Polytechnic Inst., Troy, N.Y. 12181)

Journal of the Geotechnical Engineering Division, Vol. 108, No. GT6, June, 1982, pp. 803-819

A probabilistic model for the progression of failure in a soil slope is presented. Failure progression is defined as a spatial and continuous extension of the failure zone along a potential slip surface in a statistically homogeneous medium. The local safety margin of any segment of the slip surface is assumed to follow a normal distribution; cohesive and frictional parameters of shear strength being considered as independent random variables. The joint distribution of the safety margin of any two adjacent segments of the slip surface is assumed to be bivariate normal. After defining the model and outlining the rules of transition, expressions for the probability of failure progression are derived. The model and its formulation are illustrated by a worked example and the significance of the proposed model is discussed. The suggested approach to the study of progressive failure gives insight into the interdependence of the stability of adjacent elements or sections of a soil mass. Consequently, it is potentially valuable in clarifying the real behavior of soil masses and expecially slopes.

0354 Concepts and Instruments for Improved Monitoring

Pierre Londe, Fellow, ASCE, (Technical Mgr., Coyne et Bellier, Consulting Engrs., Paris, France)

Journal of the Geotechnical Engineering Division, Vol. 108, No. GT6, June, 1982, pp. 820-834

It is shown that a good instrumentation strategy, based on reliable measurements of displacements and pressures, is necessary for monitoring the safety and the behavior of civil works during construction and operation. A significant step forward was achieved by the development of a new generation of instruments. Constant tension invar wires give best accuracy of distance measurements for minimal interference with construction works or operation. Continuous piezometers for boreholes allow precise landslide monitoring and proper remedial works implementation. Vibrating wire sensors have been used in strainmeters and in pressure cells for many years, and their reliability is proved. Recently developed self-inductance sensors allow more sensitive and accurate monitoring. Successful experience on a number of large projects show that electronic equipment can best satisfy all the conditions of a good instrumentation, and results in an improved overall safety at least cost.

0355 Permeability and Consolidation of Normally Consolidated Soils

A. Mahinda Samarasinghe, (Grad. Research Asst., Dept. of Civ. Engrg., Univ. of Kentucky, Lexington, Ky.), **Yang H. Huang**, Fellow, ASCE, (Prof. of Civ. Engrg., Univ. of Kentucky, Lexington, Ky.) and **Vincent P. Drnevich**, Member, ASCE, (Prof. and Chmn., Dept. of Civ. Engrg., Univ. of Kentucky, Lexington, Ky.)

Journal of the Geotechnical Engineering Division, Vol. 108, No. GT6, June, 1982, pp. 835-850

This paper first presents a unique relationship between permeability and void ratio. This relationship is then used for deriving the governing differential equation of consolidation for normally consolidated soils where void ratio vs. logarithm of effective stress can be approximated by a straight line. The derivation shows that the coefficient of consolidation can be related directly to the applied effective stress.

0356 K_0 - OCR Relationships in Soil

Paul W. Mayne, Assoc. Member, ASCE, (Geotechnical Engr., Law Engrg. Testing Co., Washington, D.C. 22101) and **Fred H. Kulhawy**, Member, ASCE, (Prof., School of Civ. and Environmental Engrg., Cornell Univ., Ithaca, N.Y. 14853)

Journal of the Geotechnical Engineering Division, Vol. 108, No. GT6, June, 1982, pp. 851-872

The relationships between K_0 and OCR are investigated for primary loading - unloading - reloading conditions. The study reviews laboratory data from over 170 different soils and presents an approach common to clays, silts and sands. Simple empirical methods for predicting K_0 for normally consolidated and overconsolidated soils are evaluated. The validity of the methods is supported by statistical analyses. On the basis of the findings, only the effective stress friction angle (O') and prior stress history (OCR and OCRmax) are needed to predict approximate values of K_0.

0357 Field Tests of Long-Span Aluminum Culvert

David B. Beal, Member, ASCE, (Assoc. Civ. Engr., New York State Dept. of Transportation, Albany, N.Y. 12232)

Journal of the Geotechnical Engineering Division, Vol. 108, No. GT6, June, 1982, pp. 873-890

A 28.5 ft (8.7 m) span, 17-ft 10-in. (5.4-m) high corrugated aluminum culvert was instrumented to obtain measurements of strain and displacement during backfilling and under static live load. Values of circumferential bending moment and thurst at 16 locations spaced around the structure's circumference at mid-span are reported for each 2 ft (0.6 m) of backfill from the springline to 2 ft (0.6 m) over the crown. Despite bending moments, 70 percent of the fully plastic value, and stresses which exceed the nominal yield point of the aluminum, it is concluded that the structural behavior is satisfactory. Discrepancies between measured values and design prediction are discussed.

0358 Rocking Vibrations of Footings

H. R. Sreekantiah, (Asst. Prof., Dept. of Civ. Engrg., Karnataka Regional Engrg. Coll., Surathkal, Srinvasnagar-574157, Karnataka, India)

Journal of the Geotechnical Engineering Division, Vol. 108, No. GT7, July, 1982, pp. 905-917

To predict the response of a machine foundation subjected to rocking vibrations, analytical solutions for three cases of contact pressure distributions, namely, rigid base, uniform and parabolic have been developed and are presented in the form of curves for ready use. Carefully controlled field vibration tests have been carried out on footings of various sizes and masses, resting on soil surface and subjected to horizontal vibrations which generate coupled rocking and sliding vibrations, to check the validity of the analytical solutions. The experimental investigation conducted is reported in detail. It is shown that the newly developed analytical solutions are quite satisfactory to predict the response of a machine foundation subjected to rocking vibrations, as well as predicting the rocking component of the coupled rocking and sliding vibrations, from a single field vibration test.

0359 Static Plate Loading Tests on Overconsolidated Clay

Knut H. Andersen, (Civ. Engr., Norwegian Geotechnical Inst., Oslo, Norway) and **Per Stenhamar**, (Civ. Engr., Ing. Chr. F. Groner, Sandvika, Norway)

Journal of the Geotechnical Engineering Division, Vol. 108, No. GT7, July, 1982, pp. 918-934

Three in situ static plate loading tests carried out on a medium stiff, overconsolidated clay deposit are described and interpreted. In addition to the plate loading test results, the results from undrained static laboratory tests on Haga clay are also presented. The project aimed mainly at investigating the reliability of using laboratory tests to determine the in situ static shear strength and stiffness of overconsolidated clays. It was possible to obtain reasonably good agreement between measured and calculated values of bearing capacity and displacements, provided the calculations were based on laboratory tests with the same type of load application and the same time to failure as in situ. For the displacmment calculations it was essential to account for load distributions and the very nonlinear stress-strain properties.

0360 One-Dimensional Dynamic Ground Response Analyses

Philippe P. Martin, Assoc. Member, ASCE, (Asst. Prof. of Civ. Engrg., Univ. of Illinois, Urbana, Ill.) and **H. Bolton Seed**, Fellow, ASCE, (Prof. of Civ. Engrg., Univ. of California, Berkeley, Calif.)

Journal of the Geotechnical Engineering Division, Vol. 108, No. GT7, July, 1982, pp. 935-952

The earthquake response analyses of six horizontally layered sites have been performed by means of the equivalent linear method of analysis and two non-linear methods using the method of characteristics and the finite element approach. The agreement between the equivalent linear method and the non-linear methods was found to be good in terms of the distributions of maximum accelerations and dynamic shear stresses for all the cases studied. Differences between the two types of approaches could be found for some sites between the spectral shapes of the computed surface motions.

0361 Soil Improvement with Quicklime and Filter Fabric

Toyotoshi Yamanouchi, Member, ASCE, (Prof., Dept. of Civ. Engrg., Kyushu Univ., Hakozaki, Fukuoka, Japan), **Norihiko Miura**, (Asst. Prof., Dept. of Civ. Engrg., Yamaguchi Univ., Tokiwadai, Ube, Japan), **Noboru Matsubayashi**, (Chf., Div. of Civ. Engrg., Shimonoseki City, Nanbu, Shimonoseki, Japan) and **Naozou Fukuda**, (Engr., Fukken Chosa Sekkei Co., Ltd., Higashi, Hiroshima, Japan)

Journal of the Geotechnical Engineering Division, Vol. 108, No. GT7, July, 1982, pp. 953-965

The multiple strip-sandwich method is a specially devised method using quicklime and filter fabric for embankments with difficult cohesive soils. A quicklime layer of 5 cm (2 in.) thick, sandwiched between two sheets of filter fabric of 30 cm (12 in.) wide, is placed in a triangular configuration with a horizontal pitch of about 2 m (6.6 ft) at each stage of piling the cohesive soil. Owing to combined actions resulted from the hydration of quicklime, particularly the absorption of water, the cohesive soil is effectively dewatered resulting in consolidation in success, to be able to achieve a high embankment efficiently due to an adequate increase in the shear strength of the cohesive soil by a certain depth both on and beneath the respective sandwich layers. This paper describes a case history for the construction of a high embankment, 32 m (105 ft) high, using the method with cohesive soil of a strongly weathered tuff. The work was successful and the multiple strip-sandwich method proved to be useful for constructing high embankments with difficult cohesive soils.

0362 Remedial Treatment Exploration, Wolf Creek Dam, Ky.

Marvin D. Simmons, (Chf., Geology Section, Nashville District Corps of Engrgs., Nashville, Tenn.)

Journal of the Geotechnical Engineering Division, Vol. 108, No. GT7, July, 1982, pp. 966-981

After 15 years of apparently satisfactory operation, Wolf Creek Dam developed underseepage and piping problems associated with its karstic foundation. When a sinkhole developed at the downstream toe of the earth embankment, a concentrated program of exploration was undertaken to determine both the source of the problem and possible remedial treatment. Various exploratory techniques were used; however, direct subsurface information obtained by drilling and grouting was the only method reliable enough to define the problem so that a remedy could be devised. After successful emergency grouting had arrested the piping, in all probability saving the dam from a major failure, a search for a permanent answer determined that a concrete diaphragm wall through the earth embankment into the limestone foundation would be the most acceptable method of correcting the problem. Construction of such a wall is costly; therefore, it was important to ascertain the minimal depth and extent required for it. The exploratory methods used for this purpose are reviewed in this paper.

0363 *H*-Pile Heave: A Field Test

Demetrious C. Koutsoftis, Assoc. Member, ASCE, (Assoc., Dames & Moore, Kowloon, Hong-Kong)

Journal of the Geotechnical Engineering Division, Vol. 108, No. GT8, August, 1982, pp. 999-1016

A field test was conducted to investigate the possibility of reduction in pile capacity and the potential for excessive settlements resulting from the heave of H-piles. The piles were driven through relatively softer strata to be seated into an interbedded very dense/hard deposit. A group of 38 pi(es, arranged in an 8x5 cluster, were driven as part of this test. Six previously driven piles, located outside the cluster at distances of 17 ft to 37 ft (5.2 m to 11.3 m), and two previously driven piles located within the cluster were also monitored for heave. The maximum pile heave within the cluster was 1.04 in. (26.4 mm), while 75% of the piles experienced heave between 0.4 and 0.7 in. (10.2 mm and 17.8 mm). Incremental pile heave decreases exponentially with increasing distance from the driven piles. The radius of influence of pile heave was estimated to be approximately 50 ft (15.2 m). Pile load tests showed that pile heave up to 1.04 in. (26.4 mm) had no effect on pile load capacity or on pile load deflection characteristics. Pile tip heave was only slightly less than pile butt heave.

0364 Criteria for Settlement of Tanks

W. Allen Marr, (Research Assoc., Dept. of Civ. Engrg., Rm. 1-376, Massachusetts Inst. of Tech., Cambridge, Mass. 02139), **Jose A. Ramos**, (Consulting Geotechnical Engr., Foxboro, Mass.) and **T. William Lambe**, Fellow, ASCE, (Prof. Emeritus, Dept. of Civ. Engrg., Mass. Inst. of Tech., Rm. 1-378, Cambridge, Mass. 02139)

Journal of the Geotechnical Engineering Division, Vol. 108, No. GT8, August, 1982, pp. 1017-1039

Performance criteria for the settlement of large steel tanks used to store fluids at ambient temperature and pressure are presented. Where possible, the criteria include a factor of safety, defined as the ratio of tensile stress to developed stress. Permissible values of factor of safety for safe operation are indicated. The performance criteria are developed by interpreting and expanding previous studies, making approximate analyses, and analyzing the measured performance of 90 large tanks, including the tanks at Toa Nenryo Kogyo's site at Kawasaki City, Japan. For each criterion, the mechanism of failure, the structural element to which the criterion applies, and the basis for the criterion are identified.

0365 Rock Engineering: La Grande Complex, Quebec

D. Kent Murphy, (Chf. Geologist, Rousseau, Sauve, Warren Inc., Montreal, Quebec, Canada) and **Jerry Levay**, (Deputy Head, Geology and Soil Mech. Dept., Societe d'Energy de la Baie James, 800 Blvd. de Maionneuve E., Montreal, Quebec, Canada)

Journal of the Geotechnical Engineering Division, Vol. 108, No. GT8, August, 1982, pp. 1040-1055

Rock engineering has played an important role in the design and construction management of the La Grande Complex, a 13,75O MW hydroelectric development in the James Bay area of northern Quebec. Confidence in the strength and erosion resisting properties of the precambrian granitic and metamorphic rocks of the region has led to some innovative design solutions as well as a policy of flexible response to rock support requirements. Experience gained with unlined tunnels and channels for hydraulic purposes is discussed. The importance of closely controlled drilling and blasting procedures in some areas of the work such as the crane support ledges at LG2 and a diversion conduit excavation at LG4 is also noted. The success of the flexible, generally optimistic approach to rock excavation and support activities in the Complex is due in large measure to the good communication links existing among design, cost control and construction elements within the management group, the Société d'énergie de Baie James.

0366 Volume Change Behavior of Desiccated Soils

A. Sridharan, (Assoc. Prof., Dept. of Civ. Engrg. Indian Inst. of Science, Bangalore, India) and **Mehter Mohamed Allam**, (Research Assoc., Dept. of Civ. Engrg., Indian Inst. of Science, Bangalore, India)

Journal of the Geotechnical Engineering Division, Vol. 108, No. GT8, August, 1982, pp. 1057-1071

Little information exists regarding the volume change behaviour of desiccated soils which are widespread in occurance. This paper reports a study on a few such soils and relates their behaviour with the known behaviour of pure clay minerals. The modified effective stress concept is exploited to mechanistically classify the volume change behaviour of desiccated soils. Previous experimental work indicated that repeated desiccation results in interparticle bonding which is also chemical in nature. Such bonding results in increased resistance to compression and greater permeability in such soils. Depending on the degree of bonding, an expansive soil may display either reduced expansiveness or even behave as a non-expansive soil in the undisturbed condition.

0367 Thompson Run Bridge Distresses: An Investigation

Milan Spanovich, Fellow, ASCE, (Principal Engr., Engrg. Mech., Inc., Consulting Geotechnical Engrs., Pittsburgh, Pa.)

Journal of the Geotechnical Engineering Division, Vol. 108, No. GT8, August, 1982, pp. 1072-1082

The failure of a pile-supported four lane highway bridge abutment overlying a 100-ft-high fill embankment was investigated. Subsurface data from the original bridge foundation design investigation and existing construction records were analyzed and correlated with additional subsurface explorations. The abutment movements were monitored employing standard surveying methods while the earth embankment was being monitored with slope indicators. The skin-friction loads transferred to the piles substantially exceeded the design load, resulting in the vertical displacement of those piles which did not reach bedrock. The major settlement of the abutment occurred where an existing fill of relatively high compressibility was surcharged by the newly placed embankment fill which the piles penetrated. Re-design and construction resulted in the employment of cast-in-place drilled caissons (bored piles) for support.

0368 Safety Factors for Probabilistic Slope Design

Robert A. D'Andrea, Member, ASCE, (Assoc. Prof. of Civ. Engrg., Worcester Polytechnic Inst., Worcester, Mass.) and **Dwight A. Sangrey**, Member, ASCE, (Prof. and Head, Dept. of Civ. Engrg., Carnegie-Mellon Univ., Pittsburgh, Pa.)

Journal of the Geotechnical Engineering Division, Vol. 108, No. GT9, September, 1982, pp. 1101-1118

Inconsistencies in present solution methods for slope stability problems under undrained conditions are noted and a first-order, second-moment, solution technique with a probabilistic base is suggested. The proposed procedure examines these problems with regard to the separate variables involved and utilizes partial safety factors which are proportional to the coefficient of variation of the pertinent parameters. For simplicity, a circular arc failure mechanism is assumed, and design acceptability is based on the derived value of a reliability measure, the safety index, which reflects the probability of occurrence of the assumed failure mechanism. Statistical data for the required load, resistance, and bias variables are presented. Using these, the safety index associated with current design techniques is determined, and the implications of its magnitude are examined. Sensitivity studies, performed to determine which variables have the greatest effect on design results are also described. Finally, partial safety factors are proposed for design corresponding to a desired failure probability.

0369 Use of Cyclic Element Tests to Assess Scale Models

David C. Procter, (Lect. of Civ. Engrg., Univ. of Manchester, Manchester, England) and **Jalal H. Khaffaf**, (Research Assoc., Dept. of Civ. Engrg., Univ. of Manchester, Manchester, England)

Journal of the Geotechnical Engineering Division, Vol. 108, No. GT9, September, 1982, pp. 1119-1132

Large scale laboratory model tests have been performed to aid the design of offshore gravity platforms. The paper shows that undrained conditions, necessary for correct model simulation, cannot be achieved if the full field loading program and remolded site material are used. It is theoretically possible to shorten the time of the model event by increasing cyclic load levels and reducing the number of cycles or increasing the frequency or using a dynamically weaker leaner model clay, or by a combination. Cyclic triaxial data are presented for three remolded clays of differing plasticities tested at various frequencies. The results show that for all practical purposes, even if gross model modifications are considered, similarity between model and field events is impossible. It is concluded that laboratory model tests are only indirectly helpful. They may be used to investigate a foundation's response in its weakest fully softened state or to assess a design technique by applying it to a model subjected to a brief arbitrary storm event.

0370 Housing Construction in Areas of Mine Subsidence

Felix Y. Yokel, Member, ASCE, (Leader, Geotechnical Engrg., National Bureau of Standards, Washington, D.C. 20234), **Lawrence A. Salomone**, Member, ASCE, (Research Civ. Engr., National Bureau of Standards, Washington, D.C. 20234) and **Richard E. Gray**, Member, ASCE, (Vice Pres., GAI Consultants, Inc., Monroeville, Pa. 15146)

Journal of the Geotechnical Engineering Division, Vol. 108, No. GT9, September, 1982, pp. 1133-1149

Many areas in the United States are underlain by abandoned mines and many more areas will be undermined in the future. As mine cavities collapse they cause settlement and ground distortions on the surface which may damage or destroy buildings and utilities. Many of these subsidence-prone areas are presently used or will be used in the future for residential housing development. Three problems associated with the development of mine subsidence areas are addressed: Site exploration and evaluation; site development; and housing construction in mine subsidence areas. This paper is derived from a study sponsored by the Department of Housing and Urban Development.

0371 Centrifuge Modeling of Lateral Earth Support

C. K. Shen, Member, ASCE, (Prof., Dept. of Civ. Engrg., Univ. of California, Davis, Calif. 95616), **Y. S. Kim**, (Grad. Student, Dept. of Civ. Engrg., Univ. of California, Davis, Calif. 95616), **S. Bang**, (Asst. Prof., Dept. of Civ. Engrg., Univ. of Notre Dame, Notre Dame, Ind. 46556) and **J. F. Mitchell**, (Civ. Engr., M Construction Co., Bakersfield, Calif.)

Journal of the Geotechnical Engineering Division, Vol. 108, No. GT9, September, 1982, pp. 1150-1164

Using the centrifuge modeling technique, a series of reinforced lateral earth support model tests were performed to identify the possible failure mechanism, and to verify a limit equilibrium method of analysis for the stability of the system. Detailed descriptions are given concerning the building, testing, and monitoring of the centrifuge model testing. The results of the model study show good agreement with theoretical predictions based on a plasticity solution formulated for analyzing the reinforced lateral earth system. The information obtained from this research indicates that centrifuge model testing is a useful tool for investigating the stability of earth structures particularly in the absence of prototype failure records.

0372 Initial Pore Pressure from Vertical Surface Loads

Jacobo Bielak, Member, ASCE, (Assoc. Prof., Dept. of Civ. Engrg., Carnegie-Mellon Univ., Pittsburgh, Pa. 15213)

Journal of the Geotechnical Engineering Division, Vol. 108, No. GT9, September, 1982, pp. 1167-1171

In estimating the initial settlement of footings supported on saturated clay soils, it is necessary to evaluate the instantaneous excess pore pressure developed in the soil upon application of the load. In this study, an explicit formula is derived for the instantaneous excess pore pressure beneath a corner of a uniformly loaded rectangular area. The pressure at an arbitrary point within the soil, then, is obtained by simple superposition of the basic result.

0373 Correlation of Maximum Dry Density and Grain Size

George P. Korfiatis, Assoc. Member, ASCE, (Grad. Research Asst., Dept. of Civ. and Environmental Engrg., Rutgers Univ., Piscataway, N.J.; Staff Engr., Melick-Tully and Assoc., South Bound Brook, N.J.) and **Constantine N. Manikopoulos**, (Asst. Prof., Coll. of Engrg., Rutgers Univ., Piscataway, N.J.)

Journal of the Geotechnical Engineering Division, Vol. 108, No. GT9, September, 1982, pp. 1171-1176

A predictive relationship between the maximum dry density and the grain size distribution curve for noncohesive soils has been formulated. A parametric expression has been derived and the parameters have been evaluated by fitting experimental values for 43 soil samples. The results of this evaluation have indicated that the maximum dry density of a noncohesive soil can be estimated with a reasonable degree of accuracy if the grain size distribution of the soil is known.

0374 Laterally Loaded Piles in a Two-Layer Soil System

Prabhakar J. Pise, (Asst. Prof., Dept. of Civ. Engrg., Indian Inst. of Tech., Kharagpur, India)

Journal of the Geotechnical Engineering Division, Vol. 108, No. GT9, September, 1982, pp. 1177-1181

Typical results for pile head displacement and maximum moment in nondimensional terms, for a laterally loaded free-head pile embedded in a two-layer soil system, using elastic approach, is presented. It is concluded that pile flexibility factor, thickness of surface layer, and elastic properties of the layers govern the lateral response of a pile. Depth of surface layer controlling the behavior of a pile depends significantly on pile flexibility factor and stiffness of the surface layer. Reductions in pile head displacement for flexible pile and maximum moment for relatively stiff pile, which are beneficial in practice, are greatly achieved by a stiff surface layer extending one-tenth the length of a pile. An illustrative example is solved to show the applicability of the results.

0375 Stability of Slopes by Method of Characteristics

A. Siva Reddy, (Prof. of Civ. Engrg., Indian Inst. of Science, Bangalore 560012 India) and **K. N. Venkatakrishna Rao**, (Asst. Prof. of Civ. Engrg., Government B.D.T. College of Engrg., Davangere 570004, India)

Journal of the Geotechnical Engineering Division, Vol. 108, No. GT9, September, 1982, pp. 1182-1186

The method of characteristics with some simplifying assumptions is made applicable for analyzing a given straight slope. By assuming that the mobilized shear strength varies with depth, and treating the whole soil mass as a series of layers, factors of safety of a given slope at different heights and a series of lines with different mobilized shear strength are obtained. The results show that the factors of safety obtained by the present method are lower than those obtained by friction circle method.

0376 Example Cost of 3-D FEM for Underground Openings

Charles W. Schwartz, Assoc. Member, ASCE, (Asst. Prof. of Civ. Engrg., Univ. of Maryland, College Park, Md.), **Amr S. Azzouz**, Assoc. Member, ASCE, (Asst. Prof. of Civ. Engrg., Massachusetts Institute of Technology, Cambridge, Mass.) and **Herbert H. Einstein**, Member, ASCE, (Assoc. Prof. of Civ. Engrg., Massachusetts Institute of Technology, Cambridge, Mass.)

Journal of the Geotechnical Engineering Division, Vol. 108, No. GT9, September, 1982, pp. 1186-1191

The use of three-dimensional finite element methods for the analysis of underground structures has been limited in practice, partially because of their supposed high costs. A more quantitative perspective on this issue is gained by examining the actual costs from one particular but typical study for the Peachtree Center Station cavern in Atlanta. The analysis simulated three excavation steps in high quality gneissic rock, which was modeled as a homogeneous, isotropic, elastic material. The cost of the analysis was approximately $18,000 for computer time and $20,000 for manpower, for a total of $38,000 or approximately 0.1% of the total project cost. Comments on the suitability of 3-D finite element analysis in the design process are also given.

0377 Stress Analysis of Dummy Rod Method for Sand Specimens

Han-Chin Wu, (Prof., Coll. of Engrg., Univ. of Iowa, Iowa City, Iowa 52242) and **G. S. Chang**, (Structure Design Engr., Tex-Ark Joist Co., Hope, Ariz. 71801)

Journal of the Geotechnical Engineering Division, Vol. 108, No. GT9, September, 1982, pp. 1192-1197

Stress analysis is performed for sand specimens with metal core under hydrostatic compression. It is shown that the stress state in such samples is not isotropic. The effect of membrane penetration is underestimated by such methods. The implications of this finding are examined.

0378 Energy Transfer in Standard Penetration Test

Felix Y. Yokel, Member, ASCE, (Leader, Geotechnical Engrg. Group, Center for BuiLding Tech., National Bureau of Standards, Washington, D.C. 20234)

Journal of the Geotechnical Engineering Division, Vol. 108, No. GT9, September, 1982, pp. 1197-1202

An expression is derived for the fraction of the kinetic energy transferred from the hammer to the drill rod in the SPT which reflects the energy loss attributable to drill rod length. The expression is used to study the effects of rod size, and the potential effect of reducing the hammer weight and increasing the fall height correspondingly. It is shown that 95% of the energy is transferred to the drill rod when the weight of the drill rod is 75% of that of the hammer, and that the energy transfer losses in shallow borings could be reduced considerably by using a half-weight hammer, delivering half the energy used in the standard test.

0379 Application of Modulus Degradation Model of Clays

Albert T. F. Chen, Member, ASCE, (Research Civ. Engr., U. S. Geological Survey, MS-74, 345 Middlefield Rd., Menlo Park, Calif. 94025)

Journal of the Geotechnical Engineering Division, Vol. 108, No. GT10, October, 1982, pp. 1203-1214

A degradation model is applied in conjunction with different soil models and stress-strain relations to site response analyses during earthquakes. To evaluate the effects of degradation, computations on two clay deposits subjected to both high and low-level input excitations are conducted. In addition, modulus degradation with or without shear strength reduction is discussed. In terms of surface response, degradation has little effect in the case of low-level excitation. With high-level excitatfons, however, degradation causes lower surface spectral response in the shorter period range, and the difference is on the same order as that from different soil models. Where surface response differs, the use of degradation with strength reduction is less conservative when compared to the use of degradation without strength reduction.

0380 Repetitive Load Deformation of Cohesionless Soil

Vishnu A. Diyaljee, Assoc. Member, ASCE, (Former Grad. Student, Dept. of Civ. Engrg., Queen's Univ., Kingston, Ontario, Canada K7L 3N6) and **Gerald P. Raymond**, Member, ASCE, (Prof., Dept. of Civ. Engrg., Queen's Univ., Kingston, Ontario, Canada K7L 3N6)

Journal of the Geotechnical Engineering Division, Vol. 108, No. GT10, October, 1982, pp. 1215-1229

Presented is a simplified procedure for obtaining material characterization of free draining cohesionless soils for use in calculating permanent deformations of transportation support systems. Experimental data is presented in support of the described simplified procedure, which shows that plastic strains are largely governed by the deviator stress level and, thus, the factor of safety against shear failure. Plotting logarithm of plastic strain versus logarithm of number of load repetitions resulted in a slope essentially independent of confining pressure or deviator stress level for stress levels less than 60%. Plotting logarithm of plastic strain at a given number of load cycles versus deviator stress level resulted in a slope which was essentially constant. These two findings allowed the development of the presented procedure resulting in a method of obtaining material characterization which gives a significant savings in laboratory testing time over present methods. Such material characterization is an essential requirement for predicting the rate of accumulation of permanent deformation (rutting) using the well-known layered elastic or non-elastic technique.

0381 Synthetic Fabrics as Soil Filter: A Review

David J. Hoare, (Lect., Dept. of Civ. Engrg., Univ. of Birmingham, P.O. Box 363, Birmingham B15 2TT, England)

Journal of the Geotechnical Engineering Division, Vol. 108, No. GT10, October, 1982, pp. 1230-1245

The fundamentals of fabric filters, their uses, design, longevity, strength requirements, and testing are reviewed and discussed. For unidirectional laminar flow conditions, particle retention and permeability criteria for fabric filters are well established. All the approaches currently adopted tend to give similar answers and are dependent on the fabric being the 'catalyst' in the formation of a 'self-induced' filter within the retained soil. When problems occur in practice, it is normally due to clogging of the filter, i.e. the fabric is too fine. Thin fabrics appear most suited for this use. For severe hydrodonamic and reversing flow conditions, the self-induced' filler will not form, and design criteria are not so well established or accepted. Thicker fabrics offer a more tortuous path to the migration of particles and are thus more suited than thin fabrics. However, thin fabrics have successfully been used in conjunction with (out of specification) granular filters.

0382 Conventional and Probabilistic Embankment Design

Verne McGuffey, (Assoc. Soils Engr., New York State Dept. of Transportation, Albany, N.Y. 12226), **Dimitri Athanasiou-Grivas**, Assoc. Member, ASCE, (Assoc. Prof., Dept. of Civ. Engrg., Rensselaer Polytechnic Inst., Troy, N.Y. 12181), **John Iori**, Assoc. Member, ASCE, (Asst. Soils Engr., New York State Dept. of Transportation, Albany, N.Y. 12226) and **Zenon Kyfor**, Assoc. Member, ASCE, (Asst. Soils Engr., New York State Dept. of Transportation, Albany, N.Y. 12226)

Journal of the Geotechnical Engineering Division, Vol. 108, No. GT10, October, 1982, pp. 1246-1254

The safety of a proposed highway embankment is evaluated both conventionally and probabilistically. Results from tests on undisturbed soil samples, secured during an extensive exploration program, are statistically analyzed and presented. The factor of safety is introduced as a random variable whose variability is due exclusively to the variability of the shear strength of the foundation material. Failure is defined as the event whereby the factor of safety (computer at end-of-construction) receives values smaller than or, at most, equal to unity. The probability of failure is defined as the probability of the occurrence of this event. From the three models (normal, beta, and lognormal) considered for the distribution of the factor of safety, the normal always provides the largest, and the lognormal the smallest values for the probability of failure of the embankment. This difference in the results decreases as the height or angle of the embankment increases. Results obtained using the beta model are presented in a nomography relating embankment height and slope angle to probability of failure.

0383 Short Rigid Piers in Clays

C. V. Girija Vallabhan, Member, ASCE, (Assoc. Prof. in Civ. Engrg., Texas Tech Univ., Lubbock, Tex.) and **Fariborz Alikhanlou**, (Grad. Student, Dept. of Civ. Engrg., Texas Tech Univ., Lubbock, Tex.)

Journal of the Geotechnical Engineering Division, Vol. 108, No. GT10, October, 1982, pp. 1255-1272

A simple discrete model is developed to represent the nonlinear behavior of the soil around a short rigid pier foundation, When it is subjected to large Lateral forces or large overturning moments. The soil resistances include lateral soil resistance, bottom vertical resistance, bottom frictional resistance, bottom moment resistance and frictional resistance on the sides of the pier. All these soil resistances are modelled using a nonlinear soil spring concept. Comparison of the results using the model with full-scale experimental results show good promising characteristics.

0384 Phase II Inspection of Lake Manatee Dam

W. David Carrier, III., Member, ASCE, (Executive Vice Pres. Cromwell Engrg., Inc., Lakeland, Fla. 33803), **Berg Keshian, Jr.**, Member, ASCE, (Vice Pres., Bromwell Engrg., Inc., Denver, Colo.) and **Leslie G. Bromwell**, Member, ASCE, (Pres., Bromwell Engrg., Inc., Lakeland, Fla. 33803)

Journal of the Geotechnical Engineering Division, Vol. 108, No. GT10, October, 1982, pp. 1273-1287

Lake Manatee Dam was constructed during the period 1965-1967. The reservoir is the primary water supply for a quarter of a million people in the Bradenton-Sarasota area of Florida. In early 1978, the Jacksonville District Corps of engineers performed a preliminary (Phase I) inspection of the earth dam and appurtenant structures as part of the National Dam Safety Program. The Corps found that the existing spillway did not have sufficient capacity to prevent overtopping during a large storm and that the downstream slope was potentially unstable, due to inadequate seepage control. The Corps classified the dam as large, with a high hazard potential, and recommended that a more detailed investigation be performed. Design calculations and many construction records for the dam were not available. Therefore, the subsequent Phase II investigation was considerably more extensive than would otherwise have been necessary. This investigation indicated that the embankment has a satisfactory factor of safety against slope failure and the concrete spillway has a satisfactory factor of safety against overturning. However, the drain at the toe of the dam was found to be partially plugged, and construction of a new toe drain was recommended. It was also recommended that an emergency spillway be constructed to supplement the existing spillway.

0385 Fracturing of Rock with Expansive Cement

Charles H. Dowding, Aff., (Assoc. Prof., Dept. of Civ. Engrg., Northwestern Univ., Evanston, Ill. 60201) and **Joseph F. Labuz**, Assoc. Member, ASCE, (Grad. Student, Dept. of Civ. Engrg., Northwestern Univ., Evanston, Ill. 60201)

Journal of the Geotechnical Engineering Division, Vol. 108, No. GT10, October, 1982, pp. 1288-1299

A study was conducted to determine the efficiency of a highly expansive cement (called Bristar) used to fracture rock and concrete. The 48-hour expansive pressure of the cement, calculated to be about 46 MPa (6700 psi), is independent of hole size when tested in steel, thick-walled cylinders. Because the pressure is constant, a scaling relationship of S/d emanates from elastic theory, where S is the spacing between boreholes and d is the borehole diameter. However, the rate of pressure development is a function of cement temperature; consequently thermodynamic considerations may be as important as hole spacing. At the same scaled distances, 38-mm (1.5-in.) boreholes fracture dolomite 24-48 hours sooner than 12.7-mm (0.5-in.) holes. Optimal spacings in dolomite ranged between 8 and 16 borehole diameters.

0386 Filter Criteria for Geotextiles: Relevance and Use

Chris R. Lawson, (Principal, Lawson Consulting, Sydney, Australia)

Journal of the Geotechnical Engineering Division, Vol. 108, No. GT10, October, 1982, pp. 1300-1317

Thm general requirements for filters, both granular and geotextile, in one-directional flow applications and the relevant geotextile properties affecting good filtration performance are presented. The relevance of existing one-directional geotextile filter criteria is also examined. An application involving the specification and performance evaluation of a geotextile for an internal filter in an earth dam, constructed on flyash, where large hydraulic heads were anticipated at the flyash-geotextile interface (up to 10.8 m) validated the use of existing geotextile filter criteria.

0387 Failure of an Anchored Bulkhead

David E. Daniel, Assoc. Member, ASCE, (Asst. Prof., Dept. of Civ. Engrg., Univ. of Texas, Austin, Tex. 78712) and **Roy E. Olson**, Fellow, ASCE, (Prof., Dept. of Civ. Engrg., Univ. of Texas, Austin, Tex. 78712)

Journal of the Geotechnical Engineering Division, Vol. 108, No. GT10, October, 1982, pp. 1318-1327

An 800-ft (240 m) long anchored bulkhead with a height of 65 ft (20 m) was designed more or less in accord with published recommendations by several acknowledged experts. Nevertheless, the wall failed. Post-failure analysis indicated that the source of error was a lack of understanding of soil behavior under conditions involving significant excavation near the bulkhead. The design was adequate for undrained conditions but was inadequate for fully drained conditions.

0388 Seismic Response of Earth Dams in Triangular Canyons

Faiz I. Makdisi, Assoc. Member, ASCE, (Project Engr., MIDCO, Saudi Arabia), **Takaaki Kagawa**, (Staff Engr., McClelland Engrs., Houston, Tex.) and **H. Bolton Seed**, Fellow, ASCE, (Prof. of Civ. Engrg., Univ. of California, Berkeley, Calif.)

Journal of the Geotechnical Engineering Division, Vol. 108, No. GT10, October, 1982, pp. 1328-1337

The effects of narrow and sloping canyons on the dynamic response of embankment dams is investigated by a three-dimensional finite element analysis. Analytical results for embankments with several width to height ratios are compared with the results of plane strain solutions for comparable sections and significant differences are shown to exist between the two solutions. Comparisons are presented of the natural frequencies of vibration of dams in rectangular and triangular cannons of different slopes with those of long dams for which the assumption of plane-strain rcsponse is valid. For an embankment in a triangular valley with a width to height ratio of three and subjected to transient ground motions, the computed accelerations at the mid-section were about 60 to 80% higher in a three-dimensional response analysis than in a plane-strain analysis. On the other hand the computed shear stresses for the same section in the three-dimensfonal analysis were generally about one half the values computed from plane strain analyses. It is concluded that the dynamic response of an earth embankment is significantly affected by the shape of the canyon in which it is constructed.

0389 Shallow Foundations in Cold Regions: Design

Edward C. McRoberts, (Consulting Engr., Hardy Assoc. Ltd., 4810-93 St., P. O. Box 746, Edmonton, Alberta, Canada)

Journal of the Geotechnical Engineering Division, Vol. 108, No. GT10, October, 1982, pp. 1338-1349

The techniques now avilable in the North American literature for the design of shallow foundations in permafrost are reviewed. Subjects considered are geothermal aspects, thaw settlement and consolidation effects, frost heaving, and frozen ground aspects. The major thrust is to provide an overview of the major factors that must be considered and to provide a bibliography from which more specific information can be obtained. Data pertaining to the long-term strength and creep rates of ice rich frozen ground at temperatures below but near the freezing point (0°C) is given. It is noted that the major deficiency in shallow foundation design is lack of reliable data on adfreeze frost-jacking loads and frost-heave predictions.

0390 Installation of Pile Group in Overconsolidated Clay

Michael W. O'Neill, Member, ASCE, (Assoc. Prof. of Civ. Engrg., Univ. of Houston, Houston, Tex.), **Richard A. Hawkins**, Assoc. Member, ASCE, (Project Mgr., Raymond Technical Facilities, Inc. Houston, Tex.) and **Jean M. E. Audibert**, Assoc. Member, ASCE, (Assoc. and Mgr., Ertec, Inc. Gulf Div., Houston, Tex.)

Journal of the Geotechnical Engineering Division, Vol. 108, No. GT11, November, 1982, pp. 1369-1386

Measurements made during and immediately after the installation of a full-sized group of nine driven piles and two isolated control piles are presented. Rates of pile set-up in saturated, desiccated overconsolidated clay, development of lateral pressures on the piles, ground movements and accumulation of residual loads in the piles are described. Note is made of the similarities between group and isolate isolated piles.

0391 Effective Stress Capacity Model for Piles in Clay

Leland M. Kraft, Jr., Member, ASCE, (Mgr., Special Projects Group, McClelland Engrs., Inc., Houston, Tex. 77081)

Journal of the Geotechnical Engineering Division, Vol. 108, No. GT11, November, 1982, pp. 1387-1404

Several rigorous, theoretical methods, using effective stress concepts, have been developed in the last five years to predict the ultimate shaft friction on piles supported in clays. The latest generation of these methods, which were developed as part of an industry sponsored project administered by Amoco, is described and critiqued. Computed results for four effective stress models and two conventional methods are compared with values measured on 10 piles at two field and one laboratory test sites. Predicted and measured values are in good agreement. The development of effective stress approaches to compute shaft friction provides for an improved understanding of factors that affect shaft friction and should lead in the future to improved techniques.

0392 In Situ Stress Determination by Iowa Stepped Blade

Richard L. Handy, Member, ASCE, (Prof. of Civ. Engrg., Iowa State Univ., Ames, Iowa 50011), **Bernard Remmes**, Assoc. Member, ASCE, (Geotechnical Engr., McClelland Engrg., Houston, Tex.), **Steven Moldt**, Assoc. Member, ASCE, (Geotechnical Engr., Warzyn Engrg., Inc., Madison, Wisc.), **Alan J. Lutenegger**, Assoc. Member, ASCE, (Pres., Geotechnical Test Systems, Inc., Ames, Iowa 50010) and **Gary Trott**, Assoc. Member, ASCE, (Geotechnical Engr., Weideman & Singleton, Atlanta Ga.)

Journal of the Geotechnical Engineering Division, Vol. 108, No. GT11, November, 1982, pp. 1405-1422

Lateral soil pressure is a vital element in soil mechanics theory and practice, but is very sensitive to disturbance and, therefore, difficult to measure. A step-tapered blade was developed to compensate for disturbance by measuring soil pressures on three thicknesses of blades, and extrapolating for the hypothetical pressure at zero blade thickness. The nature of the extrapolation was found to be exponential, with the slope theoretically inverse to the soil compression index C_c. In situ soil stresses measured in fine-grained soils generally agree with stresses from overburden pressures, elastic theory, and pressuremeters. Measured K_o's in natural soil deposits are generally within acceptable ranges, 0.5-1.5 depending on the consolidation state of the soil, and up to 4.2 in expansive clay. About 8-10 tests may be performed per hour; the test may not be reliable in dense sands. Directional horizontal stresses may be obtained with a three-bladed stepped vane.

0393 Predicted and Measured Resilient Response of Track

Harry E. Stewart, Assoc. Member, ASCE, (Asst. Prof. of Civ. Engrg., Univ. of Massachusetts, Amherst, Mass.) and **Ernest T. Selig**, Fellow, ASCE, (Prof. of Civ. Engrg., Univ. of Massachusetts, Amherst, Mass.)

Journal of the Geotechnical Engineering Division, Vol. 108, No. GT11, November, 1982, pp. 1423-1442

Field measurements were made at the Facility for Accelerated Service Testing (FAST) research track in Pueblo, Colorado to determine the magnitude of resilient or elastic stresses, strains and deflections in the ballast, subballast and subgrade resulting from train loading. These data were analyzed statistically to determine significant trends. A computer model, GEOTRACK, was used to predict the results measured at FAST for comparison with the experimental data. Then the influence of variations in the parameters on track response was demonstrated with GEOTRACK calculations. In general, the strains and deflections did not differ significantly with changes in ballast thickness and ballast type and tie type, compared to the variations within one test section. However, the ballast strains were much larger in the wood tie sections than in the concrete tie sections, apparently because of tie spring-up. The subgrade stress in the concrete tie section was larger than in the wood tie section. This appears to be a result of more uniform pressure distribution beneath the wood ties in the field, because the GEOTRACK model did not indicate such differences for homogeneous ballast layers. The GEOTRACK model showed that the most important parameters influencing track response were ballast depth, tie spacing and tie stiffness.

0394 Nonlocal Variational Method in Stability Analysis

Casuba G. Prabhakar Narayan, (Asst. Prof., Dept. of Civ. Engrg., Indian Inst. of Tech., New Delhi, India), **Vijay D. Bhatkar**, (Principal Scientific Officer, Electronics Commission, Government of India, New Delhi, India) and **T. Ramamurthy**, (Prof., Dept. of Civ. Engrg., Indian Inst. of Tech., New Delhi, India)

Journal of the Geotechnical Engineering Division, Vol. 108, No. GT11, November, 1982, pp. 1443-1459

A mathematical technique is developed for analyzing the stability analysis in terms of effective stresses. The stability equations are obtained based on limiting equilibrium conditions. The factor of safety defining the stable state of equilibrium is derived with respect to shear strength and follows Coulomb-Mohr failure criterion. The analysis considers the influence of effective inter-slice forces, and makes no *a priori* assumption regarding the shape of the slip surface, internal stress distribution or on the point of application of horizontal effective thrust line. In this method the stability analysis is transcribed as a minimization problem in the calculus of variations. Using this the critical slip surface associated with the minimum factor of safety is obtained. The numerical solutions obtained by the variational technique showed that the technique can be more appropriately utilised for the rigorous stability analysis of slopes. The slip-surface obtained by variational technique resembles more closely to logarithmic spiral. The analysis satisfies all equilibrium, boundary necessary and sufficient conditions. For the existence of a minimal, the functional is verified for Legendre and Weierstress conditions of positive definiteness and global extremal.

0395 Reinforced Soil: An Orthorhombic Material

Charles Gerrard, (Sr.Principal Research Sci., Commonwealth Scientific and Industrial Research Organization, Division of Building Research, P. O. Box 56, Highett 3190, Victoria, Australia)

Journal of the Geotechnical Engineering Division, Vol. 108, No. GT11, November, 1982, pp. 1460-1474

The reinforcement of a soil mass is simulated by a series of parallel, equally spaced sheets occurring in a single set or in two or three orthogonal sets. The behavior of the reinforced soil mass under working load conditions can be analyzed by applying, repeatedly in the cases where there are two or three sets of reinforcement, a theory that defines the equivalent homogeneous material that can represent a sequence of alternating orthorhombic layers. The properties of the equivalent materials can be used to calculate stress distributions from which the stresses in the soil and reinforcement components can be determined. In the case of thin, stiff reinforcements, the properties of the equivalent materials are found to depend on reinforcement terms that are products of relative thicknesses and relative moduli. When the reinforcement is in the form of meshes or strips, and the soil has pseudo-isotropic properties, the properties of the equivalent materials are simple functions of the reinforcement terms. A numerical example indicates the importance of reinforcement alignment and type, i.e. mesh or strip.

0396 Earthquake-Induced Deformations of Soft Clay Slopes

Yoshiharu Moriwaki, Member, ASCE, (Sr. Project Engr., Woodward-Clyde Consultants, Three Embarcadero Center, San Francisco, Calif. 94111), **Izzat M. Idriss**, Fellow, ASCE, (Principal, Woodward-Clyde Consultants, Three Embarcadero Center, San Francisco, Calif. 94111) and **Earl H. Doyle**, Member, ASCE, (Sr. Research Engr., Shell Development Co., Houston, Tex.)

Journal of the Geotechnical Engineering Division, Vol. 108, No. GT11, November, 1982, pp. 1475-1493

An existing one-dimensional, non-linear, degrading stress-strain soil model for horizontal sites is extended to mildly sloping sites and a practical procedure to analyze the earthquake-induced deformation of mildly sloping soft clayey sites is presented. The proposed soil mode assumes a unique relationship between the degradation index and residual strain and appears to provide satisfactory results when compared to laboratory tests results. An example analysis of a hypothetical mildly sloping site using the computer program DCHARMS is shown to illustrate the application of the procedure. The proposed procedure is similar to the stress path method as applied to the earthquake deformation analysis of an embankment. The procedure, however, assumes the decoupling of the cyclic part and the residual part of the response and incorporates omly the cyclic part of the response in the actual response calculations.

0397 Factors Affecting Stress Cell Measurements in Soil

William A. Weiler, Jr., Assoc. Member, ASCE, (Staff Engr., Haley and Aldrich, Inc., Cambridge, Mass.) and **Fred H. Kulhawy**, Member, ASCE, (Prof., School of Civ. and Environmental Engrg., Cornell Univ., Ithaca, N.Y.)

Journal of the Geotechnical Engineering Division, Vol. 108, No. GT12, December, 1982, pp. 1529-1548

The factors affecting stress cell measurements in soil are reviewed, evaluated and synthesized in this paper. Fifteen major factors are identified, and the method for taking into account each factor is described. Incorporating these factors will lead to better stress cell measurements through a more complete understanding of the interaction between the cell and the soil.

0398 Compressibility Behavior of Soft Lignite

Andreas G. Anagnostopoulos, (Reader of Foundation Engrg., Dept. of Civ. Engrg., National Technical Univ., Athens, Greece)

Journal of the Geotechnical Engineering Division, Vol. 108, No. GT12, December, 1982, pp. 1549-1566

The compressibility behavior of the Megalopolis (Greece) soft, earthy lignite is investigated by one-dimensional and triaxial consolidation tests in a wide pressure range. The material behaves as a typical preconsolidated clay and consolidates according to Terzaghi theory, exhibiting a significant secondary compression. The virgin curves from w-log p diagrams fall in the same zone as for known German and Australian lignites. The compressibility, coefficient of consolidation, and coefficient of secondary compression, investigated under various loading conditions, are influenced both by the stress history and the change in structure occuring at a pressure of about 260 psi (1800 kPa).

0399 Specimen Size Effect in Simple Shear Test

M. Vucetic, (Civil Engr., Civil Engrg. Sciences, Zagreb, Yugoslavia) and **S. Lacasse**, Member, ASCE, (Civii Engr., Norwegian Geotechnical Inst., Oslo, Norway)

Journal of the Geotechnical Engineering Division, Vol. 108, No. GT12, December, 1982, pp. 1567-1585

This paper presents the results of a laboratory investigation of the influences of height to diameter ratio and membrane stiffness on the behavior of clay in the NGI simple shear apparatus. The study included 21 strain-controlled, constant volume, monotonic loading, (static) tests on undisturbed specimens of medium-stiff clay. The clay was sampled in a special 200-mm ID research piston sampler. The testing program included use of three different height to diameter ratios and two types of reinforced rubber membranes for clay specimens at overconsolidation ratios of one and 10. The nominal height of the specimens was kept constant at 16 mm with diameters of 50 mm, 80 mm and 115 mm. The experimental results show that neither the height to diameter ratio nor the membrane reinforcement types used have any significant influence on the static-stress behavior measured in the NGI simple shear test.

0400 Dynamic Analysis of Earth Dams in Three Dimensions

Lelio H. Mejia, Assoc. Member, ASCE, (Sr. Engr., Harding-Lawson Assocs., Novato, Calif. 94948), **H. Bolton Seed**, Fellow, ASCE, (Prof., Civil Engrg., Univ. of California, Berkeley, Calif.) and **John Lysmer**, Member, ASCE, (Prof., Civil Engrg., Univ. of California, Berkeley, Calif.)

Journal of the Geotechnical Engineering Division, Vol. 108, No. GT12, December, 1982, pp. 1586-1604

The number of large dams constructed in narrow canyons, where plane strain conditions provide only an approximation to the true behavior of the structure, has increased considerably in recent years. Accordingly, the development of efficient numerical techniques for the three-dimensional dynamic analysis of earth dams has become a problem of practical importance. This paper presents a description of finite element techniques for the dynamic response analysis of earth and rockfill dams in three dimensions. Additionally, the method has been used to compute the dynamic response of Oroville Dam for the August 1, 1975 Oroville earthquake and to evaluate the dynamic material properties of the dam. The values obtained using this procedure are in reasonable agreement with the results of laboratory tests on the Oroville Dam shell material.

0401 Load Transfer Mechanisms in Piles and Pile Groups

Michael W. O'Neill, Member, ASCE, (Assoc. Prof. of Civ. Engrg., Univ. of Houston, Houston, Tex.), **Richard A. Hawkins**, Assoc. Member, ASCE, (Project Mgr., Raymond Technical Facilities, Inc., Houston, Tex.) and **Larry J. Mahar**, Assoc. Member, ASCE, (Project Engrg., ERTEC, Inc., Long Beach, Calif.)

Journal of the Geotechnical Engineering Division, Vol. 108, No. GT12, December, 1982, pp. 1605-1623

This paper describes patterns of measured load transfer in a full-sized instrumented pile group; subgroups within the main group; and in single control piles loaded in compression. The effects of residual stresses and uplift loading on load transfer are also described. Unit load transfer relationships are presented that are related to observed properties of the soil, which was a saturated overconsolidated clay. Similarities and differences between load transfer in single piles and load transfer in a pile group are also emphasized.

0402 Measured Performance of a Laboratory Culvert

Lawrence C. Rude, Member, ASCE, (Asst. Prof., Dept. of Civ. Engrg., Virginia Polytechnic Inst. and State Univ., Blacksburg, Va. 24061)

Journal of the Geotechnical Engineering Division, Vol. 108, No. GT12, December, 1982, pp. 1624-1641

This study shows that the behavior of a soil culvert system can be modeled by a linear finite element program and a mathematical solution of a hole in an elastic sheet reinforced by a concentric ring. The finite element program produced equivilant computations as the mathematical solution. The results of the theoretical calculations were compared to the measured behavior of a 6.6-in. (167.6-mm) diameter instrumented culvert installed in a laboratory testing tank. The comparison between the theoretical and actual behavior correlated well. Parameter studies were done to show the influence of variations of Young's modulus and Poisson's ratio of the soil on the culvert response.

0403 New Flat Plate Turbulent Velocity Profiles

Ben L. Sill, Member, ASCE, (Assoc. Prof., Dept. of Civ. Engrg., Clemson Univ., Clemson, S.C.)

Journal of the Hydraulics Division, Vol. 108, No. HY1, January, 1982, pp. 1-15

A new, simple equation describing the turbulent velocity profile over flat plates is developed by using the control volume statements of mass and momentum conservation. This expression accurately gives the velocity in the laminar sublayer, buffer zone, and logarithmic regions. Velocity profile results are compared to accepted data and a new expression for the eddy viscosity is also compared to measurements made near the wall. The development is then extended to include the effects of plate roughness, producing a simple but more general equation. An application which utilizes the new equations to analyze sediment transport is presented.

0404 Approximate sizing of Reservoirs for Detention Time

John S. Tapp, Assoc. Member, ASCE, (Vice-Pres.,, Commonwealth Tech., Inc., 2520 Regency Rd., Suite 104, Lexington, KY 40503), **Andrew D. Ward**, Assoc. Member, ASCE, (Research Specialist, Agr. Engrg. Dept., Univ. of Kentucky, Lexington, Ky. 40503) and **Billy J. Barfield**, (Prof., Agr. Engrg. Dept., Univ. of Kentucky, Lexington, Ky. 40503)

Journal of the Hydraulics Division, Vol. 108, No. HY1, January, 1982, pp. 17-23

To meet the detention time requirements of the Office of Surface Mining, sediment control reservoirs must be sized to give a certain detention time, as defined by the time difference between the center of mass of the inflow and outflow hydrographs. Designing a reservoir to give a certain detention time is time consuming as a principal spillway size must be selected, the routing accomplished, and the detention time calculated. Often several pipe sizes might be tried before the necessary detention time is attained. To minimize the trial and error, a procedure using triangular hydrographs is developed to give a first approximation of the peak outflow and storage volume necessary to give a desired detention time. The procedures are tested against detailed routings and are found to be good first approximations. The procedures are applicable to reservoirs with principal spillway crests at the top of the sediment storage pool.

0405 Estimating Culvert Scour in Cohesive Material

Steven R. Abt, Assoc. Member, ASCE, (Asst. Prof. of Civ. Engrg., Colorado State Univ., Fort Collins, Colo. 80523) and **James F. Ruff**, Member, ASCE, (Head, Civ. Engrg. Dept., Banner Assoc., Inc., Laramie, Wyo.)

Journal of the Hydraulics Division, Vol. 108, No. HY1, January, 1982, pp. 25-34

An investigation of scour at culvert outlets in a cohesive bed material is presented. Twelve experiments of 1000 minutes duration were conducted using a cohesive bed material. Culvert 27.3 cm (10-in.), 35.6 cm (14-in.)nd 45.7 cm (18-in.) diameters were tested with discharges ranging from 0.054 m^3/sec + 0.824 m^3/sec (1.91 cfs to 29.13 cfs). The tailwater elevation was maintained at 0.45 D ± 0.05 D above the culvert invert where D is the diameter of the culvert. The results yielded a series of empirical relationships expressing the depth, width, length and volume of scour as a function of the culvert diameter, culvert outlet velocity, soil plasticity index, soil shear strength and fluid density. General observations concerning scour hole formation, growth and stabilization are reported.

0406 Equivalent Hydraulic Pipe for Parallel Pipes

Roland W. Jeppson, Member, ASCE, (Prof. of Civ. and Environmental Engrg., College of Engrg., Utah State Univ., Logan, Utah 84322)

Journal of the Hydraulics Division, Vol. 108, No. HY1, January, 1982, pp. 35-45

Methods are defined for reducing two or more parallel pipes to a single hydraulically equivalent pipe for analysis using the Darcy-Weisbach equation that allows the friction factor to depend upon the flow conditions in the pipe, as well as the roughness of the pipe wall. The motivation for reducing groups of parallel pipes to single equivalent pipes is to reduce computer costs associated with analyses of large piping networks. Comparisons of the saving in computer execution time achieved in network analysis by utilizing the equivalent pipe technique are given for different methods of solving large networks. For the real system used in this comparison one method of solution required only one-quarter of the time if parallel pipes were formed first. The recommendation is that existing network analysis computer programs incorporate in their code the creation of hydraulic equivalent pipes for all groups of parallel pipes.

0407 Dispersion in Three Dimensions: Approximate Analytic Solutions

Budhi Sagar, Member, ASCE, (Assoc., Analytic and Computational Research, Inc., 12029 Clover Ave., Los Angeles, Calif. 90066)

Journal of the Hydraulics Division, Vol. 108, No. HY1, January, 1982, pp. 47-62

Solutions to three-dimensional convection dispersion equation under different boundary conditions are obtained by integrating the appropriate Green's functions. Two approximations are used during the integration. First, error function differences are approximated by exponential functions through Taylor expansion. Second, the time integral is approximated by Laplace method. The error introduced due to these approximations is analysed and discussed. The pollution source is assumed to be an arbitrary function of time and of finite size in space. The boundary conditions considered are of a stream in a horizontal flow field and a water table in a vertical flow field. Such solutions have potential use in quick preliminary analysis of groundwater pollution problems.

0408 Flood Probability Estimation

Samuel O. Russell, Member, ASCE, (Assoc. Prof., Dept. of Civ. Engrg., Univ. of British Columbia, Vancouver, B.C., Canada)

Journal of the Hydraulics Division, Vol. 108, No. HY1, January, 1982, pp. 63-73

A procedure is developed for estimating design floods in situations where data are insufficient for a conventional frequency analysis. It involves the use of a compound distribution which is a weighted combination of individual two parameter distributions. Initial values are assigned to the parameters of the component distributions and their weights on the basis of subjective estimates of the mean and standard deviation of the flood peaks. The weights of the component distributions are updated with Bayes' theorem in the light of any additional information such as recorded floods, the largest flood in a number of years or a flow which has not been exceeded in a given number of years. Numerical methods are used. A computer program is developed to carry out the necessary computations and is described to illustrate how the method could be used by practicing hydrologists. Examples are presented to illustrate the procedure.

0409 Spectrum of Shallow Water Waves

Angel N. Menendez, (Research Asst., Iowa Inst. of Hydr. Research, The Univ. of Iowa, Iowa City, Iowa 52242) and **Ruben Norscini**, (Physicist, Centro de Informatica Hydrica, Instituto Nacional de Ciencia y Technica Hidricas)

Journal of the Hydraulics Division, Vol. 108, No. HY1, January, 1982, pp. 75-94

A complete and coherent formulation for defining different spectral shallow water wave bands is stated. It takes into account the attenuation and dispersion characteristics of the waves and the predominant physical mechanisms. It is shown that the relevant parameter which accounts for the importance of attenuation and dispersion effects is the phase difference between the "velocity-wave" and the "depth-wave". Practical criteria for defining the limits of the spectral bands are introduced for any required accuracy. The results presented can be used to obtain a rough and quick determination of the main characteristics of a flood wave from its dimensionless wave number and the Froude number of the mean-flow.

0410 Incipient Motion of Sand Particles on Side Slopes

Syunsuke Ikeda, Member, ASCE, (Assoc. Prof., Dept. of Foundation Engrg., Saitama Univ., Urawa, Saitama, Japan)

Journal of the Hydraulics Division, Vol. 108, No. HY1, January, 1982, pp. 95-114

Several criteria have been theoretically proposed for the critical shear stress of noncohesive sediment on side slopes. However, no laboratory work has ever been done on this problem. A theoretical study of the incipient motion of sediment particles on side slopes is presented, including the effects of viscous sublayer, turbulence and lift force. As a special case of level beds, the theory describes the experimental values reported by several workers. A specially designed rotary wind tunnel permits the evaluation of the critical shear stress on side slopes, and the experiment supports the theory. These works reveal that the lift force can be neglected in estimating the critical shear stress using the proposed theory if the averaged experimental values are concerned, but inclusion of the lift force in the theory is recommended for safety and for describing the lower limit of the possibility of incipient motion.

0411 Asymptotic Solution for Dam-Break Problem

Bruce Hunt, (Reader, Dept. of Civ. Engrg., Univ. of Canterbury, Christchurch, New Zealand)

Journal of the Hydraulics Division, Vol. 108, No. HY1, January, 1982, pp. 115-126

The kinematic wave approximation is used to obtain a simple, closed-form solution for the catastrophic failure of a dam upon a dry, sloping channel. The solution validity is investigated by computing orders of magnitude for neglected terms in the equations of motion and by comparing the solution with the results of previously calculated numerical solutions and with experimental results. These calculations and comparisons suggest that the kinematic wave solution becomes asymptotically valid after the flood wave has advanced about four reservoir lengths downstream.

0412 Tapered Inlet Design Using Specific Energy Curves

Ronald L. Rossmiller, Member, ASCE, (Asst. Prof., Dept. of Civ. Engrg. and Engrg. Extension, Iowa State Univ., Ames, Iowa) and **Merwin D. Dougal**, Member, ASCE, (Prof. of Civ. Engrg. and Dir., Dept. of Civ. Engrg., Iowa State Water Resources Research Inst., Iowa State Univ., Ames, Iowa)

Journal of the Hydraulics Division, Vol. 108, No. HY1, January, 1982, pp. 127-135

Design charts to simplify the hydraulic design of tapered inlet culverts for both rectangular and circular sections are presented. The use of these design charts greatly reduces the opportunity for error and the amount of time needed to design a tapered inlet. The use of these tapered inlet culverts in the state of Iowa has resulted in the savings of millions of dollars in construction costs over a period of years. In addition, many of these raised inlet culverts have enabled the highway department to provide ponds for recreation or for stock watering, and have helped control channel degradation and bank erosion.

0413 Obtaining the Layout of Water Distribution Systems

William F. Rowell, (Captain, United States Air Force, Pentagon, Washington, D.C.) and **J. Wesley Barnes**, (Assoc. Prof., Dept. of Mech. Engrg., Univ. of Texas at Austin)

Journal of the Hydraulics Division, Vol. 108, No. HY1, January, 1982, pp. 137-148

An essential first step in the design of a municipal water distribution system is the determination of the locational placement or layout of the links of pipe that will form the system. A two-level hierarchically integrated system of models is developed for the layout of both single and multiple source water distribution systems. The first level, a nonlinear programming model, selects an economical tree layout for the major pipe links. The second level, an integer programming model, chooses the loop- forming links to add to the first level tree layout in order to minimize the cost of providing a specified level of reliability in case of failure of the larger first level links. The system of models is applied to an example two-source water distribution system layout problem.

0414 Resistance of Beauharnois Canal in Winter

Gee Tsang, (Research Sci., Environmental Hydr. Div., National Water Research Inst., Burlington, Ontario, Canada)

Journal of the Hydraulics Division, Vol. 108, No. HY2, February, 1982, pp. 167-186

The flow in an ice-affected river can be calculated using the open-water Chezy equation and ignoring the presence of ice if a corrective coefficient for Chezy's C is introduced to take into account the ice effects. Analysis of the data from the Beauharnois Canal shows that the corrective coefficient follows a well-defined trend during the ice season. Frazil presence is found to be the most important factor in affecting the resistance of the river. A small amount of frazil ice in the water can reduce the flow by 50 percent. The approach proposed in this study has the advantage over the currently popular approach of using the Manning equation and the composite Manning's n because it can deal with situations when a river is only partially ice covered or is frazil affected. It also avoids the situation of having to choose one from many formulae for evaluating the composite Manning's n. By using the corrective coefficient curve derived from the study, an alternative ice management procedure seems possible which may lead to higher power production.

0415 Circular Wall Jets in Coflowing Streams

Nallamuthu Rajaratnam, Member, ASCE, (Prof. of Civ. Engrg., Univ. of Alberta, Edmonton, Alberta, Canada) and **Michael J. Stalker**, (Water Resources Engr., Stanley Assocs. Engrg. Ltd., Edmonton, Alberta, Canada)

Journal of the Hydraulics Division, Vol. 108, No. HY2, February, 1982, pp. 187-198

The mixing and diffusion of circular turbulent wall jets in coflowing open channel streams is discussed with the ratio of the jet to channel velocity varying from 2 to 30 and the depth of flow being 10 to 30 times the jet diameter. The longitudinal velocity profiles with respect to the channel velocity have been found to be similar. Using the excess momentum thickness as the length scale, it has been possible to develop practically useful correlations for the length, relative velocity and shear velocity scales, using the previous results of circular free jets in coflowing streams as a guide.

0416 Bed Load Fluctuation: Applications

Han Qiwei, (Engr., Water Conservancy and Hydroelectric Power Research Inst., Beijing, Chia.) and **He Mingmin**, (Engr., Water Conservancy and Hydroelectric Power Research Inst., Beijing, China)

Journal of the Hydraulics Division, Vol. 108, No. HY2, February, 1982, pp. 199-210

Based on the distribution of transport rate, the fluctuation of bed load and its application to hydrological measurement have been studied. The variance and coefficient of variation of transport rate for uniform sand and nonuniform sand with determinate and random size distribution have been derived. The relationship between the coefficient of variation of transport rate measured for the whole cross section and that of the local points has been obtained. All this will make it possible to determine the sampling duration with the desired accuracy or to estimate the accuracy available from the sampling duration performed.

0417 Stochastic Model of Incipient Sediment Motion

He Mingmin, (Water Conservancy and Hydroelectric Power Scientific Research Inst., Beijing, China) and **Han Qiwei**, (Water Conservancy and Hydroelectric Power Scientific Research Inst., Beijing, China)

Journal of the Hydraulics Division, Vol. 108, No. HY2, February, 1982, pp. 211-224

A stochastic model of incipient sediment motion is developed. The probability of incipient motion, the life distribution of stationary particles, the number distribution of particles in incipient motion, the intensity of the incipient motion as well as the distribution of transport rate and the expression of mean transport rate in the case of low transport load are all derived. The analytical relationships between the velocity of bottom flow and each of the parameters describing the stage of incipient sediment, such as probability of incipient motion, intensity of incipient motion, period of incipient motion and mean transport rate are obtained. This research discloses the nature of incipient motion and gives quantitative relationships between various criteria of incipient motion. It gives a theoretical basis for determining a uniform and reliable criterion of incipient motion in hydraulic engineering.

0418 Hypolimnion Oxygenation Studies in Clark Hill Lake

Richard E. Speece, Member, ASCE, (Drexel Univ., Philadelphia, Pa.), **Charles Givler**, Member, ASCE, (Lafayette Coll., Easton, Pa.), **Raymond Aubert**, Member, ASCE, (Drexel Univ., Philadelphia, Pa.), **Jeffrey Crate**, Member, ASCE, (Virginia Polytechnical Inst. and State Univ., Blacksburg, Va.), **Raphael Caire**, Member, ASCE, (Virginia Polytechnical Inst. and State Univ., Blackburg, Va.) and **Rashid H. Siddiqi**, Member, ASCE, (Coll. of Petroleum and Minerals, Dharan, Saudi Arabia)

Journal of the Hydraulics Division, Vol. 108, No. HY2, February, 1982, pp. 225-244

Discharges from stratified hydroelectric generating impoundments are frequently drawn from the hypolimnin due to the low elevatin of the penstocks. The hypolimnion contains the colder water and thus provides a cold water fishery below the dam. Oxygen depletion within the hypolimnion results in very low D.O. levels in the waters being discharged. Economic analysis of supplemental areation techniques found hypolimnion oxygenation with commercial oxygen had the least cost. The seasonal oxygen deficit was 7500 tons to provide 6 mg/ D.O. downstream. The maximum daily requirement was 110 tons per day. On-site production of cyrogenic oxygen was estimated to cost about $30 per ton. Pulsed oxygen injection at the dam with the oxygen injection rate matched to the peaking power discharge rate added 6 mg/ D.O. with 85% absorption efficiency. Continuous injection of 100 tons per day of oxygen at a location one mile above the dam resulted in 50% withdrawal of the injected oxygen.

0419 Three-Dimensional Slot Jets

Misganaw Demissie, Assoc. Member, ASCE, (Sci., Ill. Water Survey, Champaign, Ill.) and **W. Hall C. Maxwell**, Fellow, ASCE, (Assoc. Prof., Dept. of Civ. Engrg., Univ. of Illinois, Urbana, Ill.)

Journal of the Hydraulics Division, Vol. 108, No. HY2, February, 1982, pp. 247-251

A comparison between a momentum flux analysis and data for slot jet of limited aspect ratio is described for one set of consistent data. It is shown that an analytic model, based on the synthesis of momentum flux profiles, gives good agreement with experimental data when non-uniformity of exit velocity is taken into account, and the spread coefficient and origin correction are carefully selected.

0420 Thermal Front Formation from Buoyant Surface Jets

J. P. Gosink, Assoc. Member, ASCE, (Postdoctoral Fellow, Geophysical Inst., Univ. of Alaska, Fairbanks, Alaska 99701)

Journal of the Hydraulics Division, Vol. 108, No. HY2, February, 1982, pp. 252-257

An explanation is given for the formation of thermal fronts for outfall Froude numbers in the range of $F \leq 3.3$. It is shown that the homogeneous round jet has a characteristic natural frequency in the range of $0.3 < St < 0.5$ which is damped out beyond $x/D \geq 5$. It is suggested that further investigation of the phenomena of thermal fronts is necessary.

0421 Nature of Wave Attenuation in Open Channel Flow

Victor Miguel Ponce, Member, ASCE, (Assoc. Prof. of Civ. Engrg., San Diego State Univ., San Diego, Calif. 92182)

Journal of the Hydraulics Division, Vol. 108, No. HY2, February, 1982, pp. 257-262

The identification of the term or group of terms which will cause a free-surface flow to dissipate is investigated. It is shown that wave attenuation is caused by the interaction of the kinematic term (friction and bottom slope) with the local acceleration term. When the latter is absent or negligible, wave attenuation is caused by the interaction of the kinematic term with the pressure gradient, or conventive acceleration terms, or both.

0422 Erosion by Submerged Circular Jets

Nallamuthu Rajaratnam, Member, ASCE, (Prof., Dept. of Civ. Engrg., Univ. of Alberta, Edmonton, Canada)

Journal of the Hydraulics Division, Vol. 108, No. HY2, February, 1982, pp. 262-267

An experimental study on the erosion of sand beds by impinging submerged circular water jets shows similarities between air-sand system and water-sand system. It is shown that there are significant differences between the two systems regarding the depth and radial extent of the scour hole.

0423 Hot-Film Response and Digital Analysis in Bubly Flows

Sahim Tekeli, (Asst. Prof., Dept. of Civ. Engrg., Bogazici Univ., Istanbul, Turkey) and **W. Hall C. Maxwell**, Fellow, ASCE, (Assoc. Prof., Dept. of Civ. Engrg., Univ. of Illinois at Urbana-Champaign, Urbana, Ill. 61801)

Journal of the Hydraulics Division, Vol. 108, No. HY2, February, 1982, pp. 267-272

Hot film responses in bubbly flows are analyzed for the development of a digital analysis technique. It is shown that hot-film anenometry and digital data analysis techniques can be excellent tools for investigations in bubbly flows. In addition to fluid velocity, the distributions of void ratios, bubble size, and velocities are also accessible.

0424 Hydrothermal Analysis of Lakes and Reservoirs

Donald R. F. Harleman, Member, ASCE, (Ford Prof. of Engrg. and Dir., Ralph M. Parsons Lab., Dept. of Civ. Engrg., Massachusetts Inst. of Tech., Cambridge, Mass. 02139)

Journal of the Hydraulics Division, Vol. 108, No. HY3, March, 1982, pp. 301-325

The impact of Hunter Rouse on the application of fluid mechanics to hydraulic engineering practice in the United States is reviewed. The development of physical and mathematical techniques for predicting the hydrothermal structure of lakes and reservoirs during the past fifteen years is discussed. Laboratory models have played a key role in the development and testing of early mathematical models based on the vertical diffusion concept. The sensitivity of these models to variations in turbulent diffusivity, internal heat absorption and vertical advection due to inflows and outflows is demonstrated. A wind-mixing model is proposed as an alternative to specifying temporal and spatial variations in the vertical diffusivity. The wind-mixing algorithm converts turbulent kinetic energy input by wind shear to potential energy through entrainment of hypolimnetic water across the thermocline. An iterative technique is used to achieve a heat balance based on meteorological inputs averaged over hourly or daily time steps. The temperature predictions of the numrical model are compared with field data for a number of lakes and reservoirs.

0425 Dilution Analysis for Undirectional Diffusers

E. Eric. Adams, Assoc. Member, ASCE, (Lect., Principal Research Engr., Dept. of Civ. Engrg., Mass. Inst. of Tech., Cambridge, Mass., 02139)

Journal of the Hydraulics Division, Vol. 108, No. HY3, March, 1982, pp. 327-342

The paper discusses the mixing mechanics of submerged, multi-port diffusers used to discharge heated water from steam-electric power plants into shallow receiving waters. An analysis for co-flowing diffusers is presented. The analysis is compared with similar fluid flow situations and predicted dilutions are compared with available laboratory measurements. The behavior of a co-flowing diffuser is then contrasted with that of a tee diffuser. A semi-empirical theory is derived for the tee diffuser and is compared with available measurements.

0426 Lower Flow Regime Bedforms: Rational Classification

Timothy R. H. Davies, (Sr. Lect., Dept. of Agr. Engrg., Univ. of Canterbury, Lincoln College, Canterbury, New Zealand)

Journal of the Hydraulics Division, Vol. 108, No. HY3, March, 1982, pp. 343-360

A classification of lower flow regime bedforms is proposed, based upon consideration of the effect of water flow on the shape and size of bedforms, which is more rational than other systems presently in use. It is suggested that a more quantitative system of bedform classification may lead to a better understanding of the occurrence and behaviour of bedforms. The assumptions on which the proposed classification is based are shown to agree with published data relating bedform characteristics to mean flow depth and velocity. The proposed classification offers a rational explanation for several hitherto puzzling aspects of bedform behaviour in laboratory and field situations. Since much presently available data is unsuited to testing these proposals, further extensive experimental work on bedforms is required.

0427 Pump-Sump Design Experience: Summary

Charles E. Sweeney, Assoc. Member, ASCE, (Sr. Engr. Mgr., Engrg. Hydr. Inc., Longmont, Colo.), **Rex A. Elder**, Fellow, ASCE, (Engrg. Mgr., Hydr. and Hydrology, Bechtel, Inc., San Francisco, Calif.) and **Duncan Hay**, Assoc. Member, ASCE, (Pres., Western Canada Hydr. Lab., Ltd., Port Coquitlam, British Columbia, Canada)

Journal of the Hydraulics Division, Vol. 108, No. HY3, March, 1982, pp. 361-377

The designs of pump sumps for seven different projects, as developed through hydraulic model testing, are compared with published design standards. Modifications to the sump designs, in addition to the basic sump dimensions recommended by the published standards, are required to achieve satisfactory hydraulic performance. It is theorized that performance criteria applied in developing the standards are not as stringent as those required for present day large capacity pumps. Typical sump modifications and dimensions presented may provide a basis for developing conceptual designs for sumps for large capacity pumps. Final designs should be proven through hydraulic model testing.

0428 Sewer Network Routing by Dynamic Wave Characteristics

A. Suha Sevuk, (Assoc. Prof., Dept. of Civ. Engrg., Middle East Technical Univ., Ankara, Turkey) and **Ben Chie Yen**, Fellow, ASCE, (Prof. of Civ. Engrg., Univ. of Illinois at Urbana-Champaign, Ill.)

Journal of the Hydraulics Division, Vol. 108, No. HY3, March, 1982, pp. 379-398

Hydraulics of unsteady flow in storm sewer networks is investigated by using a dynamic wave simulation model. The Illinois Storm Sewer (ISS) Model solves the complete dynamic equations using the method of characteristics. Two examples are presented. One is a 48-pipe sewer system to show that for unsteady open channel flow in sewer networks, the backwater effect of the junctions cannot be ignored, under certain circumstances reversal flow may occur, the depth-discharge relationship is not unique and a loop-type rating curve exists, and the occurrence of maximum discharge in a sewer need not coincide with that of maximum depth. The second example shows how a dynamic wave routing model can be used to evaluate and improve a sewer system.

0429 Sediment Analyses for Urbanizing Watersheds

Gary R. Dyhouse, Member, ASCE, (Chf. Hydr., Engrg. Section, Hydr. and Hydraulics Branch, St. Louis District, Corps of Engrs., St. Louis, MO)

Journal of the Hydraulics Division, Vol. 108, No. HY3, March, 1982, pp. 399-418

An analysis of channel modifications for Harding Ditch, the main drainage canal for two urbanizing watersheds near St. Louis, Missouri, required an estimate of depositional changes between existing and future land use conditions, and the frequency and quantity of dredging required to maintain a modified channel configuration. A grid cell data bank was used to breakdown each watershed by land use category and the percent of the watershed in each category. Sediment washoff parameters were estimated for each land use category and input to the STORM computer program. Hourly values of discharge and tons of sediment washoff per event from each watershed were computed for 20 years of rainfall data for existing and future land use. STORM output was manipulated through an interface program and supplied to an

HEC-6 sediment transport model of 11.5 miles (18.5 km) of Harding Ditch to evaluate depositional patterns. Comparison of 1982 and 2020 thalweg profiles showed less deposition under future land use conditions. Dredging analysis indicated a 10-year excavation interval was sufficient to maintain a with-project rating curve relationship.

0430 Stochastic Generation of Monthly Streamflows

Ratnasingham Srikanthan, (Postgrad. Student, Dept. of Civ. Engrg., Monash Univ., Clayton, Vic. 3168, Australia) and **Thomas A. McMahon**, (Assoc. Prof., Dept. of Civ. Engrg., Monash Univ., Clayton, Vic. 3168, Australia)

Journal of the Hydraulics Division, Vol. 108, No. HY3, March, 1982, pp. 419-441

Two streamflow generation models are used to generate monthly flows for Australian streams. In one model, seasonalities and periodicities in the monthly flows are removed and the resulting weekly stationary series is modelled. The second approach uses the Thomas-Fiering monthly model. The results of these models are compared. A modified two-tier model is recommended for less variable streams and the method of fragments is recommended for highly variable streams.

0431 Carrying Capacity of Flood Plains

Nani G. Bhowmik, Member, ASCE, (Principal Sci., Illinois State Water Survey, P.O. Box 5050, Station A, Champaign, Ill. 61820) and **Misganaw Demissie**, Assoc. Member, ASCE, (Assoc. Professional Sci., Illinois State Water Survey, P.O. Box 5050, Station A, Champaign, Ill. 61820)

Journal of the Hydraulics Division, Vol. 108, No. HY3, March, 1982, pp. 443-452

The field data of flood flows for several streams were analyzed to determine the distribution of flow in the main channel and in the flood plains. The discharge carrying capacity of flood plains were found to vary from a few percent to more than 80 percent of the total flow. The carrying capacity of flood plains depends upon many factors including the nature of the flood plain and the main channel, and the flood frequency. In general, the carrying capacity of flood plains increases with the return period of the flood. When the return period of the flood is about 40 years or more, the flood plain and the main channel appear to behave as a single unit.

0432 Flow Resistance in Wide Rectangular Channels

W. R. C. Myers, (Principal Lect., School of Civ. Engrg., Ulster Polytechnic, Shore Road, Newtownabbey, County, Antrim, Northern Ireland)

Journal of the Hydraulics Division, Vol. 108, No. HY4, April, 1982, pp. 471-482

Friction coefficients are determined for rectangular channels over a wide range of Reynolds number and a range of width-to-depth ratios of 35 to 1. Comparison with other investigators shows general agreement that the relationship of $\sqrt{f}$ to $R_e \sqrt{f}$ is different from the pipeflow equation,but not on the values of the numerical constants. The influence of width-to-depth ratio on friction factor is examined. It is shown to be complex, probably influenced by both secondary circulation and non-uniformity of shear patterns, in agreement with other work. Two methods of adjusting the channel data to conform to the pipeflow friction factor equation using shape factors are applied. Using these methods, the average divergence between the channel friction factor equation and that applicable to pipeflow can be reduced from 8% to under 2%.

0433 Outfall Diffuser Behavior in Stratified Ambient Fluid

Stephen J. Wright, Assoc. Member, ASCE, (Asst. Prof., Dept. of Civ. Engrg., Univ. of Michigan, Ann Arbor, Mich.), **Derek R. Wong**, Assoc. Member, ASCE, (Grad. Research Asst., Dept.of Civ. Engrg., Univ. of Michigan, Ann Arbor, Mich.), **Karen E. Zimmerman**, Assoc. Member, ASCE, (Grad. Research Asst., Dept. of Civ. Engrg., Univ. of Michigan, Ann Arbor, Mich.) and **Roger B. Wallace**, Member, ASCE, (Grad. Research Asst., Dept. of Civ. Engrg., Univ. of Michigan, Ann Arbor, Mich.)

Journal of the Hydraulics Division, Vol. 108, No. HY4, April, 1982, pp. 483-501

A method for the preparation of design charts for estimating the mixing behavior from outfall diffusers in stagnant stratified ambient fluids is presented. Numerical analyses by traditional integral methods which consider the interaction and merging of individual port charges are used to predict the diffuser behavior. Predictions for the maximum height of plume rise and minimum dilution at that point are presented in the form of design curves. Experimental data for the cases of a single and opposing rows of horizontal ports are presented and compared to the numerical results. Additional data of the thickness and vertical location of the horizontally spreading layer developed after the maximum reached are presented. The numerical model was found to be inaccurate for cases of discharges with significant initial momentum. A criterion is developed to indicate when this effect will be important.

0434 Gumbel's Extreme Value I Distribution: A New Look

Dennis P. Lettenmaier, Assoc. Member, ASCE, (Research Assoc. Prof., Dept. of Civ. Engrg., Univ. of Washington, Seattle, Wash. 98195) and **Stephen J. Burges**, Member, ASCE, (Prof., Dept. of Civ. Engrg., Univ. of Washington, seattle, Wash. 98195)

Journal of the Hydraulics Division, Vol. 108, No. HY4, April, 1982, pp. 502-514

Many basic hydrologic textbooks, following Gumbel's original work, advocate use of frequency factors dependent on an integer m for estimation of event levels using the Extreme Value I distribution. Although Gumbel recommended use of m = n, the sample size, it is shown that this choice results in poor estimates, which are much improved by use of $m = \infty$. This is equivalent to use of the population values. Thus, use of m-dependent tables of frequency factors is misleading and should be avoided. It is also shown that a biased variance estimate (n divisor) yields slightly better estimates of extremes, both in terms of variability and bias, than does the unbiased (n-1 divisor) estimate usually advocated. Finally, modest improvements in estimates, most significant for large return periods and small sample sample sizes, are shown to result from use of maximum likelihood parameter estimators in favor of the best moment estimators.

0435 Revised Rainfall Frequency Analysis for Houston

Tracey C. Bouvette, (Grad. Student, Dept. of Environmental Sci. and Engrg., Rice Univ., Houston, Tex.), **Jeff L. Lambert**, (Staff Engr., Bovay Engrg., Houston, Tex.) and **Philip B. Bedient**, Assoc. Member, ASCE, (Assoc. Prof., Dept. of Environmental Sci. and Engrg., Rice Univ., Houston, Tex.)

Journal of the Hydraulics Division, Vol. 108, No. HY4, April, 1982, pp. 515-528

A statistical analysis of rainfall data for four Houston area stations is presented. A similar analysis was performed on the same stations in 1961 by the U.S. Weather Bureau. Both methods used the Gumbel Type I distribution to predict design storms for particular durations (1 hr. to 24 hrs.) and for particular return periods (2 to 100 yrs.). A Comparison of the design storms calculated in each of the studies identified rainfall rate changes which have occurred over the past two decades. The largest changes found were associated with the 24 hr. event. The downtown station recorded a 13 decrease in the 24 hr. 100-yr. storm over the past 20 years, while Conroe, Galveston, and Hobby stations (outside of Houston) recorded an increase of 15 percent for the same storm. These changes can be linked to the heavy urban development in the Downtown area, and to changing climatic patterns along the Gulf Coast. Drainage designs for the affected areas should be adjusted upward to account for these changes.

0436 Flow Under Weir on Scoured Bed

Devendra Kumar, (Exec. Engr., Tehri Dam Division, 14, Rishikesh, U.P., India), **Satish Chandra**, (Prof., School of Hydrology, Univ. of Roorkee, Roorkee, India) and **G. C. Mishra**, (Reader, School of Hydrology, Univ. of Roorkee, Roorkee, India)

Journal of the Hydraulics Division, Vol. 108, No. HY4, April, 1982, pp. 529-543

An approximate analysis for seepage flow under a weir, resting on isotropic porous medium of infinite depth, with a vertical sheet pile at the toe and a segmental circular scour, is obtained using conformal mapping techniques. An exact analysis for semi-circular scour, using conformal mapping, is also given. The results of approximate and exact analysis for semi-circular scour are compared. From the analysis it is shown that the maximum exit gradient is increased by about 75 percent in case of semi-circular scour. For semi-circular scour profile, as s/b increases, where s is the depth of sheet pile and b is width of the floor, the point of maximum exit gradient shifts away from the sheet pile. The exit gradient distribution obtained from approximate analysis is very close to the one given by the exact analysis, but the potentials at salient points under the apron obtained from the two analysis compared closely for values of s/b up to 0.1. The maximum reduction in potential for a semi-circular scour of l/b = 1/8 is about 48 percent, where 2l is the length of scour.

0437 Bedload and Size Distribution in Paved Gravel-Bed Streams

Gary Parker, Member, ASCE, (Assoc. Prof., St. Anthony Falls Hydraulic Lab., Univ. of Minnesota, Minneapolis, Minn. 55414), **Peter C. Klingeman**, Member, ASCE, (Prof., Dept. of Civ. Engrg., Oregon State Univ., Corvallis, Oreg.) and **David G. McLean**, (Research Officer, Research Council of Alberta, Edmonton, Alberta, Canada)

Journal of the Hydraulics Division, Vol. 108, No. HY4, April, 1982, pp. 544-571

Field data are used to study the size distribution of bedload in paved gravel-bed streams. Similarity analysis yields the results that all grain size ranges are of approximately equal transportability when the critical condition for breaking the pavement is exceeded. This result is only approximately correct due to deviations from similarity. However, it is adequate to justify development of a method for calculating total bedload, which requires only the subpavement median grain size rather than the size distribution. A method for calculating bedload size distribution that accounts for deviation from similarity is also developed.

Errata: HY Oct. '82, pp. 1249.

0439 Groundwater Contamination due to Continuous Sources

Anand Prakash, Member, ASCE, (Chf. Water Resources Engr., Dames & Moore, Golden, Colo.)

Journal of the Hydraulics Division, Vol. 108, No. HY4, April, 1982, pp. 572-90

Analytical methods are described to predict the concentrations of effluents released in a two- or three-dimensional groundwater flowfield from a point, line, plane or parallel piped source. The effects of the upper and lower confining boundaries in an artesian aquifer and those of the bedrock and water table in an unconfined aquifer are accounted for by the method of images. The same technique is used to model the contribution of a constant concentration boundary in the flowfield like the one provided by a fully penetrating perennial stream. Using the results of a sensitivity analysis, the number of images at which computations can be truncated without significantly affecting the accuracy of the results is indicated. Examples are provided to illustrate the effect of source configuration, e.g., point, line or plane, and the presence or absence of a fully penetrating perennial stream in the flowfield. To validate the model, concentration isopleths are developed for the release of effluents from an ashpond in a two-dimensional stream-aquifer system an the results are compared with those contained from a finite-element study.

0440 Dispersion in Tumbling Flow

S. Beltaos, (Research Sci., National Water Research Inst., Canada Centre for Inland Waters, Burlington, Ontario, Canada)

Journal of the Hydraulics Division, Vol. 108, No. HY4, April, 1982, pp. 591-612

Steep mountain streams are often in the tumbling flow regime, that is alternating sub- and super-critical flow with energy dissipation caused mainly by sudden changes in slope and cross section. Though longitudinal dispersion characteristics of tumbling flows are important in time-of-travel, stream gauging, and contaminant spread considerations, very little pertinent information is available. The results of a laboratory study intended to provide preliminary data on tumbling flow dispersion are discussed. To simulate tumbling flow, a 1 m wide flume was divided into 11 pools by placing identical triangular weirs at 3 m intervals. Slugs of fluorescent dye were injected at the first weir and the resulting concentration-time variations measured downstream at every second weir. Test results are analyzed according to existing theoretical models and corresponding parameters evaluated. It is concluded that conventional river dispersion models are possible means of describing tumbling flow dispersion but it is unclear at this time whether storage dispersion models are relistic.

0441 Vadose Zone Permeability Tests: Summary

Daniel B. Stephens, (Asst. Prof. of Hydr., Dept. of Geoscience, New Mexico Inst. of Mining and Tech., Socorro, N.M. 87801) and **Shlomo P. Neuman**, (Prof. of Hydr., Dept. of Hydr. and Water Research, Univ. of Arizona, Tucson, Ariz. 85721)

Journal of the Hydraulics Division, Vol. 108, No. HY5, May, 1982, pp. 623-639

Solutions to determine saturated hydraulic above a water table by means of constant head borehole infiltration tests are summarized and critiqued. Analytical solutions are available for open and partly cased boreholes and deep and shallow water table conditions. The formulas are based on various simplifying assumptions pertaining to soil properties and boundary conditions. None of the solutions account for capillary effects. For identical test conditions values of hydraulic conductivity calculated from these formulas may differ by as much as a few hundred percent.

0442 Vadose Zone Permeability Tests: Steady State Results

Daniel B. Stephens, (Asst. Prof. of Hydr., Dept. of Geoscience, New Mexico Tech., Socorro, N.M.) and **Shlomo P. Neuman**, (Prof. of Hydro., Dept. of Hydro. & Water Resources, Univ. of Arizona, Tucson, Ariz.)

Journal of the Hydraulics Division, Vol. 108, No. HY5, May, 1982, pp. 640-659

Free surface and saturated-unsaturated flow models are applied to the problem of constant head borehole infiltratlon tests for determining saturated hydraulic conductivity above a water table. The free surface model follows the same general approach employed to develop analytical solutions, but this model does not suffer from some of the many simplifying assumptions regarding boundary conditions. Nonetheless, the results of steady state simulations with the free surface model agree rather well with the analytical solutions. The free surface approach provides a distorted view of the infiltration process. Saturated-unsaturated models used to simulate quasi-steady state conditions indicated that only a small volume of soil in the vicinity of the borehole is completely saturated with water. Based on saturated-unsaturated simulations in four different homogeneous, isotropic soils, an empirical solution is developed for deep water table conditions to estimate saturated hydraulic conductivity with improved accuracy. Due to the effects of borehole geometry unsaturated soil characteristics, water table depth, heterogeneity, and anisotropy, it is extremely difficult to develop a completely general formula.

0443 Vadose Zone Permeability Tests: Unsteady Flow

Daniel B. Stephens, (Asst. Prof. of Hydr., Dept. of Geoscience, New Mexico Inst. of Mining and Tech., Socorro, N.M.) and **Shlomo P. Neuman**, (Prof. of Hydr., Dept. of Hydr. and Water Research, Univ. of Arizona, Tucson, Ariz.)

Journal of the Hydraulics Division, Vol. 108, No. HY5, May, 1982, pp. 660-677

Saturated-unsaturated flow models are used to simulate transient conditions in four soils during constant head borehole infiltration tests in the vadose zone. Infiltration rate varies directly with saturated hydraulic conductivity and is also influenced by unsaturated soil properties, antecedent moisture conditions, anisotropy and borehole conditions. The method recommended by the U.S. Bureau of Reclamation to determine when to terminate a test may lead to overestimating saturated hydraulic conductivity. Simple graphical procedures are developed to overcome field problems associated with large water volume requirements or long time required to reach steady state by using early time date to estimate the steady infiltration rate.

0444 Mathematical Model for Erodible Channels

Howard H. Chang, Member, ASCE, (Prof. of Civ. Engrg., San Diego State Univ., San Diego, Calif. 92182)

Journal of the Hydraulics Division, Vol. 108, No. HY5, May, 1982, pp. 678-689

A mathematical model for erodible channels is formulated, developed and tested with field data. A significant feature of the model is its ability to account for changes in both channel width and channel-bed profile of aggrading and degrading streams. Since changes in channel width and channel-bed elevation are closely interrelated, this model represents an improvement in this direction. This model employs the techniQues for water and sediment routing; an additional condition based upon the concept of minimum stream power is used as the width predictor. Analysis using this model shows that the total stream power of a reach is reduced if the widths at all cross-sections are so adjusted as to reduce the variation in energy gradients at these sections. When this model is applied, a varied flow will eventually develop into a uniform flow for which the stream power is an eventual minimum.

0445 Minimum Energy and Energy Dissipation Rate

Charles C. S. Song, (Prof., St. Anthony Falls Hydraulic Lab., Dept. of Civ. and Mineral Engrg., Univ. of Minnesota, Minneapolis, Minn. 55414) and **Chin Ted Yang**, Member, ASCE, (Civ. Engr., U.S. Dept. of the Interior, Bureau of Reclamation, Engrg. and Research Center, Denver, Colo. 80225)

Journal of the Hydraulics Division, Vol. 108, No. HY5, May, 1982, pp. 690-706

The theory of minimum energy and the minimum rate of energy dissipation have been applied to flow around bluff bodies, stability of falling bodies, statics and dynamics of gas bubbles, generation of ripples and dunes, and drag reduction by suspended load. These examples are used to illustrate how the minimization principles (variational principles) can be used to solve the problems of dissipative mechanical systems in static or dynamic equilibrium conditions. The minimization principle can be used independently or in tandem with the equations of motion to solve problems of a large degree of freedom. The equilibrium solution thus obtained has been shown to contain certain implicit information on the stability characteristics of the equilibrium state. For this reason, the method is not only a viable alternative to the vector mechanics method, but it also provides a relatively simple way of determining certain stability criteria.

0446 Thermal Fronts in Heated Water Discharges

Vahid Alavian, Assoc. Member, ASCE, (Assistant Prof., Dept. of Civ. Engrg., Univ. of Illinois, Urbana-Champaign, Ill.) and **John A. Hoopes**, M, (Prof., Dept. of Civ. and Environmental Engrg., Univ. of Wisconsin, Madison, Wis.)

Journal of the Hydraulics Division, Vol. 108, No. HY5, May, 1982, pp. 707-725

Large scale, periodic motions known as thermal fronts, which are observed in surface, heated water discharges were investigated in a laboratory model. Characteristic length scales corresponding to various regions of flow in a heated water discharge are evolved from the kinematic mass, momentum, and buoyancy fluxes and the ambient velocity considerations. A one-dimensional, wave type model, describing the velocity of propagation of fully-developed thermal fronts, is developed. Experiments show that thermal fronts become fully developed in the far field region of the discharge mixing zone where the horizontal motions due to the discharge buoyancy and to the ambient water velocity dominate the discharge behavior. The mean behavior of the heated discharges with thermal fronts is found to be similar to heated discharges without thermal fronts. The measured velocities are in good agreement with the one-dimensional, wave model predictions. Criteria for the discharge densimetric Froude number and discharge-ambient velocity ratio are developed for the occurrence of thermal fronts.

0447 Experiments on Coaxial Heater Water Discharges

Wolfgang Rodi, (Prof., Sonderforschungsbereich 80, Univ. of Karlsruhe, 75 Karlsruhe, West Germany) and **Klaus Weiss**, (Research Engr., Univ. of Karlsruhe, Karlsruhe, West Germany)

Journal of the Hydraulics Division, Vol. 108, No. HY6, June, 1982, pp. 737-746

The paper describes an experimental investigation in which heated water was discharged coaxially into rectangular open channel flow from a slot placed at the channel center. Discharge and channel velocity were the same, and discharges with excess temperatures of 9 °C and 20 °C were studied. The temperature distribution in the near field was measured in detail with thermistors, and the surface motion was recorded by photographing surface floats. The photographs show how the warm water spreads at the surface and that it forms a straight front whose angle with the center line increases with increasing density difference. Isotherm pictures are given for various cross-sections down to 16 water depths from the discharge, and these show the rise to and the spreading at the surface of the warm water, the formation of warm-water pools off the centre line at the surface, the inward motion of cold channel water near the bottom, and the eventual detachment of the warm water from the bottom.

0448 Accuracy Criteria in Diffusion Routing

Victor Miguel Ponce, Member, ASCE, (Assoc. Prof., Dept. of Civ. Engrg., Coll. of Engrg., San Diego State Univ., San Diego, Calif.) and **Fred D. Theurer**, Member, ASCE, (Civ. Engr., Soil Conservation Service, Cooperative Instream Flow Service Group, Fort Collins, Colo.)

Journal of the Hydraulics Division, Vol. 108, No. HY6, June, 1982, pp. 747-757

An analysis of Muskingum diffusion model, in which routing parameters are calculated based on channel and grid characteristics, is carried out. Numerical experiments are performed to throw additional light onto the criterion setting an upper limit to the space step if accuracy is to be preserved. Computational experience indicates that for very large values of the space step, there is a tendency for physically unrealistic negative outflows to occur. The results of the numerical experiments point to the condition $C_2 \geq \zeta$ if accuracy is to be preserved, in which C_2 is one of the routing coefficients and ζ is a real number. A value of $\zeta = 0.33$ is recommended for practical applications. The spatial resolution criterion is expressed in terms of Courant and cell Reynolds numbers.

0449 Origin of Step-Pool Systems in Mountain Streams

Jeffrey G. Whittaker, (Research Fellow, Dept. of Agric. Engrg., Univ. of Canterbury, Lincoln Coll., Canterbury, New Zealand) and **Martin N. R. Jaeggi**, (Sr. Research Engr., Lab. of Hydr., Hydrology, and Glaciology, Federal Inst. of Tech., Zurich, Switzerland)

Journal of the Hydraulics Division, Vol. 108, No. HY6, June, 1982, pp. 758-773

Mountain streams very often present a step-pool aspect. Antidune formation alone; dispersion and sorting theory or velocity reversal as advocated for the origin of riffle-pool sequences do not explain how steps and pools form. Experiments have been conducted to clarify their origin. It was found that the step-pool aspect appears after high flows had deformed the bed. During the presented tests a self armoring process occurred which resulted in a stabilization of the bed. When coarsening of the top layer was little classic antidunes resulted. An important coarsening was accompanied by a structuring of the bed into roughness elements whose spacing did correspond to maximum flow resistance. The substantial increase in resistance was an essential factor in bed-stabilization.

0450 Sediment Transport and Unit Stream Power Function

Chih Ted. Yang, Member, ASCE, (Civ. Engr., U.S. Dept. of the Interior, Bureau of Reclamation, Engrg. and Research Center, Denver Federal Center, Denver, Colo.) and **Albert Molinas**, Assoc. Member, ASCE, (Research Assoc., Dept. of Civ. Engrg., Colorado State Univ., Ft. Collins, Colo.)

Journal of the Hydraulics Division, Vol. 108, No. HY6, June, 1982, pp. 774-793

Basic fluid mechanics and turbulence theories are applied to show that suspended sediment concentration at a given depth of an open channel flow is a function of the turbulence energy production rate at that depth. Depth-averaged suspended sediment concentration can be obtained by integrating a function of the turbulence energy production rate over the depth of flow which in turn is a function of unit stream power. Total sediment concentration can also be expressed as a function of unit stream power following the foresaid and a procedure similar to that proposed by Einstein. Comparisons of seven total load equations indicate that equations derived from the concept that the rate of sediment transport should be related to the rate of energy dissipation of the flow are more accurate than other equations for both laboratory flumes and natural rivers.

0451 Improvement of Nonlinear Rainfall-Runoff Model

Otto J. Helweg, Member, ASCE, (Assoc. Prof. and Prof., Dept. of Civ. Engrg., Univ. of California, Davis, Calif.), **Jaime Amorocho**, Member, ASCE, (Prof., Dept. of Civ. Engrg., Univ. of California, Davis, Calif.) and **Ralph H. Finch**, (Research Asst., Mexico Research Inst., Mexico City, Mexico)

Journal of the Hydraulics Division, Vol. 108, No. HY7, July, 1982, pp. 813-822

Water resources engineers have long been aware that the linear assumption of unit hydrograph models produces poor results in some basins. The Amorocho-Branstetter nonlinear model is an alternative to unit hydrograph models in these instances, but has not been as popular, perhaps because the user had to determine the parameters by trial-and-error, thereby needing to understand more of the mathematical background of the model than desirable. Moreover, in many instances, the trial-and-error procedure was expensive and tedious. This paper presents a way to efficiently calculate the three main model parameters, memory, U, order, IORD, and truncation, M1 and M2. These calculations are programmed into the model so the user need not supply them.

0452 Infiltration Formulas by Curve Number Procedure

Cheng-Lung Chen, Member, ASCE, (Hydro., U. S. Geological Survey, Gulf Coast Hydroscience Center, NSTL Station, Miss. 39529)

Journal of the Hydraulics Division, Vol. 108, No. HY7, July, 1982, pp. 823-829

The Soil Conservation Service (SCS) curve number procedure for estimating runoff volume is examined in terms of the validity and applicability of the derived infiltration equations. Two different sets of the runoff and infiltration equations can be formulated from the water balance equation and one of the two SCS hypotheses postulating the basic proportionality among the relevant variables. The infiltration rate formula derived based on the SCS hypothesis with the potential maximum retention or infiltration defined to include the initial abstraction proves to be as general as, if not more general than, the Holtan-Overton equation. If the initial abstraction results mainly from infiltration that occurs before runoff begins, the amount of initial abstraction can be reasonably estimated from the derived infiltration rate formula; therefore, the SCS curve number procedure can be generalized without sacrificing its simplicity. The proposed method, however, may not be valid for large watersheds, where the surface interception and depression storage dominate the initial abstraction.

0453 The IUH of General Hydrologic System Model

Ven Te Chow, Fellow, ASCE, (Prof. of Civ. Hydrosystems Engrg., Univ. of Illinois, Urbana-Champaign, Urbana, Ill.) and **V. C. Kulandaiswamy**, (UNESCO Adviser and expert in Civ. Engrg. , (Hydrology and Water Resources), UNDP and UNESCO Mission, Colombo, Sri Lanka)

Journal of the Hydraulics Division, Vol. 108, No. HY7, July, 1982, pp. 830-844

This paper describes two simple and practical forms of the general hydrologic system (GHS) model, including a 5-parameter model and a 3-parameter model, and derives and compares their instantaneous unit hydrographs (IUH) which are useful to hydrologic analysis and design. It is shown that the 3-parameter model can be derived from the 5-parameter model by modifying the assumption on the boundary conditions in solving the differential equation that represents the model. It is found that the 5-parameter model is more realistic to represent the real-world system and fit the hydrologic data consistently better than the 3-parameter model. On the other hand, the 3-parameter model is slightly simpler in form and produces IUHs always of positive ordinates.

0454 Modeling Hydraulic Phenomena: A Glossary of Terms

ASCE Task Committee on Glossary of Hydraulic Modeling Terms of the Committee on Research of the Hydraulics Division, J. W. Kamphuis, chmn.

Journal of the Hydraulics Division, Vol. 108, No. HY7, July, 1982, pp. 845-852

A glossary is presented for the purposes of study by researchers with a view to generating a common terminology among those involved with modeling hydraulic phenomena. The glossary is divided into five categories: Types of models in general; scaling and design of hydraulic models; types of hydraulic models; types of mathematical models; and terms referring to numerical models.

0455 Vertical Circulation in Density-Stratified Reservoir

Zygmunt Meyer, (Sr. Lect., Technical Univ. of Szczecin, Water Engrg. Inst., Piastow St. 50, 70-311 Szczecin, Poland)

Journal of the Hydraulics Division, Vol. 108, No. HY7, July, 1982, pp. 853-873

The paper presents a mathematical model of vertical circulation in density-stratified reservoirs. An analytical solution of the formation of circulation regions is presented. The solutions includes two shapes of reservoirs, rectangular and semi-infinite strip, inflow-outflow exchange and wind action. The method allows to take into account additional horizontal water density changes in the reservoir which appear to be significant in calculating the extent of the circulation region.

0456 Mathematical Description of Turbulent Flows

Angelos N. Findikakis, Assoc. Member, ASCE, (Sr. Engr., Hydr./Hydrology Grp., Bechtel Civ. and Minerals Inc., P.O. Box 3965, San Francisco, Calif. 94119) and **Robert L. Street**, Member, ASCE, (Prof., Dept. of Civ. Engrg., Stanford Univ., Stanford, Calif. 94305)

Journal of the Hydraulics Division, Vol. 108, No. HY8, August, 1982, pp. 887-903

The mathematical description of uniform density and thermally stratified flows in terms of space averages is reviewed. The governing equations are presented and the use of alternative space filters is discussed. Different models for Sub-Grid-Scale (SGS) turbulent transport are examined. A simple algebraic model for the SGS Reynolds stresses and turbulent heat flumes in thermally stratified flows is presented. The model is cast in eddy viscosity - eddy diffusivity form. Its derivation is based on making appropriate closure approximations in the differential equations for the Reynolds stresses and the turbulent heat fluxes and introducing simplifying assumptions to reduce these equations to a system of algebraic equations. The solution of this system yields eddy coefficients which are functions of a local SGS turbulence length scale and the local structure of the velocity and the temperature fields. The variation of these coefficients with a local Richardson number is examined. Potential numerical problems which can be caused by improper formulation of the model are discussed.

0457 Finite Element Simulation of Stratified Turbulent Flows

Angelos N. Findikakis, Assoc. Member, ASCE, (Sr. Engr., Hydr./Hydrology Grp., Bechtel Civ. and Minerals Inc., Box 3965, San Francisco, Calif. 94119) and **Robert L. Street**, Member, ASCE, (Prof., Dept. of Civ. Engrg., Stanford Univ., Stanford, Calif. 94305)

Journal of the Hydraulics Division, Vol. 108, No. HY8, August, 1982, pp. 904-920

A two-dimensional, time dependent Finite Element (FE) code for simulating uniform density or thermally stratified flows is developed and applied to cavity flows. The choices of the schemes used for time stepping and for solving the nonlinear equations are discussed. Smagorinsky's model and the vorticity gradient model for Sub-Grid-Scale (SGS) turbulence in uniform density flows are incorporated in the FE code to simulate cavity flows driven by a surface shear stress with a randomly varying component. The general structure of the flow field and its spectral characteristics, as predicted with the two models, are compared. An algebraic model for turbulence in thermally stratified flows is incorporated in the FE code to simulate a shear driven flow in a stratified cavity. The transient development of the circulation under the action of a constant shear and the response to a shear stress with a randomly varying component are simulated. Velocity, and isotherm displacement spectra are presented. The spectra show a sharp drop in the spectral density function near the buoyancy frequency of the thermocline.

0458 Boundary Integral Solutions of Water Wave Problems

Gerard P. Lennon, Assoc. Member, ASCE, (Asst. Prof., Civ. Engrg. Dept., Lehigh Univ., Bethlehem, Pa.), **Philip L-F. Liu**, Assoc. Member, ASCE, (Assoc. Prof., Civ. and Environmental Engrg. Dept., Cornell Univ., Ithaca, N.Y.) and **James A. Liggett**, Member, ASCE, (Prof., Civ. and Environmental Engrg. Dept., Cornell Univ., Ithaca, N.Y.)

Journal of the Hydraulics Division, Vol. 108, No. HY8, August, 1982, pp. 921-931

The Boundary Integral Equation Method (BIEM) is extended to solve three-dimensional, linearized, transient water wave problems. Only the radiation problems concerning long waves generated by an upthrust block of the sea bed are considered in this paper. Due to the linearization, free-surface boundary conditions are applied on the still water level rather than the actual location of the free surface. Thus the problem becomes a fixed boundary problem. A radiation boundary condition is used to truncate the computational domain. Reasonable agreement between present numerical results and previous analytical solutions is observed.

0459 Shape and Dimensions of Stationary Dunes in Rivers

Jorgen Fredsoe, (Assoc. Prof., Inst. of Hydrodynamics and Hydraulic Engrg., Technical Univ. of Denmark, Lyngby, Denmark)

Journal of the Hydraulics Division, Vol. 108, No. HY8, August, 1982, pp. 932-947

The shape and dimensions (length, height) of sand dunes in rivers are calculated by use of the measured bed shear stress distribution downstream a rearward-facing step. The transport of sediment is split up into bed load and suspended load, which makes it possible to explain the transition to plane bed. The model can predict flow-resistance curves in alluvial streams.

0460 Calculation of Secondary Currents in Channel Flow

Dan Naot, (Visiting Sci., Univ. of Karlsruhe, Karlsruhe, West Germany) and **Wolfgang Rodi**, Member, ASCE, (Prof., Univ. of Karlsruhe, West Germany)

Journal of the Hydraulics Division, Vol. 108, No. HY8, August, 1982, pp. 948-968

A model is presented for calculating the flow in channels with turbulence-driven secondary motion, with an emphasis on open channel flow. Algebraic expressions are derived for the Reynolds stresses in the momentum equations for the secondary motion by simplifying modelled Reynolds stress equations. A simple eddy viscosity model is used for the shear stresses in the logitudinal momentum equation. The kinetic energy k, and the dissipation rate ϵ of the turbulent motion appearing in the algebraic and eddy viscosity expressions are determined from transport equations for these quantities. The restricting influence of a free surface on the length scale of turbulence is accounted for by a special free surface boundary condition for ϵ. The resulting set of equations is solved with a marching forward numerical procedure for three-dimensional boundary layers. The model is tested by application to developed two-dimensional closed and open channel flow, closed square duct flow, and flow in open channels of various width-to-depth-ratios. Most features of these flows are simulated well by the model, including the reduction of the eddy viscosity near the free surface and the depression of the velocity maximum below the surface.

0461 Discharge Assessment in Compound Channel Flow

Peter R. Wormleaton, (Lect., Queen Mary Coll., London Univ., London, England), **John Allen**, (Sr. Lect., Queen Mary Coll., London Univ., London, England) and **Panos Hadjipanos**, (Research Student, Queen Mary Coll., London Univ., London, England)

Journal of the Hydraulics Division, Vol. 108, No. HY9, September, 1982, pp. 975-994

A series of laboratory tests were performed to measure the boundary shear stresses in a compound channel consisting of a main channel and associated shallow floodplains. Four different floodplain roughnesses were used. From these results apparent shear stresses across three different assumed main channel/floodplain interface planes (vertical, horizontal and diagonal) were calculated. The data was used with data from previous studies to derive a statistical relationship between the apparent shear stress on the vertical interface and several easily calculated parameters of the channel geometry and hydraulic characteristics. The accuracies of different methods of discharge calculation were shown to be dependent upon the assumptions they make regarding the magnitude of the apparent shear stress on the particular interface chosen. An apparent shear stress ratio was proposed which was found to be a useful yardstick in selecting the best method of discharge calculation for particular channel and flow conditions. Generally it was found that at low floodplain depths using the horizontal or diagonal interfaces gave better discharge results than the vertical.

0462 Vertical Turbulent Diffusion in Air-Driven Water Flows

Louis H. J. Goossens, (Scientific Officer, Safety Sci., Delft Univ. of Tech., Kanaalweg 2b, Delft, The Netherlands), **Hans J. A. van Pagee**, (Project Engr., Delft Hydr. Lab., Rotterdamseweg 185, Delft, The Netherlands) and **Piet J. Tessel**, (Process Engr., Zuiveringsschap Veluwe, Brinklaan 268, Apeldoorn, The Netherlands)

Journal of the Hydraulics Division, Vol. 108, No. HY9, September, 1982, pp. 995-1009

Turbulence intensities and Reynolds stresses produced in a water channel by an air flow over the surface have been measured. An expression for the vertical eddy diffusion coefficient is derived from these results. The drag necessary for the creation of waves and the development of a boundary layer, including the acceleration of the water mass, have been taken into account and are shown to be significant. A linear shear stress distribution has been measured within the boundary layer, and the Prandtl mixing length theory and Boussinesq hypothesis are shown to provide a good basis for describing the diffusion process.

0463 Resistance to Flow in Alluvial Channels

Gerd Holtorff, (Water Resources Engr., Ministry of Agr. and Water, Riyadh, Saudi Arabia)

Journal of the Hydraulics Division, Vol. 108, No. HY9, September, 1982, pp. 1010-1028

Resistance to flow in alluvial channels depends on the geometry of the bed patterns that are formed by the flow, the properties of the sediment, and the transport rate of sediment. A theory is presented to predict the geometry of the bed patterns (ripples, dunes, antidunes, and bars) and to evaluate the influence of the transport rate of sediment on the flow resistance. The theory is found to be in good agreement with experimental results.

0464 Self-aerated Flow in Steep, Partially Filled Pipes

Peter U. Volkart, (Sr. Research Engr., Lab. of Hydraulics, Hydrology, and Glaciology annexed to the Federal Inst. of Tech., 8091 Zurich, Switzerland)

Journal of the Hydraulics Division, Vol. 108, No. HY9, September, 1982, pp. 1029-1046

The self-aerated flow in steep and partially filled conduits of circular shape is investigated. Prototype measurements are carried out on pipelines with slopes up to 1.0 (45°) and diameters from 110–900 mm. Instruments for measuring the air concentration, the velocity of the mixture and the mixture level are described. For the mixture flow, the isotachs, the contour lines of air concentration, and the shape of the free surface are analyzed. Formulae are presented for the mean air concentration, the mean mixture velocity as well as for the critical slope that initiates a water-air flow (air entrainment). The results are compared with known formulae for open rectangular cross-sections. The calculation of the pipe diameter that guarantees partially filled flow (also the width for not-too-wide rectangular channels) is illustrated by examples.

0465 Stability of Armour Layers in Rivers

Arved J. Raudkivi, (Prof., Dept. of Civ. Engrg., Univ. of Auckland, Auckland, New Zealand) and **Robert Ettema**, (Research Engr., Iowa Inst. of Hydr. Research, Iowa City, Iowa 52242)

Journal of the Hydraulics Division, Vol. 108, No. HY9, September, 1982, pp. 1047-1057

The importance of both the overpassing and the embedding of river bed particles is demonstrated by experiments. When a single layer thick armouring layer of uniform d_1 size particles on a bed of uniform finer d_2 size particles is broken at one location the stability of the remainder of the armour layer depends on the tendency of the d_1 particles to embed into the surface of d_2 particles. If a d_1 size particle is seated in the exposed position on the surface of the d_2 sized bed sediment it may, by virtue of its exposure to the flow, be preferentially entrained and it may then overpass the finer bed sediment. Alternatively, the d_1 particle may be too large for overpassing and it will become embedded. Armour layers composed of particles which embed, tend to be more stable.

0466 Urban Flood Damage Estimating Curves

Thomas N. Debo, Member, ASCE, (Asst. Prof., City Planning Program, Georgia Inst. of Tech., Atlanta, Ga. 30332)

Journal of the Hydraulics Division, Vol. 108, No. HY10, October, 1982, pp. 1059-1069

Because of increased flood damages resulting from development in flood plain areas many local and state politicians in Georgia have expressed interest in adopting state legislation which would encourage or mandate the development of local stormwater management programs. Before such legislation is drafted or programs developed for specific areas, the legislature wanted some documentation to estimate the magnitude of present and future flood damages for urban areas throughout the state. Thus, one of the objectives of a recently completed research project was to develop general flood damage curves to roughly estimate flood damages for urban flood plain areas. The development of the flood damage curves was accomplished by using the results from an economic computer model and its application to generalized data in Georgia. The results of applying curves to five case study areas in Georgia are given. Uses and limitations of the curves are outlined.

0467 Tabular Representation of Cross-Sectional Elements

Delbert D. Franz, Member, ASCE, (Vice Pres., Linsley, Kraeger Assoc., Ltd., Mountain View, Calif. 94043)

Journal of the Hydraulics Division, Vol. 108, No. HY10, October, 1982, pp. 1070-1081

A convenient and consistent method for representing the top width, area, first moment of area, and second moment of area for an irregular stream cross section is presented. The variation of top width is taken to be piecewise linear and continuous. The top width is then integrated exactly to obtain the values for the other elements. The unique features of the method is the use of integration to not only define the tabulated values, but also for interpolating for values not tabulated. This procedure yields values consistent with the mathematical definition of the elements. The errors introduced by the operational assumptions and the errors inherent in the basic top width data are separately estimated in an example. A comparison of these two errors is used to evaluate the adequacy of the assumption made for top width.

0468 Asymmetric Flow in Symmetric Expansions

S. David Graber, Member, ASCE, (Consulting Engr., 118 Larson Road, Stoughton, Mass.)

Journal of the Hydraulics Division, Vol. 108, No. HY10, October, 1982, pp. 1082-1101

Asymmetric flow patterns may occur in perfectly symmetric abrupt expansions, in which the main flow deflects and attaches arbitrarily to one wall of the expansion. With the notable exception of "fluidic" devices, asymmetric flow patterns are detrimental in most civil engineering applications. The adverse effects include lengthening of the distance to full-width flow in expansions, inefficient operation of screens in screen channels, and increasing the potential for vortexing in pump intakes. No good explanation or prediction of the onset of asymmetric flow patterns is known to exist. This paper attributes the asymmetric behavior to a static instability of the system occurring under certain conditions of expansion geometry. The stability is investigated by analyzing the forces acting on the system when it undergoes a small deflection from its symmetric position. A mathematical stability criterion is developed and applied to subcritical, horizontal, rectangular expansions. The results are in excellent agreement with experimental observations (limited to small Froude Numbers). The predictive method is extended theoretically to the analysis of certain preventive and corrective measures. Means of extending the analyses to more complex two-dimensional and three-dimensional configurations are examined.

0469 Point Slope Approach for Reservoir Flood Routing

Stanley S. Butler, Fellow, ASCE, (Assoc. Prof., Dept. of Civ. Engrg., Univ. of Southern California, Los Angeles, Calif. 90007)

Journal of the Hydraulics Division, Vol. 108, No. HY10, October, 1982, pp. 1102-1113

The point-slope method of routing floods through reservoirs uses equations and criteria for determining the outflow hydrograph slope and the inflection point of the rising limb of the outflow hydrograph on the basis of the inflow hydrograph, the topography of the reservoir, and the hydraulic characteristics of the outflow structure. The point-slope method is less broadly applicable than the traditional incremental-time methods, but within its limitations (simple single-peaked inflow hydrographs), it is fast, accurate, and adaptable for investigating alternate designs. A check and adjustment procedure provides assurance that the result is valid. This procedure and the criteria for determining the inflection point of the outflow hydrography are of general applicability with reservoir flood routing methods.

0470 Distributed Model for Small Semiarid Watersheds

Leonard J. Lane, Assoc. Member, ASCE, (Hydro., United States Dept. of Agriculture-ARS, Southwest Watershed Research Center, 442 East Seventh St., Tucson, Ariz. 85705)

Journal of the Hydraulics Division, Vol. 108, No. HY10, October, 1982, pp. 1114-1131

A distributed model for estimation of runoff volumes and peak rates of flow from small semiarid watersheds is shown to produce reasonable estimates for mean runoff and flood frequency distributions. The model is simplified and constructed to require a minimum of observed data for calibration. The model simulates runoff volume and peak discharge rates for individual storm events. It can also be used to estimate water yield and a surface water balance incorporating transmission losses in ephemeral stream channels. Experimental data are compared with the simulation model predictions. Individual components of the model could be improved with further research and data. However, based on available information contained in soils and topographic maps and published sources, the distributed model can be used to estimate runoff rates and amounts from ungaged watersheds in semiarid regions.

0471 Model Study of Low Drop Grade Control Structures

W. C. Little, Member, ASCE, (Research Hydr. Engr., Erosion and Channels Research Unit., U. S. Dept. of Agric. Sedimentation Lab., Oxford, Miss.) and **J. B. Murphey**, (Geologist, Erosion and Channels Research Unit, U. S. Dept. of Agric. Sedimentation Lab., Oxford, Miss.)

Journal of the Hydraulics Division, Vol. 108, No. HY10, October, 1982, pp. 1132-1146

A generalized model study was conducted on low drop, grade control structures. A low drop was defined from this study as an hydraulic drop with a difference in elevation between the upstream and downstream channel beds, H; a discharge, Q; and a corresponding critical depth, Y_c, such that the relative drop height, H/Y_c, is equal to or less than 1.0. Conversely a high drop is defined as one with a relative drop height, H/Y_c, greater than 1.0. Tentative design criteria, in the form of dimensionless equations, are given to proportion and hydraulically design the stilling basin for low drops with energy dissipation devices. A method is given to determine the size and placement of a baffle pier or baffle plate to achieve optimum flow conditions in the downstream channel.

0472 Time Scale for Scour Downstream of Hydraulic Jump

Javad Farhoudi, (Lect. in Hydr. Struct., Univ. of Urmia, Urmia, Iran) and **Kenneth V. H. Smith**, Member, ASCE, (Lect. in Hydr. Engrg., Univ. of Southampton, Southampton, England)

Journal of the Hydraulics Division, Vol. 108, No. HY10, October, 1982, pp. 1147-1162

Scour downstream a hydraulic structure takes time to develop. Therefore, for a river in which peak flows are of short duration, the time history of scour development is of major importance. The paper briefly describes the work of Breusers on the time scale of local scour downstream of estuary closure structures and then extends it by reference to a program of experiments to the case of local scour downstream a hydraulic jump. The paper includes an example calculation of the determination of the time scale between a model and prototype overflow weir with plunge-type dissipator, subject to scour downstream.

0473 Optimal Design of Stilling Basins for Overflow Spillways

Yeou-Koung Tung, (Asst. Prof., Dept. of Civ. Engrg., Univ. of Nevada, Reno, Nev. 89557) and **Larry W. Mays**, Assoc. Member, ASCE, (Assoc. Prof., Dept. of Civ. Engrg., Univ. of Texas, Austin, Tex. 78712)

Journal of the Hydraulics Division, Vol. 108, No. HY10, October, 1982, pp. 1163-1178

A model has been developed for use in determining the optimal dimensions of stilling basins for overflow dam spillways and their appurtenances that minimize cost and satisfy hydraulic performance. The decision variables which are the width and elevation of the stilling basins are determined such that the hydraulic performance is satisfied for a wide range of possible spillway discharges. The model is based upon an optimization procedure with two phases. The first phase is used to optimize the hydraulic performance of a stilling basin and the second phase is to use the first phase results to minimize the cost satisfying required basin performance. Both phases are based upon a gradient search technique. Application of the optimization model was made using data for the spillway and stilling basin designs for three dams in Texas.

0474 Analytical Approach to River Regime

William Rodney White, (Principal Scientific Officer, Hydraulics Research Station, Wallingford, United Kingdom), **Roger Bettess**, (Sr. Scientific Officer, Hydraulics Research Station, Wallingford, United Kingdom) and **Enio Paris**, (Lect., Instituto di Ingegnere Civile, Universitya di Firenze, Italy)

Journal of the Hydraulics Division, Vol. 108, No. HY10, October, 1982, pp. 1179-1193

Rational regime relationships for the width, depth and slope of a channel in equilibrium are developed using the Ackers and White sediment transport formulae and the White, Paris and Bettess friction relationships, together with the principle of maximum sediment transporting capacity. Favorable comparisons are made with data from sand and gravel channels and with existing empirical regime relationships.

0475 Recovery in Large Diameter Wells

P. Basak, (Head, Ground Water Division, Centre for Water Resources Development and Management, Calicut 673 571, Kerala, India)

Journal of the Hydraulics Division, Vol. 108, No. HY10, October, 1982, pp. 1194-1207

A closed-form approximate analytical solution for recovery in fully penetrating large diameter wells tapping a confined aquifer is presented. The solutions are also applicable for unconfined flow under conditions of small drawndown. The analysis predicts exponential decreases of inflow rate, a transient logarithmic recovery response, and an asymptotic increase of comulative recovery with time. The solution is verified against the field recovery response in a wide range of geological aquifer formations, e.g., fissured rock, lateritic as well as sandy alluvial deposits. The predicted response and the actual field response for all the types of aquifers mentioned are found to match extremely well. The available theoretical time-recovery relationship for unconfined flow conditions approaches to the present theory of confined aquifer under smaller drawdown.

0476 Envelope Curves for Extreme Flood Events

John R. Crippen, (Supervisory Hydro., U.S. Geological Survey, 855 Oak Ave., Menlo Park, Calif. 94025)

Journal of the Hydraulics Division, Vol. 108, No. HY10, October, 1982, pp. 1208-1212

Maximum flood experience in a region can be depicted by a graph on which maximum known floods are plotted against drainage area. A smooth curve enveloping the plotted points provides the most concise description. such curves have been developed for 17 regions in the conterminous United States and are described by equations. The curves are not associated with specific probabilities or frequencies, but, within the region to which they apply, they give evidence as to the magnitude of flow that was occurred. Given similar basin characteristics, a peak lying close to the envelope curve might occur at other basins in the same region.

0477 Equivalent Roughness of Alluvial Bed

Leo C. van Rijn, (Project Engr., Delft Hydr. Lab., Delft, The Netherlands)

Journal of the Hydraulics Division, Vol. 108, No. HY10, October, 1982, pp. 1215-1218

The eguivalent roughness of a movable bed is considered. Relationships are given which can be used to determine the equivalent bed roughness from the bed material size and the bed-form dimensions. In the case of a plane bed the equivalent roughness is related to the D_{90} of the bed material. Based on flume and field data, the equivalent roughness may vary from 1-10 D_{90}. Similar values were reported by other research-workers. In the case of bed forms the equivalent roughness is related to the average height and length of the bed forms. Both flume and field data were used to determine a functional relationship. The proposed relationship yields values which are considerably smaller than other existing relationships.

0478 Simplified Testing of Hydrologic Regression Regions

Gary D. Tasker, Member, ASCE, (Hydrologist, U.S. Geological Survey, National Center, Reston, Va. 22090)

Journal of the Hydraulics Division, Vol. 108, No. HY10, October, 1982, pp. 1218-1222

The Wilsoxon Signed Ranks Test is suggested to test the statistical significance of "homogenous hydrologic regions" defined by subjectively drawing regional boundaries around regression residuals plotted on a map. Such a test provides the regression analyst and user with helpful insight into what extent the definition of hydrologic region boundaries are supported by the regression data, and to what extent the boundaries are the subjective judgment of the data analyst.

0479 Design of Dual Drainage Systems Using SWMM

J. H. Ellis, (Grad. Student, Dept. of Civ. Engrg., Univ. of Waterloo, Ontario, Canada N2L 3G1), **Edward A. McBean**, (Prof., Dept. of Civ. Engrg., Univ. of Waterloo, Waterloo, Ontario, Canada N2L 3G1) and **G. Mulamoottil**, (Assoc. Prof., School of Urban and Regional Planning, Univ. of Waterloo, Waterloo, Ontario, Canada N2L 3G1)

Journal of the Hydraulics Division, Vol. 108, No. HY10, October, 1982, pp. 1222-1227

In the absence of a proper calibration procedure, traditional methods of applying the Storm Water Management Model, (SWMM), to dual stormwater drainage systems, results in considerable underprediction of system flow and overprediction of pollutant mass flux. These errors can be of considerable importance when designing stormwater management facilities. A methodology is presented which consists of manipulation of selected SWMM input parameters in order to calibrate a dual drainage system response to a singular system reference. Until such time as calibration against reliable field data is possible, such a methodology constitutes a reasonable first step toward improving SWMM modeling accuracy of dual drainage systems.

0480 Bed Material Sorting in Pools and Riffles

Nani G. Bhowmik, Member, ASCE, (Principal Sci., Illinois State Water Survey, P.O. Box 5050, Station A., Champaign, Ill. 61820) and **Misganaw Demissie**, Member, ASCE, (Assoc. Prof. Sci., Illinois State Water Survey, P.O. Box 5050, Station A, Champaign, Ill. 61820)

Journal of the Hydraulics Division, Vol. 108, No. HY10, October, 1982, pp. 1227-1231

When rivers flow in a series of riffles and pools during periods of low flow, bed material sizes change from coarser to finer in the riffle-pool sequence. The Froude Number also changes from high to low from riffle to pool. Larger particle sizes are therefore associated with higher Froude Number and vice versa. The sorting of bed materials, with coarser material at the riffles and finer materials at the pools, is mainly due to the high bed shear stress at the riffles, where the fine materials unable to withstand the high shear stress are washed away and deposited in the pools downstream where the bed shear stress is lower. The aforementioned process of sorting takes place during low discharges when the difference in velocity between that of a riffle and a pool are the greatest.

0481 Shore Line Effects on Mixing of a Tee Diffuser

E. Eric Adams, Member, ASCE, (Lect., Principal Research Engr., Dept. of Civ. Engrg., Massachusetts Inst. of Tech., Cambridge, Mass.), **William C. Skamarock**, (Undergraduate Student, Dept. of Civ. Engrg., Massachusetts Inst. of Tech., Cambridge, Mass.) and **Rita U. Nothaft**, (Undergraduate Student, Dept. of Civ. Engrg., Massachusetts Inst. of Tech., Cambridge, Mass.)

Journal of the Hydraulics Division, Vol. 108, No. HY10, October, 1982, pp. 1232-1238

Submerged multi-port diffusers are often used to discharge waste heat resulting from once-through cooling at coastal power plants. For shallow receiving waters characterized by weak ambient currents (e.g., large lakes), a tee diffuser, with diffuser axis parallel to shore and ports oriented offshore, provides the most effective mixing. However, if the diffuser is sufficiently close to shore, dilution is reduced because of the limited availability of ambient dilution water reaching ports from behind the diffuser. A semi-empirical theory, with laboratory calibration, accounts for the reduced dilution for a tee diffuser in shallow, quiescent water of constant depth near a parallel shoreline. Dilution is reduced by 10% when the ratio of distance offshore to diffuser length is about 0.8 while dilution drops by 30% when the raio is about 0.35. For a sloping bottom, similar results are expected if the distance offshore is taken as one-half of the actual distance.

0482 Saltation in Water Dynamics

Peter J. Murphy, Assoc. Member, ASCE, (Asst. Prof., Dept. of Civ. Engrg., Univ. of Massachusetts, Amherst, Mass. 01003) and **Hamid Hooshiari**, (Grad. Student, Dept. of Civ. Engrg., Univ. of Massachusetts, Amherst, Mass.)

Journal of the Hydraulics Division, Vol. 108, No. HY11, November, 1982, pp. 1251-1267

An evaluation of the fluid forces exerted on a bed load particle is presented. The study used strobe photography to measure the kinematics of a sphere moving via saltation along the bed of an open channel flow. The experiment showed that the forces of bouyancy, added mass, and drag could predict the particle motion. The added mass force included an unusual component due to the mean shear of the open channel flow.

0483 Bed Characteristics in Alluvial Channel Bends

A. Jacob Odgaard, (Asst. Prof., Inst. of Hydr. Research, The Univ.of Iowa, Iowa City, Iowa)

Journal of the Hydraulics Division, Vol. 108, No. HY11, November, 1982, pp. 1268-1281

An approach to the description of the steady-state distributions of flow and bed characteristics in an alluvial channel bend is presented. The approach, which is based on a combination of the principle of moment-of-momentum flux and a critical-shear-stress analysis, is used to explain observed transverse variations of depth, depth-averaged velocity, and mean-grain size in a bend of the Sacramento River, California. The local transverse bed slope is found to vary linearly with d/r (the ratio of depth, d, to radius of curvature, r) and almost linearly with the bed-surface particle Froude number. It is established that the distributions of depth, depth-averaged velocity, and bed-surface sediment size depend on the variety of grain sizes in the bed sediment. Equations for the transverse variation of depth, depth-averaged velocity, and mean-grain size are developed, which can provide a guide to the depth of bed erosion and depth-averaged velocity near the outer banks of river bends.

0484 Fluvial Hydraulics of Deltas and Alluvial Fans

Howard H. Chang, Member, ASCE, (Prof. of Civ. Engrg., San Diego State Univ., San Diego, Calif. 92182)

Journal of the Hydraulics Division, Vol. 108, No. HY11, November, 1982, pp. 1282-1295

Delta formation and its associated steam channel hydraulics have been studied using data obtained in a large laboratory flume. The findings for deltas are generally applicable to alluvial fans. Delta or fan shape is directly related to the pattern of stream flow, its loci, and mode of deposition. In response to a change in water discharge, sediment feeding rate or base level, stream channel adjustments on the delta are characterized by the formation of a single channel or multiple distributary channels. The changes in channel pattern are associated with large changes in channel width and stream power expenditure. Stream channel changes are related to the stream's tendency to seek equal power expenditure along the channel; they also provide delicate adjustments in velocity and sediment load to favor restoration of channel's equilibrium.

0485 Computation of Flow Patterns in Rivers

C. B. Vreugdenhil, (Specialist in Computational Hydr., Delft Hydraulics Lab., Delft, the Netherlands) and **J. H. A. Wijbenga**, (Project Engr., Delft Hydr. Lab., Delft, the Netherlands)

Journal of the Hydraulics Division, Vol. 108, No. HY11, November, 1982, pp. 1296-1310

Quasisteady flow in a river with flood plains is considered with the objective of determining flood levels and flow patterns. The two-dimensional hydrodynamic equations are solved using a time-dependent, finite-difference method. Bottom and lateral friction are found to have an important influence. The mathematical model is calibrated and verified, using detailed laboratory measurements. A fair agreement is found, but the need of calibrating frictional parameters with field data is stressed.

0486 Well Hydraulics in Vertically Heterogeneous Formations

Tatiana D. Streltsova, (Sr. Research Specialist, Reservoir Div., Exxon Production Research Co., P.O. Box 2189, Houston, Tex. 77001)

Journal of the Hydraulics Division, Vol. 108, No. HY11, November, 1982, pp. 1311-1327

Comparative analysis of time-dependent drawdown alterations caused by the heterogeneous properties of stratified formations is given based on solutions developed under different assumptions on interlayered crossflow. The assumption of proportionality of leakage to the differential flow depletion, which is widely used in leaky aquifers developments, severely modifies the calculated drawdown values for both the aquifer and the aquitard as compared to those values obtained by assuming the diffusivity distribution process for the flow within each layer. The drawdown pattern is particularly affected at the early times. It is shown that if the compressibility of the low-permeability layer is taken into account, the shape of transient drawdown curves is a function not only of the ratio of permeabilities of the permeable and the semi-permeable layers but of the ratio of their storages as well.

0487 Relationships between Morphology of Small Streams and Sediment Yield

Task Committee on Relations Between Morphology of Small Streams and Sediment Yield of the Committee on Sedimentation of the Hydraulics Division, Leonard J. Lane, chmn.

Journal of the Hydraulics Division, Vol. 108, No. HY11, November, 1982, pp. 1328-1365

The state of the art in the relations between morphology of small streams and sediment yield is assessed. Research findings and recommendations for additional research are presented. Topics include systems and interactions, simulation models, channel forms and processes, transport of sediment in small streams, and aspects of channel morphology. Selected topics for additional research are also included.

0488 Lateral Bed Load Transport on Side Slopes

Syunsuke Ikeda, Member, ASCE, (Assoc. Prof., Dept. of Foundation Engrg., Saitama Univ., Urawa, Saitama, Japan)

Journal of the Hydraulics Division, Vol. 108, No. HY11, November, 1982, pp. 1369-1373

Lateral bed load transport is studied in laboratory flume and wind tunnel. Introduction of laterally inclinable wind tunnel permits accurate measurements of shear stress on side slope and of lateral bed load. The result of wind tunnel test was examined by an open channel experiment. A universal formula for the lateral bed load q^*_l is presented as a function of inclination of side slope, θ, and dimensionless shear stress τ^*/τ^*_c; and the expression is $q^*_l/\tan\theta = \alpha(\tau^*/\tau^*_c)^\beta$. The numerical constants α and β απε φουνδ το βε 0.0039 ανδ 1.5, πεσρεϝτιςελυ, φοπ $\tau^*/t^*_c > 1.07$. The formula may add some information to bank protection engineering, prediction of river meander, and to widening of braided river.

0489 Flow Computation Using Extrapolation Procedures

Chintu Lai, Member, ASCE, (Hydro., Dept. of the Interior, U.S. Geological Survey, Reston, Va. 22092)

Journal of the Hydraulics Division, Vol. 108, No. HY11, November, 1982, pp. 1374-1380

The extrapolation procedure is a powerful numerical technique by which to raise the accuracy of computational results with little extra computational efforts. The procedure assumes the computational error as a polynomial function of grid interval, $\triangle t$; which permits elimination of one or more terms of lower powers of $\triangle t$ through a linear combination of a set of numerical solutions using different sizes of $\triangle t$. Because it is often difficult to express the numerical error in an analytical or explicit form of powers of $\triangle t$, numerical experiments can be used with advantages to test the applicability and the effectiveness of the extrapolation procedure in flow computations. A series of numerical experiments conducted for a Sacramento River reach has demonstrated that the technique is a remarkably powerful and economical tool in attaining a second-order result from simple linear computations.

0490 On Zero-Inertia and Kinematic Waves

Nikolaos D. Katopodes, Member, ASCE, (Asst. Prof., Dept. of Civ. Engrg., Univ. of Michigan, Ann Arbor, Mich.)

Journal of the Hydraulics Division, Vol. 108, No. HY11, November, 1982, pp. 1380-1387

Small perturbation analysis of the linearized open-channel flow equations leads to the conclusion that zero-inertia waves propagate in the downstream direction only with a speed equal to that of kinematic waves. In addition, such an analysis indicates that kinematic waves travel with no attentuation in amplitude. In this work, it is shown both by theory and nonlinear numerical experiments that zero-inertia waves propagate both in the downstream and upstream direction. In fact, the zero-inertia wave characteristics are horizontal lines, and the problem requires two-point boundary data, which extends the applicability of zero-inertia models to cases of great practical interest. It is also shown the nonlinear kinematic-wave characteristics are inclined relative to each other and that they may intersect to create a kinematic shock. This shock is always present in wave motion on a dry bed and leads to significant subsidence of the depth and discharge in the flow behind it, a behavior which is not due to amplitude attentuation, as in the case of dynamic waves. Merely it is due to a consequence of the presence of the kinematic-shock front.

0491 Muskingum-Like Approximations for Water Routing

William C. Peterson, (Mathematician, Space Systems Div., General Electric Co., Valley Forge, Pa.) and **Frank H. Verhoff**, Member, ASCE, (Prof., Dept. of Chemical Engrg., West Virginia Univ., Morgantown, W. Va. 26506)

Journal of the Hydraulics Division, Vol. 108, No. HY11, November, 1982, pp. 1387-1393

An integral approximation is obtained for the continuity equation as applied to stream flow during storms. The resulting general algebraic equation is similar in form to Muskingum approximations, and it is of cubic order in distance. The approximation parameters can be calculated from the graph of the flow rate, Q, as a function of the cross-sectional flow area, A.

0492 Celerity of Transient Bed Profiles

Victor Miguel Ponce, Member, ASCE, (Assoc. Prof. of Civ. Engrg., San Diego State Univ., San Diego, Calif.)

Journal of the Hydraulics Division, Vol. 108, No. HY11, November, 1982, pp. 1393-1397

The applicability of a formula for the calculation of celerity of alluvial channel bed transients has been verified using data from flume experiments. The formula is easy to use; the celerity is expressed in terms of readily identified hydraulics and sediment transport parameters. The results obtained are encouraging, and show the promise of this approach in the qualitative analysis of unsteady alluvial channel hydraulics. The need for additional research in this area is recognized.

0493 Multiport Diffusers for Heat Disposal: A Summary

Gerhard H. Jirka, Member, ASCE, (Assoc. Prof., School of Civ. and Environmental Engrg., Cornell Univ., Ithaca, N.Y. 14853)

Journal of the Hydraulics Division, Vol. 108, No. HY12, December, 1982, pp. 1423-1468

This paper reviews the fluid mechanical characteristics - as derived from theoretical and experimental studies during the past dozen years - of submerged multiport diffusers used for heat disposal from thermal power plants into the water environment. Foremost among these characteristics is the near field instability produced by such thermal diffusers in typical receiving water conditions. Rather than forming a distinct buoyant plume as is the case for the traditional sewage diffuser, the high discharge momentum of thermal diffusers leads to a flow breakdown with local recirculation zones and full vertical mixing. Stability diagrams for both stagnant and flowing ambient conditions are presented. The flow and temperature fields at larger distances, in the intermediate field, are, in turn, critically dependent upon how the discharge momentum is introduced into the ambient fluid layer. Out of a spectrum of possible diffuser designs three major types have evolved. The unidirectional and staged diffusers are designs which result in concentrated vertically mixed plume motions. The alternating diffuser with appropriate nozzle control generates a stratified flow field outside the unstable near field. Predictive techniques for these basic types are summarized. A typical case comparison illustrates their differences in engineering design and environmental impact.

0494 Ground Water Pumping from Shallow Axisymmetric Ponds

Mustafa M. Aral, Member, ASCE, (Assoc. Prof., School of Civ. Engrg., Georgia Institute of Technology, Atlanta, Ga.) and **Terry W. Sturm**, Assoc. Member, ASCE, (Asst. Prof., School of Civ. Engrg., Georgia Institute of Technology, Atlanta, Ga.)

Journal of the Hydraulics Division, Vol. 108, No. HY12, December, 1982, pp. 1469-1485

The hydraulic aspects of pumping from axisymmetric ponds, or large diameter wells, as a means of developing shallow, unconfined aquifers are investigated, and it is shown that pond storage allows groundwater pumping at higher rates during short time periods than would be possible for continuous pumping. A finite element model that considers the effects of pond storage, partial penetration, pond size, and pumping time on the pumping discharge from the pond is developed. Numerical results are presented in terms of generalized dimensionless variables, and are applied to a design problem of selecting the pond diameter, pumping rate, and pumping schedule for given aquifer properties.

0495 Routing Graded Sediments in Streams: Formulations

Deva K. Borah, Member, ASCE, (Sr. Civ. Engr., Tippetts-Abbett-McCarthy-Stratton, The TAMS Building, 655 Third Ave., New York, N.Y. 10017), **Carlos V. Alonso**, Member, ASCE, (Research Hydr. Engr., U.S. Dept. of Agr. Sedimentation Lab., Oxford, Miss.) and **Shyam N. Prasad**, (Prof., Dept. of Civ. Engrg., The Univ. of Mississippi, University, Miss.)

Journal of the Hydraulics Division, Vol. 108, No. HY12, December, 1982, pp. 1486-1503

A one-dimensional numerical model is developed for simulating the movement of well graded sediments through a stream network. The model is based on the physical processes governing the mechanics of sediment movement in alluvial channels. It incorporates the effect of sediment gradation on the stream's potential transport capacity. It also accounts for the effect of particle exchange between the sediment load and the bed, while simulating aggradation, degradation and armoring of the bed. The applicability of the model is restricted to noncohesive materials and relatively stable channel geometries. Any valid water routing algorithm can be used in simulating the movement of water.

0496 Routing Graded Sediments in Streams: Applications

Deva K. Borah, Member, ASCE, (Sr. Civ. Engr., Tippetts-Abbett-McCarthy-Stratton, The TAMS Building, 655 Third Ave., New York, N.Y. 10017), **Carlos V. Alonso**, Member, ASCE, (Research Hydr. Engr., U.S. Dept. of Agr. Sedimentation Lab., Oxford, Miss.) and **Shyam N. Prasad**, (Prof., Dept. of Civ. Engrg., The Univ. of Mississippi, University, Miss.)

Journal of the Hydraulics Division, Vol. 108, No. HY12, December, 1982, pp. 1504-1517

A one-dimensional numerical model has been tested in simulating the movement of well graded sediments through a stream reach. The model is applied to two laboratory flumes and two alluvial channel reaches, found in the literature. Three of these examples are under steady state flow conditions and the other is under unsteady flow conditions. In the unsteady flow situation, the flow computations are done by using a water routing scheme based on the kinematic wave equations. This scheme needs adjustment of one parameter. The sediment model also needs adjustment of only one parameter which is the erodibility parameter. The model predicts water and sediment discharges coming out of each reach, size distributions of these transported materials, size distributions of the bed material or the armor coat, and changes in bed elevations. Comparisons of these predictions are made with the measurements available from these examples, and are found to be promising.

0497 Sedimentology of Erect Vegetal Filters

Ernest W. Tollner, (Asst. Prof., Agr. Engrg., Univ. of Georgia, Experiment, Ga.), **Billy J. Barfield**, (Prof., Agr. Engrg., Univ. of Kentucky, Lexington, Ky.) and **John C. Hayes**, (Asst. Prof., Agr. and Biological Engrg., Mississippi State Univ., Mississippi State, Miss.)

Journal of the Hydraulics Division, Vol. 108, No. HY12, December, 1982, pp. 1518-1531

In order to analyze sediment deposition in vegetated areas, it is necessary to develop relationships that define the steady-state sediment transport capacity of a given filter media. The development of these relationships was greatly facilitated by drawing an analogy between the flow in a narrow, deep channel and flow through closely spaced erect bodies. Based on the analogy, a dimensionless shear parameter is calculated using a "spacing hydraulic radius" as the characteristic length. Utilizing data collected with artificial media, this dimensionless shear is correlated to the bed load and total load parameters of Graf and Einstein with r^2 values ranging from 0.81 - 0.94. Neil's relationship was also analyzed and yielded a high r^2 value. Relationships based on the analogy were further examined by incorporating them into a model that predicted the sediment deposition wedge advance distance in actual vegetation. The agreement between the predicted and observed values was excellent. Thus, the sedimentology of grass filters presented herein is considered useful for design and research applications.

0498 Separation of Hydraulic Flow over Topography

Herbert E. Huppert, (Lect., Dept. of Applied Mechanics and Theoretical Physics, Univ. of Cambridge, Silver Street, Cambridge CB3 9EW, England) and **Rex E. Britter**, (Lect., Dept. of Engrg., Univ. of Cambridge, Cambridge, CB3 9EW, England)

Journal of the Hydraulics Division, Vol. 108, No. HY12, December, 1982, pp. 1532-1539

The hydraulic flow of either a single layer or multiple layers of fluid over a topographic feature is frequently employed in many areas of civil engineering, meteorology, and oceanography. It is usually tacitly assumed that such flows remain attached as they flow over the topography. This paper determines the conditions under which flow separation occurs. Hydraulic equations for the horizontal velocity, and the criterion that separation occurs when the flow is locally decelerating are used to determine that for a single-layer separation occurs whenever the flow is everywhere subcritical. The results of experiments (in an open channel flume) are in agreement with this theoretical criterion. The analysis is extended to determine the condition under which a two-layer hydraulic flow separates.

0499 Volume-Accuracy Relationship in Soil Moisture Sampling

Mark E. Hawley, (Faculty Research Asst., Dept. of Civ. Engrg., Univ. of Maryland, College Park, Md. 20742), **Richard H. McCuen**, Aff., (Assoc. Dean, Coll. of Engrg., Univ. of Maryland, College Park, Md. 20742) and **Thomas J. Jackson**, Assoc. Member, ASCE, (Hydro., Hydr. Lab., USDA-SEA-AR, Beltsville, Md.)

Journal of the Irrigation and Drainage Division, Vol. 108, No. IR1, March, 1982, pp. 1-11

The number of soil samples required to achieve a given level of accuracy of the mean value is directly proportional to the variance of the samples. Analysis of a set of soil samples of different sample volumes showed that there is a relationship between the variance of the samples and the sample volume. Generally, a larger sample volume results in a lower variance because the larger sample includes more of the overall spatial variability than a smaller sample. Samples of eight different sizes were collected; sample volumes ranged from about 7 to 825 cm^3 (1/2 to 50 in^3). Analysis of variance techniques were used to determine the relationships between both bias and sample volume and precision and sample volume. This analysis showed that when sampling to a depth of 10 cm (4 in), the minimum desirable sampling volume is about 50 cm^3 (3 in^3); smaller volumes will result in lower precision of the mean value. There does not seem to be any direct relationship between bias and sample volume. The minimum desirable sampling volume is a function of the homogeneity of the area being sampled and the elapsed time since the last significant rainfall.

0500 Application of Penman Equation Wind Function

Richard H. Cuenca, Member, ASCE, (Asst. Prof., Dept. of Agricultural Engrg., Oregon State Univ., Corvallis, Oreg.) and **Martin T. Nicholson**, (Agricultural Engr., CH2M Hill Consulting Engrs., Yakima, Wash.)

Journal of the Irrigation and Drainage Division, Vol. 108, No. IR1, March, 1982, pp. 13-23

The Penman equation for computing evapotranspiration is frequently applied as the basis of irrigation system design. It has seen widespread application when sufficient data are available because of its sound theoretical basis. This equation combines an energy balance and an aerodynamic term. The aerodyamic portion of the equation contains a measure of vapor pressure deficit multiplied by an empirically derived wind function. Uncertainties and errors have arisen in the application of this equation because of misconceptions regarding calculation of the vapor pressure deficit. Derivation of the original Penman aerodynamic term is indicated along with recent modifications to the wind function. Correct methods of computing the vapor pressure deficit corresponding to various wind functions are shown. The importance of using wind function coefficients consistent with the method of computing the vapor pressure deficit for design applications is stressed.

0501 Water Yield Estimation in Western United States

Mark E. Hawley, (Faculty Research Asst., Dept. of Civ. Engrg., Univ. of Maryland, College Park, Md. 20742) and **Richard H. McCuen**, Member, ASCE, (Assoc. Dean, College of Engrg., Univ. of Maryland, College Park, Md.)

Journal of the Irrigation and Drainage Division, Vol. 108, No. IR1, March, 1982, pp. 25-34

The mean annual water yield of a watershed is used in the planning, design, and management of some watershed projects, especially those involving irrigation. A statistical analysis of water yield data from 605 watersheds in the western United States resulted in five equations, one for each of five regions of hydrologic similarity. The separation of the study area into these five regions improved the accuracy of estimateion significantly. The equations explained as much as 95 percent of the variance in mean annual water yield.

0502 Advance and Runoff-Surge Flow Furrow Irrigation

Phillip S. Coolidge, (Research Assoc., Agr. & Irrigation Engrg. Dept., Utah State Univ., Logan, Utah), **Wynn R. Walker**, Member, ASCE, (Assoc. Prof., Agr. & Irrigation Engrg. Dept., Utah State Univ., Logan, Utah.) and **A. Alvin Bishop**, Member, ASCE, (Prof. Emeritus, Agr. & Irrigation Engrg. Dept., Utah State Univ., Logan, Utah.)

Journal of the Irrigation and Drainage Division, Vol. 108, No. IR1, March, 1982, pp. 35-42

Furrow advance and runoff evaluations were conducted on a field having a silt loam soil using pulsed water applications. The irrigation method, defined as "surge flow," was applied in two experiments, each covering 100 m furrow sections. The first experiment consisted of 13 combinations of on- and off-times, replicated five times. The second experiment included 10 combinations of cycle period and cycle ratio with two replications. The results indicate that surge flow treatments significantly accelerated furrow advance per unit of applied water and reduced the temporal and spatial variability among furrows in the field. It is theorized that furrow drainage between surges accelerates the formation of thin surface seals, thereby reducing intake rates.

0503 Managing Salinity in Irrigated Agriculture

Irfan A. Kahn, Member, ASCE, (Asst. Prof., Dept. of Civ. Engrg., Youngstown State Univ., Youngstown, Ohio 44555)

Journal of the Irrigation and Drainage Division, Vol. 108, No. IR1, March, 1982, pp. 43-56

Excessive accumulation of salts in soils and ground water has plagued the irrigated semiarid and subhumid regions of the world for centuries. The structural measures of salinity control are effective but extremely capital intensive. A non-structural management model is presented which may help maintain a permanent irrigated agriculture at a reasonable cost. The model is based on the theory that an irrigated system consists of a number of components and that the good health of each component is vital to the overall good health of the entire system, and that ignoring even a single component could lead to disastrous consequences. A simplified limited version of the model is applied to a case study and it is concluded that the implementation of the model devised strategies could reduce the normal rate of degradation of the ground water by about 80 percent.

0504 New Evapotranspiration Crop Coefficients

James L. Wright, (Soil Scientist, U.S. Dept. of Agriculture, Snake River Conservation Research Center, Route 1, Box 186, Kimberly, Idaho, 83341)

Journal of the Irrigation and Drainage Division, Vol. 108, No. IR1, March, 1982, pp. 57-74

Improved crop coefficients for various Pacific Northwest irrigated crops were developed for estimating crop evapotranspiration (ET) from estimates or measurements of reference ET. Reference ET was based on that for well watered, actively growing alfalfa with sufficient growth for near maximum ET in arid, irrigated regions. ET for the alfalfa reference crop and other crops was measured with sensitive weighing lysimeters at the field site near Kimberly, Idaho. The new crop coefficients are basal or minimal coefficients for conditions when soil evaporation is minimal but root-zone soil moisture is adequate. When combined with improved estimates of evaporation from wet soils, they should permit more accurate estimates of daily crop ET, more accurate irrigation scheduling, and more reliable estimates of crop water requirements. Curves were developed for alfalfa, potatoes, snap beans, sugarbeets, peas, sweet and field corn and winter and spring cereals.

0505 Optimum Depth and Spacing of Subsurface Drains

Haridas S. Acharya, (Asst. Prof. of Mathematics, Marathwada Agr. Univ., Parbhani, India) and **Dnyanoba G. Holsambre**, (Asst. Prof. of Agr. Engrg., Marathwada Agr. Univ., Parbhani, India)

Journal of the Irrigation and Drainage Division, Vol. 108, No. IR1, March, 1982, pp. 77-80

Optimization theory is used to minimize the cost of installation of subsurface drainage systems. The optimization problem is presented and constraints are identified. The analytic solution yields formulas for direct determination of optimum depth and spacing of subsurface drains.

0506 Designing for Deficit Irrigation

Marshall J. English, Member, ASCE, (Asst. Prof., Degt. of Agricultural Engrg., Oregon State Univ., Corvallis, Oreg.) and **Gary S. Nuss**, (Grad. Research Asst., Dept. of Agricultural Engrg., Oregon State Univ., Corvallis, Oreg.)

Journal of the Irrigation and Drainage Division, Vol. 108, No. IR2, June, 1982, pp. 91-106

Deficit irrigation is the practice of deliberately under-irrigating a crop. This practice may be preferable to full irrigation when water supplies are limited or irrigation costs are high. This paper investigates the merits of deficit irrigation for a farm in eastern Oregon. Two distinctly different irrigation systems were designed, one for full irrigation, the other for deficit irrigation. The costs and performances of these two systems are compared. It is determined that the system designed for deficit irrigation could lead to increased farm income while substantially reducing energy, water and capital requirements. These advantages would be gained by designing the deficit irrigation system for long intervals between irrigations and for low soil moisture uniformity. It is concluded that deficit irrigation can offer significant benefits under some circumstances, these benefits may be largely dependent on system design, and a comprehensive economics analysis is required to accurately assess the merits of a deficit irrigation scheme.

0507 Antecedent Moisture Condition Probabilities

Donald D. Gray, Member, ASCE, (Asst. Prof. of Hydromechanics, School of Civ. Engrg., Purdue Univ., West Lafayette, Ind. 47907), **Peter G. Katz**, (Engr., Snowy Mountains Engrg. Corp., Cooma, Australia), **Sharon M. deMonsabert**, (Grad. Research Asst., School of Civ. Engrg., Purdue Univ., West Lafayette, Ind. 47907) and **Neroli P. Cogo**, (Grad. Student, Dept. of Agronomy, Purdue Univ., West Lafayette, Ind. 47907)

Journal of the Irrigation and Drainage Division, Vol. 108, No. IR2, June, 1982, pp. 107-114

The problem of selecting an appropriately conservative Antecedent Moisture Condition (AMC) when using the Soil Conservation Service curve number method or the Illinois Urban program for prediction is considered. Daily rainfall data from 17 stations in Indiana, Kentucky, and Tennessee are analyzed to find the probability of the various Antecedent Moisture Conditions during the growing season at each station. Regression equations are proposed relating these probabilities to average annual precipitation in areas having similar rainfall amounts and distributions.

0508 Effluent in Trickle Irrigation of Cotton in Arid Zones

Gideon Oron, (Research Engr. and Agronomist, Jacob Blaustein Inst. for Desert Research, Ben-Gurion Univ. of the Negev, 84990 Kiryat Sde-Boker, Israel), **Jiftah Ben-Asher**, (Agronomist, Jacob Blaustein Inst. for Desert Research, Ben-Gurion Univ. of the Negev, 84990 Kiryat Sde-Boker, Israel) and **Yoel DeMalach**, (Researcher, Ramat-Negev Agric. Field Station , (Sde-Chail), D.N. Chalutsa, 85415 Revivim, Israel)

Journal of the Irrigation and Drainage Division, Vol. 108, No. IR2, June, 1982, pp. 115-126

A trickle system was installed in a cotton field to study the effects of effluent quality, emitter discharge and irrigation regime on the yield of cotton. The experiments were conducted in a typical arid zone and the cotton was grown on loess soil. The results show that a high cotton hield of more than 6000 kg/hectare can be obtained with the use of a high frequency (every 2 days) trickle irrigation system. A total of approximately 5900 m^3 of effluent per hectare was applied, with no additional fertilization required. With proper screen filtration no emitter clogging was observed.

0509 Limits for Practical Level-Basin Design

Albert J. Clemmens, Assoc. Member, ASCE, (Research Hydr. Engr. and Agric. Engr., U.S. Water Conservation Lab., Phoenix, Ariz. 85040) and **Allen R. Dedrick**, (Research Hydr. Engr. and Agric. Engr., U.S. Water Conservation Lab., Phoenix, Ariz. 85040)

Journal of the Irrigation and Drainage Division, Vol. 108, No. IR2, June, 1982, pp. 127-141

A previous solution for level basin design is expanded to cover a wider range of infiltration conditions. An analysis is performed to determine an approximate limit on basin length which depends upon application depth, infiltration conditions, roughness and the desired uniformity. Once a basin length is determined, the minimum unit flow rate necessary to achieve the desired uniformity is found. Basin area and width are then based on the available stream size. The solution is developed from the zero-inertia border irrigation model and is displayed in dimensionless form.

0510 The V-Shaped Broad-Crested Weir

Wubbo Boiten, (Research Engr., Delft Hydraulics Lab., Wageningen Branch, Nieuwe Kanaal 11, 6709 PA Wageningen, The Netherlands) and **Remmet H. Pitlo**, (Research Engr., Hydraulics Lab., Agricultural Univ., Wageningen, The Netherlands)

Journal of the Irrigation and Drainage Division, Vol. 108, No. IR2, June, 1982, pp. 142-160

A discharge measurement structure with a triangular control section and a broad crest provides a rather wide measuring range while the head-discharge relation remains unaffected by tailwater up to a high submergence ratio. Experiments with the V-shaped broad-crested weirs have been carried out in order to investigate the influence of a boundary layer above the crest. The boundary displacement thickness is replaced by a correction height which includes the effects of a non-uniform velocity distribution, non-parallel streamlines and lateral pressure gradients. The experiments have results in a large collection of laboratory data with the aim to derive the head-discharge relations for different shapes under free flow as well as submerged flow conditions. The weir can be constructed with a fixed crest or a movable crest in vertical runways. The maximum discharge depends on the chosen dimensions. The structure is a suitable device for small and moderate discharges. Results of investigation, design rules and limits of application are given.

0511 Uniform Flow over Skew Side-Weir

Subhash C. Jain, Member, ASCE, (Research Engr., Inst. of Hydr. Research, The Univ. of Iowa, Iowa City, Iowa) and **Edward E. Fischer**, Assoc. Member, ASCE, (Research Assoc., Inst. of Hydr. Research, The Univ. of Iowa, Iowa City, Iowa)

Journal of the Irrigation and Drainage Division, Vol. 108, No. IR2, June, 1982, pp. 163-166

A rectangular channel with a skew side-weir is designed to obtain a uniform discharge distribution over the side-weir. Numerical and experimental results are presented. Geometric and flow characteristics of the side weir and the channel are given. It is shown that the theoretical results are in good agreement with the experimental data.

0512 Estimating Soil Water Retention from Soil Properties

Walter J. Rawles, (Hydro., USDA-ARS, Hydr. Lab., Beltsville, Md. 20705) and **D. L. Brakensiek**, Member, ASCE, (Research Hydr. Engr., USDA-ARS, Northwest Watershed Research Center, Boise, Idaho 83701)

Journal of the Irrigation and Drainage Division, Vol. 108, No. IR2, June, 1982, pp. 166-171

Soil water retention characteristics are needed to describe the movement and storage of water in soils. Since conventional laboratory methods of obtaining these are expensive and time consuming, a method of estimating soil water retention at various matric potentials from readily available soil properties was investigated. Data from 2543 soil horizons representing many agricultural areas of the country were utilized. The soil properties investigated were the percent sand, silt and clay, the percent organic matter, the bulk density, median particle size, particle size index, and the -0.33 and -15 bar soil water contents. A series of regression equations for predicting soil water retention at 12 matric potentials ranging from -0.04 to -15 bar were developed utilizing different combinations of soils data.

0513 Zero-Inertia Modeling of Furrow Irrigation Advance

Ronald L. Elliott, (Asst. Prof., Agr. Engrg. Dept., Oklahoma State Univ., Stillwater, Okla.), **Wynn R. Walker**, Member, ASCE, (Assoc. Prof., Agr. and Irrigation Engrg. Dept., Utah State Univ., Logan, Utah) and **Gaylord V. Skogerboe**, Member, ASCE, (Prof., Agr. and Chemical Engrg. Dept., Colorado State Univ., Fort Collins, Colo.)

Journal of the Irrigation and Drainage Division, Vol. 108, No. IR3, September, 1982, pp. 179-195

The rate of advance of the water front in furrows has been mathematically modeled using a zero-inertia approach, in which the surface water hydraulics are simplified by neglecting accelerations. Power curve relationships are used to relate both the flow depth and wetted perimeter to the cross-sectional flow area. Infiltration is described with the modified Kostiakov equation, which has a constant term that accounts for a soil's basic intake rate. Field evaluations from three Colorado farms were used in testing the model. It was shown that the zero-inertia model can effectively simulate the hydraulics of the advance phase of furrow irrigation.

0514 Drainage Systems for Land Treatment of Wastewater

R. W. Skaggs, (Prof. and Research Assoc., Dept. of Biological and Agricultural Engrg., North Carolina State Univ., Raleigh, N.C.) and **A. Nassehzadeh-Rabrizi**

Journal of the Irrigation and Drainage Division, Vol. 108, No. IR3, September, 1982, pp. 196-211

Simulation methods for design and evaluation of drainage systems for wastewater land treatment sites are presented. The methods are based on the water management model, DRAINMOD, for describing the performance of an artificially drained land treatment system over a long period of climatological record. The methods are demonstrated by considering an example of wastewater application on a poorly drained soil area near Wilmington, N.C. Several drainage designs and application strategies are evaluated to determine their effects on the amount of wastewater that could be treated and how that amount changes from year to year with weather conditions. Effects on water table depth outflow volumes and required wastewater storage are also analyzed. When wastewater application is limited by drainage conditions of the site, more water can be treated by applying small amounts frequently than with larger amounts on a less frequent basis. Results showed that DRAINMOD can be used to optimize the drainage system design for a given land application site.

0515 Models of Alfalfa Yield and Evapotranspiration

Johannes C. Guitjens, Member, ASCE, (Prof. of Irrigation Engrg., Dept. of Plant, Soil, and Water Sci., Univ. of Nevada, Reno, Nev. 89557)

Journal of the Irrigation and Drainage Division, Vol. 108, No. IR3, September, 1982, pp. 212-222

Data of per-cutting yields and annual yields from four cuttings were collected over six consecutive years for alfalfa grown on three lysimeters. Corresponding evapotranspiration was obtained from weekly water balances. Differences in water management created a range of conditions from water for annual yields of 10 ton/acre (22.4 ko m^{-2}) to 50% yield reductions caused by inadequate soil moisture required to satisfy evapotranspiration demands. Yield and water use efficiency, both annual and per-cutting data, were compared in correlation tests with their corresponding amounts of evapotranspiration. Annual and per-cutting yields were found to be statistically proportional to their evapotranspiration. Annual water use efficiency was not statistically significant with annual evapotranspiration. Although a questionable percentage of the data explained the variance, water use efficiency per cutting and evapotranspiration per cutting had a statistically significant negative correlation.

0516 Estimating Potential Evapotranspiration

George H. Hargreaves, Fellow, ASCE, (Research Dir., Dept. of Agr. and Irrigation Engrg., Utah State Univ., Logan, Utah 84322) and **Zohrab A. Samani**, (Ph.D. Candidate, Dept. of Agr. and Irrigation Engrg., Utah State Univ., Logan, Utah 84322)

Journal of the Irrigation and Drainage Division, Vol. 108, No. IR3, September, 1982, pp. 225-230

Increasing population and needs for an augmented food supply give greater importance to improved procedures for estimating agricultural water requirements both for irrigation and for rain-fed agriculture. Four methods for estimating potential evapotranspiration are compared and evaluated. These are the Class A evaporation pan located in an irrigated pasture area, the Hargreaves equation, the Jensen-Haise eguation, and the Blaney-Criddle method. The evaporation pan is rated as superior to the other methods. However, the difference in reliability between the pan and the Hargreaves method are not considered to be very significant. Both the Jensen-Haise and the Hargreaves methods require either measured or estimated solar radiation. Methods are presented for estimating solar radiation from the difference between maximum and minimum temperature, from the percentage of possible sunshine, and from relative humidity. These procedures have some limitations, but provide improved reliability and make the estimates more universal.

0517 Two-Dimensional Exponential Infiltration Equations

Yu-Si Fok, Member, ASCE, (Prof., Dept. of Civ. Engrg., Univ. of Hawaii at Manoa, Honolulu, Hawaii 96822), **Sang-Ok Chung**, (Grad. Student, Dept. of Civ. Engrg., Univ. of Hawaii at Manoa, Honolulu, Hawaii 96822) and **Clark C. K. Liu**, (Asst. Prof., Dept. of Civ. Engrg., Univ. of Hawaii at Manoa, Honolulu, Hawaii 96822)

Journal of the Irrigation and Drainage Division, Vol. 108, No. IR4, December, 1982, pp. 231-241

Physical two-dimensional infiltration equations expressed in exponential form were developed based upon the four time intervals derived in an earlier study for one-dimensional infiltration. Assumptions that the loci of the wetting pattern for the two-dimensional infiltration are half-ellipses, and that their vertical and horizontal flow components can be described as one-dimensional infiltration. The developed equations were checked by laboratory experiments using the soil-box method at two ("dry" and "wetter") soil-moisture levels. These experiments were conducted at the two controlled soil-moisture levels using aa clay-loam as a soil medium. For the dry soil moisture level, the computed and the measured infiltration rates were considered in good agreement (-1% - 11%). However, in the wetter soil-moisture level, the computed infiltration rates were smaller, by as much as -56%, than the measured infiltration rates until 150 minutes of the infiltration time. The former were as much as 14% larger than the latter, ranging from 150 min. - 330 min. of the infiltration time. The results, in general, show good agreement in both soil samples considering only the arithmetic differences between the measured infiltration rates and the computed ones. The assumptions made in the derivation of the two-dimensional infiltration equations were found to be responsible to some extent. However, the discrepancy warrants further investigations and a table of experimental data is provided for such further work.

0518 Channel Design to Minimize Lining Material Costs

Thomas J. Trout, Member, ASCE, (Agricultural Engr., U.S. Dept. of Agr., Agricultural Research Service, Snake River Conservation Research Center, Kimberly, Idaho 83341)

Journal of the Irrigation and Drainage Division, Vol. 108, No. IR4, December, 1982, pp. 242-249

A direct algebraic technique is developed to determine open channel cross-sectional designs which minimize lining material costs when base and side wall unit costs are different. Solution graphs, which indicate both the optimal parameter combination and the costs of deviating from the optimal design, are presented. The graphs or analytical technique are also effective in designing any trapezoidal channels. Use of the technique shows that moderate deviations from optimal designs are not costly.

0519 Optimal Design of Small Reservoir Systems

Prem N. Sharma, (Irrigation Specialist, InterAmerican Inst. for Cooperation on Agriculture at CPATSA, Petrolina, PE, Brazil) and **Otto J. Helweg**, Member, ASCE, (Assoc. Prof., Civil Engrg., Univ. of California, Davis, Calif. 95616)

Journal of the Irrigation and Drainage Division, Vol. 108, No. IR4, December, 1982, pp. 250-264

Tanks, small reservoirs to catch rain during monsoon seasons, are traditional water harvesting methods in India. Though major investment is being planned for more tank construction, no one has investigated the optimal tank design, size, and placement. This paper combines the tools of systems analysis with tank methodology to design optimal systems thereby saving considerable construction costs and improving water use efficiency. Problems of semi-arid tropics are examined. Traditional tank construction is also covered. The optimization problems in designing a tank system and covered in this paper are volume, location, and tank dimensions.

0520 Defining Cost-Effective Salinity Control Programs

Robert G. Evans, Member, ASCE, (Asst. Agricultural Engr., Washington State Univ., Irrigation Agr. Research and Extension Center, Prosser, Wash. 99350), **Wynn R. Walker**, Member, ASCE, (Assoc. Prof., Agricultural and Irrigation Engrg. Dept., Utah State Univ., Logan, Utah 84322) and **Gaylord V. Skogerboe**, Member, ASCE, (Prof., Dept. of Agricultural and Chemical Engrg., Colorado State Univ., Ft. Collins, Colo. 80523)

Journal of the Irrigation and Drainage Division, Vol. 108, No. IR4, December, 1982, pp. 265-272

Marginal cost analysis based on current damage estimates indicate that the optimal salinity control program involving PL 93-320 projects in the Upper Colorado River Basin would cost about $30 million annually and prevent about 1.2 million metric tons of salt per year from entering the Colorado River System. This optimal program is based on equating the annual costs of salinity control in the Upper Basin with annual salinity related damages in the Lower Basin. The results show that maintenance of the 1972 salinity concentration levels at Imperial Dam, dictated by PL 93-320, cannot be achieved cost-effectively with PL 93-320 salinity control projects. Although it is perhaps politically unfeasible, salinity concentration goals at Imperial Dam could be allowed to rise over the 1972 figures by as much as 180 mg/L in order to make the benefits of salinity control in the basin more in line with the anticipated costs.

0521 Fatigue Reliability: Introduction

Committee on Fatigue and Fracture Reliability of the Committee on Structural Safety and Reliability of the Structural Division

Journal of the Structural Division, Vol. 108, No. ST1, January, 1982, pp. 3-23

A four-part state of the art summary on fatigue reliability is presented by the ASCE Committee on Fatigue and Fracture Reliability. This first part reviews fatigue models commonly used by designers, e.g., the characteristic S-N curve for high cycle fatigue and the general strain life relationship for low cycle fatigue as well as the fracture mechanics approach. Statistical models (i.e., the Weibull and lognormal) used to describe the distribution of cycles to failure fatigue data, are summarized. An example of analysis of S-N data is presented.

0522 Fatigue Reliability: Quality Assurance and Maintainability

Committee on Fatigue and Fracture Reliability of the Committee on Structural Safety and Reliability of the Structural Division

Journal of the Structural Division, Vol. 108, No. ST1, January, 1982, pp. 25-46

Periodic reevaluation of a structure by inspection and proof loading with repair when needed is an alternative to complete dependence on initial quality for obtaining structural reliability. The theory of reliability of fatigue sensitive structures in service is developed. Changes in the probability density functions for strength and fatigue crack length under programs of proof loading or inspection are illustrated. Proof loading truncates the lower tail of the strength distribution. Inspection shifts the distribution upward. The state of the art of inspection for cracks is presented in a review of nondestructive methods and a detailed discussion of visual inspection. Maintenance principles are discussed and several successful repair procedures noted. The economic analysis of inspection and repair programs is outlined.

0523 Fatigue Reliability: Variable Amplitude Loading

Committee on Fatigue and Fracture Reliability of the Committee on Structural Safety and Reliability of the Structural Division

Journal of the Structural Division, Vol. 108, No. ST1, January, 1982, pp. 47-69

Models which designers use to predict fatigue under variable amplitude and random loading are summarized. Basic terminology associated with random process theory is presented. Models for computing fatigue damage associated with various statistical distributrions of stress ranges are reviewed. The Palmgren-Miner rule is described and statistical summaries which provide a description of the performance of Miner's rule is presented. A closed form expression for fatigue damage under wide band stress spectra is given. Finally, methods of predicting fatigue crack growth under variable amplitude stresses are reviewed.

0524 Fatigue Reliability: Development of Criteria for Design

Committee on Fatigue and Fracture Reliability of the Committee on Structural Safety and Reliability of the Structural Division

Journal of the Structural Division, Vol. 108, No. ST1, January, 1982, pp. 71-88

This is the fourth in a series of four articles prepared by the ASCE Committee on Fatigue and Fracture Reliability. This paper summarizes how criteria for assuring integrity against fatigue and fracture may be developed using principles of engineering reliability analysis as a basis. The probabilistic treatment allows specification-writing groups to consider explicitly uncertainties that affect fatigue performance and fracture susceptibility when setting practical criteria for design. However, this treatment need not lead to more complex provisions at the design level. Several illustrations of these concepts are given.

0525 Structural Response of Skew RC Box Girder Bridge

Alexander C. Scordelis, Fellow, ASCE, (Prof. of Civ. Engrg., Univ. of California, Berkeley, Calif), **S. Tanvir Wasti**, (Prof. of Civ. Engrg., Middle East Technical Univ., Ankara, Turkey) and **Frieder Seible**, (Research Asst., Univ. of California, Berkeley, Calif.)

Journal of the Structural Division, Vol. 108, No. ST1, January, 1982, pp. 89-104

Analytical and experimental results obtained in a study of a large-scale 45° skew two span four cell reinforced concrete box girder bridge model are presented. The model, which was a 1:2.82 scale replica of a typical California highway system prototype, had overall dimensions of 72 ft (21 m) long by 12 ft (3.7 m) wide, with two spans and a skew center bent diaphragm. Results

are given for reactions, deflections, strains and moments due to working stress point loads applied before and after overload stress levels in the bridge. The adequacy of analytical methods for the assessment of the structural response is discussed.
Errata: ST July '82, pp. 1695.

0527 Ultimate Strength of Skew RC Box Girder Bridge

Alexander C. Scordelis, Fellow, ASCE, (Prof. of Civ. Engrg., Univ. of California, Berkeley, Calif.), **Jack G. Boukamp**, Member, ASCE, (Prof. of Civ. Engrg., Univ. of California, Berkeley, Calif.), **S. Tanvir Wasti**, (Prof. of Civ. Engrg., Middle East Tech. Univ. Ankara, Turkey) and **Frieder Seible**, (Research Asst., Univ. of California, Berkeley, Calif.)

Journal of the Structural Division, Vol. 108, No. ST1, January, 1982, pp. 105-121

Analytical and experimental results obtained in a study of a large-scale 45° skew two span four cell reinforced concrete box girder bridge model are presented. The model, which was a 1:2.82 scale replica of a typical California highway system prototype, had overall dimensions of 72 ft (21 m) long by 12 ft (3.7 m) wide, with two spans and a skew center bent diaphragm. Experimental and theoretical results are considered for reactions, steel and concrete strains, deflections, and moments due to conditioning overloads producing stress values as high as 2.5 times the nominal design stress. The loading to failure is described in detail. The ultimate strength behavior is examined on the basis of postulated collapse mechanisms and the bridge is observed to possess an excellent overload capacity.
Errata: ST July '82, pp. 1694.

0529 Fatigue of Two Year Weathered A588 Stiffeners and Attachments

Ian M. Friedland, Assoc. Member, ASCE, (Engr., Stone and Webster Engrg. Co., Cherry Hill, N.J.), **Pedro Albrecht**, Member, ASCE, (Assoc. Prof., Dept. of Civ. Engrg., Univ. of Maryland, College Park, Md.) and **George R. Irwin**, Member, ASCE, (Visiting Prof., Dept. of Mech. Engrg., Univ. of Maryland, College Park, Md.)

Journal of the Structural Division, Vol. 108, No. ST1, January, 1982, pp. 125-144

An experimental program examined the effect of A588 steel specimens with transverse stiffeners and attachments. It was found that two years of continuous weathering, prior to stress cycling to failure, reduced the fatigue life of the specimens with stiffeners and attachments by 19 and 6 percent, respectively, as compared to the non-weathered control specimens. Two years of alternating between 6-month weathering and one-quarter-life stress cycling had no significant effect on the fatigue life. The present data, in conjunction with earlier results, suggest a larger weathering-induced loss in fatigue life the more fatigue strength the specimen had in a non-weathered state. Since weathering prior to cycling affects only the crack initiation phase, whereas unpainted A588 bridges are continuously exposed to the environment, it is prudent to regard the fatigue strength of the weathered specimens as an upper bound on the fatigue strength such details would exhibit on actual structures.

0530 Tests of Composite Beams with Web Openings

William C. Clawson, Assoc. Member, ASCE, (Research Engr., Exxon Production Research Co., Houston, Tex.) and **David Darwin**, Member, ASCE, (Assoc. Prof. of Civ. Engrg., Univ. of Kansas, Lawrence, Kans.)

Journal of the Structural Division, Vol. 108, No. ST1, January, 1982, pp. 145-162

An experimental investigation to study composite beams with rectangular web openings is presented. Six composite beams and one steel beam were tested. W14 × 34, W18 × 45 and W18 × 46 sections were used. Opening sizes were fixed, with depths equal to 60 percent of the steel beam depth and lengths equal to twice the opening depth. Concrete slab dimensions were held constant. Opening locations were varied to investigate moment-shear ratios ranging from 3 to 33 feet (0.9 to 10 m). Web openings reduce the strength of composite beams. Beams with high moment-shear ratios fail by general yielding in the steel below the neutral axis and crushing in the concrete. Beams with medium to low moment-shear ratios fail by the formation of plastic hinges in the steel below the opening, accompanied by a diagonal tension failure in the concrete slab. The concrete in composite beams contributes significantly not only to the flexural strength, but also to the shear strength of the beams at web openings.

0531 Bridge Rehabilitation: An Alternate Approach

John M. Yadlosky, Assoc. Member, ASCE, (Engr. I/II, Richardson, Gordon, and Assoc., Inc., Pittsburgh, Pa.), **Robert J. Brungraber**, Member, ASCE, (Presidential Prof., Dept. of Civ. Engrg., Bucknell Univ., Lewisburg, Pa.) and **Jai B. Kim**, Member, ASCE, (Chmn., Dept. of Civ. Engrg. Buckness Univ., Lewisburg, Pa.)

Journal of the Structural Division, Vol. 108, No. ST1, January, 1982, pp. 163-176

The presentation of rehabilitation plans for bridge structures requires engineering analyses of various strengthening schemes. The use of scale models as an alternate approach to conceiving and evaluating these rehabilitation schemes is examined. These schemes would be applied to the bridge in a manner that would affect the entire structure, not simply localize repairs. Schemes conceived make use of steel rods, or cables, or both. Data is presented for the experimental, and theoretical, performance of a one-seventh scale unreinforced model, in addition to the experimental results of two preliminary reinforcing schemes. Also presented are questions that arise from rehabilitation work of this type that only model analysis, supported by theoretical analysis, can answer. Emphasis is placed upon the design of the model since accurate modeling is required for accurate results.

0532 External Pressure and Sectional Behavior of Fabricated Tubes

Shouji Toma, (Asst. Mgr., Office of Welding Research Lab., Kawasaki Heavy Industries Ltd., Chiba-Prefecture, Japan), **Wai F. Chen**, Member, ASCE, (Prof. and Head of Struct. Engrg., School of Civ. Engrg., West LaFayette, Ind.) and **Lyle D. Finn**, (Sr. Research Specialist, Offshore Structures Div., Exxon Production Research Co., Houston, Tex.)

Journal of the Structural Division, Vol. 108, No. ST1, January, 1982, pp. 177-194

The effect of external hydrostatic pressure on the cross sectional behavior of fabricated tubular columns is studied analytically using tangent stiffness approach. The moment-curvature behavior of a short column is of prime importance in the analysis of any long beam-column. The results will be applied to study the effect of external pressure on the strength and behavior of long beam-columns. The fabricated tubular column contains imperfections such as out-of-roundness and longitudinal and transverse residual stresses. These imperfections combined with the effect of external pressure are studied in the present analysis. Tresca yield criterion is used for elements in biaxial stress state, which is induced by the interaction of axial stress and longitudinal and transverse residual stresses and magnified by external pressure. A computer program has been developed to obtain numerical results from which practical design formulas are derived for cross-sectional capacity of fabricated tubular columns subjected to axial force, bending moment and external hydrostatic pressure.

0533 In-Plane Stability of Parabolic Arches

Howard B. Harrison, Member, ASCE, (Assoc. Prof. of Civ. Engrg., Univ. of Sidney, Australia, 2006)

Journal of the Structural Division, Vol. 108, No. ST1, January, 1982, pp. 195-205

Studies are reported of the response to uniformly distributed loading of parabolic arches of rise equal to a fourth of the span. Base conditions are both fixed and pinned. The distributed loading varies between covering less than half the span to covering the complete span. The numerical studies use a discrete element, large deformation method that enables the prediction of the complete load-deformation history of an elastic arch up to full inversion if necessary. Account is taken of axial as well as flexural strains and, under full span loading published data are confirmed for the load intensities at which equilibrium bifurcation occurs. The significant result is that the intensity of load that can be supported by a fully elastic arch is significantly less than the bifurcation value when approximately three-fourths of the span is loaded. The reduction in load intensity is appreciable for both fixed and pinned arches.

0534 Cracks in Concrete Subject to Shear

Hans W. Reinhardt, (Prof. of Concrete Structures, Stevin Lab., Delft Univ. of Technology, Delft, The Netherlands) and **Joost C. Walraven**, (Sr. Research Asst., Stevin Lab., Delft Univ. of Technology, Delft, The Netherlands)

Journal of the Structural Division, Vol. 108, No. ST1, January, 1982, pp. 207-224

Shear tests on push-off specimens are carried out in order to study the strength and deformation behaviour of normal and lightweight concrete. Two types of specimen are used, one with embedded reinforcing bars and another with external restraint bars. In both series the experimental results show the dilatancy of cracked concrete under shear loading. The variables studied are the concrete quality, concrete composition, the reinforcement ratio, the bar diameter, the initial crack width and the influence of repeated loading. The concrete quality and the reinforcement ratio have the most pronounced influence on the behaviour. For the case of the external restraining force, diagrams are established to show the relation between shear stress, normal stress, crack width and shear displacement.

0535 Alongwind Response Estimation: Closed Form Solution

Giovanni Solari, (Research Asst., Instituto di Scienza delle Costruzioni, Univ. of Genova, Genova, Italy)

Journal of the Structural Division, Vol. 108, No. ST1, January, 1982, pp. 225-244

The problem of dynamic alongwind response of structures to forces induced by atmospheric turbulence is treated in this paper. Starting from the classical formulation, the study analyses the behaviour of two structural standard models, called "point-like" and "three-dimensional", respectively. The treatment of the problem presented in the paper leads to a closed form expression of the alongwind response. The remarkable simplicity and the very high precision of the proposed method is pointed out in general terms and illustrated by two examples. In conclusion, some prospects for possible future applications referred to this solution are outlined and briefly discussed.

0536 Uniform Element Modeling of Tapered Frame Members

Eryk Kosko, (Engrg. Consultant, 2106 Woodcrest Road, Ottawa, K1H 6K8, Canada)

Journal of the Structural Division, Vol. 108, No. ST1, January, 1982, pp. 245-264

Prismatic members of structural frameworks are conventionally analysed by using a simplified model supplied by the engineering beam theory. The paper considers the possibility of extending that model to members of varying cross sections; such extension is warranted provided the variation is smooth and not too large. The flexibility parameters which represent the elastic behavior of such members are calculated by means of simple quadratures. Explicit expressions and charts are presented for an important class of tapered beams. As an alternative to the usual representation of a member in terms of a stiffness of flexibility matrix, an equivalent model, consisting of two or three uniform segments, is suggested. Simple formulas are derived for that purpose and their application is numerically demonstrated.

0537 Performance of Cable Trusses under Static Loads

Prem Krishna, Member, ASCE, (Prof. of Civ. Engrg., Univ. of Roorkee, Roorkee, India), **Vijaya K. Gupta**, (Lect. of Civ. Engrg., Univ. of Roorkee, Roorkee, India), **Ashok K. Ahuja**, (Research Assoc., Civ. Engrg. Dept., Univ. of Roorkee, Roorkee, India) and **Ashok K. Mittal**, (Executive Engr., Central Public Works Dept., New Dehli, India)

Journal of the Structural Division, Vol. 108, No. ST1, January, 1982, pp. 265-282

Tensioned cable structures are becoming increasingly popular for covering large column free areas. Among the various forms of cable roof systems used, the truss form is the most popular because of its simplicity and versatility. These structures exhibit a non-linear structural response and thus pose a special problem of analysis. A study of the comparative behaviour of different types of cable truss configurations is presented. The influence of design parameters, such as sag to span ratio, cable sizes and prestressing forces, upon the response of the structure has been investigated for various truss forms with unyielding supports subjected to both symmetrical and unsymmetrical loading conditions.

0538 Lapped Tensile Reinforcement Splices

Ralejs Tepfers, (Assoc. Prof. in Building Tech. and Building Materials, Chalmers Univ. of Tech., Gothenburg, Sweden)

Journal of the Structural Division, Vol. 108, No. ST1, January, 1982, pp. 283-301

The bond action between concrete and deformed bars in lap splices is analysed in its different working stages. The principal tensile stress generated by bond stress cracks the concrete after which bond gives rise to splitting forces. These forces cause longitudinal cracking of the cover. A method for determination of this cracking load is proposed. The stress distribution along the lap length is analysed with modulus of displacement theory for different bond stages. The final bond failure by splitting of the cover according to minimum stressed failure pattern is analysed. The effect of confining reinforcement is discussed. Methods for calculation of bond resistance in lap splices are proposed and compared with tests.

Errata: ST Dec. '82, pp. 2861.

0540 Beam Web Shear Connections: An Experimental Study

Joseph A. Yura, Member, ASCE, (Prof., Dept. of Civ. Engrg., The Univ. of Texas at Austin, Austin, Tex.), **Peter C. Birkemoe**, Member, ASCE, (Prof., Dept. of Civ. Engrg., Univ. of Toronto, Ontario, Canada) and **James M. Ricles**, Assoc. Member, ASCE, (Grad Student Research Asst., The Univ. of Texas at Austin, Austin, Tex)

Journal of the Structural Division, Vol. 108, No. ST2, February, 1982, pp. 311-325

Bolted shear web connections were tested to evaluate the 1978 AISC Specification provisions concerning edge distance, end distance, and bearing stresses. Tests wee performed on beams with both standard holes and also slotted holes; single row and double rows of bolts were considered. Both coped and uncoped beams were used in the test. The results of each test are compared with the allowable service loads. For each test a plot of the connection reaction versus the deformation is presented along with a description of the behavior. Results of the tests indicate that design methods of connections which develop the full depth of the web are conservative; connections with two rows of bolts which were concentrated in the upper half of the web gave lower factors of safety than desired. Also, connections with slotted holes had less capacity than those with standard holes when edge distance was critical.

0541 Finite Element Method for Interface Problems

Santosh K. Arya, Assoc. Member, ASCE, (Principal Development Engr., Dept. of Applied Mechanics and Engrg. Sci., Univ. of California at San Diego, LaJolla, Calif. 92093) and **Gilbert A. Hegemier**, (Prof., Dept. of Applied Mechanics and Engrg. Sci., Univ. of California at San Diego, La Jolla, Calif. 92093)

Journal of the Structural Division, Vol. 108, No. ST2, February, 1982, pp. 327-342

A method for incorporating interface discontinuities in a finite element analysis of discontinuous media is presented. For computational purposes the procedure is implemented into the nonlinear structural analysis program NONSAP. Several example problems concerning the modeling of concrete masonry mortar joints are presented, and demonstrate the accuracy and efficiency of the algorithm.

0542 Beam-to-Column Connection Research in Japan

Ben Kato, (Prof. of Structural Engrg., Dept. of Architecture, Univ. of Tokyo, Tokyo, Japan)

Journal of the Structural Division, Vol. 108, No. ST2, February, 1982, pp. 343-360

Present status of the research on the beam-to-column connection in Japan is reviewed. The design philosophy of beam-to-column moment connections of steel building frames under the unique situation that the earthquake loading dominates the structural design is discussed. The importance of the contribution of the excellent shear deformation capacity of joint-panels to the overall ductility of the frames which should be developed against the extreme earthquake is emphasized. Some design criteria for the beam-to-column moment connections suitable for seismic design are suggested.

0543 Columns with Semirigid Joints

Stephen W. Jones, (Research Student, Dept. of Civ. and Struct. Engrg., Univ. of Sheffield, Sheffield, S1 3JD, England), **Patrick A. Kirby**, (Lect., Dept. of Civ. and Struct. Engrg., Univ. of Sheffield, Sheffield, S1 3JD, England) and **David A. Nethercot**, (Sr. Lect., Dept. of Civ. and Struct. Engrg., Univ. of Sheffield, Sheffield, S1 3JD, England)

Journal of the Structural Division, Vol. 108, No. ST2, February, 1982, pp. 361-372

The influence of realistic semi-rigid end restraint on the strength and behaviour of real steel columns is investigated. An analytical procedure is developed which includes the effects of initial out-of-straightness, spread of yield and initial residual stresses. The procedure also permits the inclusion of the nonlinear restraining effect of the column's end connections as determined by tests on actual connections. Comparisons of initial results with available experimental data for end-restrained columns are made to verify the validity of the procedure. Agreement is shown to be acceptable. Load-deflection curves and column curves are produced for a series of column assemblies including both initially stress-free and residual stress conditions. Effective length factors calculated from these results indicate possible column design economies due to the actual end restraint present. The influence of end restraint on the distribution of internal stresses and hence the initiation of column yield is also discussed.

0544 Seismic Behavior of Moment Connections and Joints

Helmut Krawinkler, Member, ASCE, (Assoc. Prof. of Civ. Engrg., Stanford Univ., Stanford, Calif. 94305) and **Egor P. Popov**, Fellow, ASCE, (Prof. of Civ. Engrg., Univ. of California, Berkeley, Calif.)

Journal of the Structural Division, Vol. 108, No. ST2, February, 1982, pp. 373-391

The behavior of moment connections and beam-column joints in moment resisting steel frames subjected to severe earthquakes is studied. The cyclic inelastic deformation capacities of different types of connections are discussed and design recommendations are presented that are intended to assure the development of plastic hinges in beams at the column faces. A conceptual approach to the prediction of the low cycle fatigue life of welded connections under random loading is outlined, utilizing concepts of elastic-plastic fracture mechanics. The shear behavior of beam-column joints subjected to large cyclic beam moment reversals is summarized.

Based on experimental evidence and a simplified mathematical model for shear strength, a method for the shear design of joints is proposed that should permit a reduction in the demand for shear stiffeners in joints.

0545 Beam-To-Column Web Connection Details

G. P. Rentschler, Assoc. Member, ASCE, (Engr., Gilbert Assoc., Reading, Pa.), **W. F. Chen**, Member, ASCE, (Prof. and Head of Structural Engrg., School of Civ. Engrg. Purdue Univ., West Lafayette, Ind.) and **G. C. Driscoll**, Member, ASCE, (Prof. and Assoc. Head, Civ. Engrg. and Fritz Engrg. Lab., Lehigh Univ., Bethleham, Pa.)

Journal of the Structural Division, Vol. 108, No. ST2, February, 1982, pp. 393-409

For a beam-to-column web connection where the beam is attached to the column perpendicular to the plane of the column web, the theoretical maximum strength of this beam and column assemblage is reached when plastic hinges are formed in the column or in the beam. Other factors that limit this maximum strength include the development of yield line mechanism in column web, fracture or shear punch of column web, and shear yielding of beam flange plate. These could preclude obtaining the connection load level required to produce plastic hinges. The results of eight simulated tests on web connection details are presented. In the simulations, pairs of steel plates are welded to column sections using four different attachment details. These plates represent the tension and compression flanges of a beam attached to the column. The pair of flange plates are loaded by tension and compression forces to simulate the bending moment of a beam acting upon the column. Observations are made of stresses and deformations as well as phenomena that prevented the connections from obtaining this theoretical maximum load level.

0546 Ultimate Strength of Gapped Joints in RHS Trusses

Jeffrey A. Packer, (Asst. Prof., Dept. of Civ. Engrg., Univ. of Toronto, Toronto, Canada), **Gwynne Davies**, (Sr. Lect., Dept. of Civ. Engrg., Univ. of Nottingham, England) and **Martin G. Coutie**, (Lect., Dept. of Civ. Engrg., Univ. of Nottingham, England)

Journal of the Structural Division, Vol. 108, No. ST2, February, 1982, pp. 411-421

A theoretical and empirical analysis is presented for the ultimate static strength of gapped joints in welded Rectangular Hollow Section trusses having one compression bracing member and one tension bracing member. Ultimate joint strength is determined from a set of observed failure modes which are studied with the yield line method, adapted for large joint deformations, as the main analytical tool. The results of 84 tests on isolated and truss joints from three different countries have been used to successfully verify the analyses.

0547 Welded and Bolted Beam-to-Column Connections

Jelle Witteveen, (Prof. of Struct. Mechanics, Dept. of Civ. Engrg., Delft Univ. of Tech., The Netherlands), **Jan W. B. Stark**, (Head, Dept. of Steel Structures of the Inst. TNO for Building Structures, Delft, The Netherlands), **Frans S.K. Bijlaard**, (Research Officer, Dept. of Steel Structures of the Inst. TNO for Building Materials and Building Structures, Delft. The Netherlands) and **Piet Zoetemeijer**, (Research Officer, Stecin Lab., Dept. of Civ. Engrg., Delft Univ. of Tech. The Netherlands)

Journal of the Structural Division, Vol. 108, No. ST2, February, 1982, pp. 433-455

For many years research on welded and bolted beam-to-column connections has been carried out in The Netherlands, with particular emphasis on those connections which cannot transmit the full plastic moment. This concept allows for a wide range of moment distributions in non-sway frames, thus giving the possibility of optimizing the cost of the construction in balancing the material cost of the members against the labour cost of the connections. It will be evident that for unstiffened connections special precautions should be taken in order to avoid lack of rotation capacity and stiffness of the connections. The appropriate design rules for unstiffened welded connections are bolted connections in non-sway frames are discussed.

0548 Modified Analysis of Reinforced Concrete Beams

Theodor Krauthammer, Assoc. Member, ASCE, (Sr. Engr., Applied Research Assocs., Inc., 2601 Wyoming Blvd., N.E., Suite H-1, Albuquerque, N.M. 87112) and **William J. Hall**, Fellow, ASCE, (Prof. of Civ. Engrg., Univ. of Illinois at Urbana-Champaign, Urbana, Ill.)

Journal of the Structural Division, Vol. 108, No. ST2, February, 1982, pp. 457-475

A method for analyzing reinforced concrete beams and beam-columns is developed, and the analytical results are compared to experimental data obtained by other investigators. The present method employs the stress-strain relationships for reinforced concrete and steel bars, and considers the combined effects of flexure, shear, and axial forces. The ultimate flexural moment and rotation capacities of reinforced concrete beams is associated with the amount of shear reinforcement provided. As a result, a relationship is proposed between the ultimate flexural moment capacity, the amount of shear reinforcement, and the inclination of cracks.

0549 Restrained Crooked Aluminum Columns

Jacques Chapuis, Assoc. Member, ASCE, (Consulting Engr., Lausanne, Switzerland) and **Theodore V. Galambos**, Fellow, ASCE, (Prof., Civ. and Mineral Engrg. Dept., Univ. of Minnesota, Minneapolis, Minn.)

Journal of the Structural Division, Vol. 108, No. ST3, March, 1982, pp. 511-524

The strength of aluminum pinned crooked columns has been derived as the basis of a design column curve. The effective length factor concept has then been presented and a value for crooked columns introduced ($K_{crooked}$). This factor can only be numerically determined and has no direct theoretical relation with the classical effective length factor (K_{el}) derived from elastic buckling analyses. Many examples have, however, shown that $K_{crooked}$ is less than K_{el} and that it is a conservative to base a design on K_{el} and a column curve derived for pinned crooked columns. It has also been shown that the amount of restraint necessary to compensate the effect of initial imperfection is strongly dependent on the slenderness ratio of the member.

0550 Local and Overall Buckling of Welded Box Columns

Tsutomu Usami, Member, ASCE, (Assoc. Prof. of Civ. Engrg., Nagoya Univ., Nagoya, Japan) and **Yushsi Fukumoto**, Member, ASCE, (Prof. of Civ. Engrg., Nagoya Univ., Nagoya, Japan)

Journal of the Structural Division, Vol. 108, No. ST3, March, 1982, pp. 525-542

An experimental study is presented on the local and overall interaction buckling strength of welded built-up box columns made of high strength steel. A total of twenty-seven box columns with large width-thickness ratios were tested to failure, of which twenty-four columns were centrally loaded and the rest eccentrically loaded. Based on the test results, an empirical design formula is presented to predict the interaction buckling strength. Further, a discussion of an economical way of proportioning box sections is presented.

0551 Method for Analysis for Guyed Towers

Felix Rosenthal, (Head, Ocean Structures Section, Marine Tech. Div., Naval Research Lab., Wash., D.C.) and **Richard A. Skop**, (Head, Applied Mechanics Branch, Marine Tech. Div., Naval Research Lab., Wash., D.C.)

Journal of the Structural Division, Vol. 108, No. ST3, March, 1982, pp. 543-558

A method for static analysis of guyed towers is extended and applied to problems in three dimensions. The tower and guy cables may be loaded by arbitrary distributed or concentrated forces, or both, and there are no apriori assumptions about cable shape (such as the usual catenary assumption). The effect of axial forces as well as shear and twist deformations in the tower are taken into account. The base of the tower may be in any support condition between clamped and hinged including either extreme in each of three rotational degrees of freedom. Cables are handled by the Method of Imaginary Reactions. For the tower, highly accurate integration algorithms for the three dimensional beam column equations are presented. Results are applied to a 1500 foot tower bearing a 15 million pound platform and other loads, and supported by 20 cables attached in four clusters at the tope of the tower.

0552 Generalized Stiffness Matrix for Thin Walled Beams

Anil B. Chaudhary, (Research Asst., Civ. Engrg. Dept., Massachusetts Inst. of Tech., Cambridge, Mass.)

Journal of the Structural Division, Vol. 108, No. ST3, March, 1982, pp. 559-577

Stiffness matrix for a structural member is obtained using the differential equations derived by V.Z. Vlasov. Presence of additional off-diagonal elements over the conventional tridiagonal form is observed. A number example is shown in order to compare the relative magnitudes of elements in the stiffness matrix and some of the new off-diagonal elements are found to be of considerable importance. This stiffness matrix conclusively demonstrates the inter-relationship between the three rotational degrees of freedom. This coupling is found to reduce the value of the critical axial force below all three individual buckling loads, viz: lateral buckling about major and minor axes and torsional buckling. All these effects are attributed to an internal force called as "bimoment'.

0553 Prestressed Steel Grid Erected with Bent Forms

Oren Vilnay, (Lect., Dept. of Civ. and Struct. Engrg., Univ. College, Cardiff, Wales), **Avraham Ben-Arroyo**, (Sr. Lect., Faculty of Civ. Engrg., Technion-Israel Inst. of Tech., Haifa, Israel) and **Eliahn Racah**, (Research Officer, School of Architecture and Building Engrg., Univ. of Bath, Bath, England)

Journal of the Structural Division, Vol. 108, No. ST3, March, 1982, pp. 579-590

A new method for constructing one-layer steel grids erected by prestressing with the aid of recoverable bending forms is presented, followed by an approximate structural analysis. The grid elements are joined together by cables passing through them. By stressing other cables the flat grid erects itself to a vault, the shape of which is determined by the recoverable bending forms. The preliminary analysis covers the erection stages and dead, snow and wind loads during service. Using the results of a design example it is predicted that this new method, which dispenses with special joints on the one hand, and with high scaffolding and erection cranes on the other, will not require excessive amounts of steel.

0554 Post-Buckling of Plates in Compression and Bending

Tsutomu Usami, Member, ASCE, (Assoc. Prof. of Civ. Engrg., Nagoya Univ., Nagoya, Japan)

Journal of the Structural Division, Vol. 108, No. ST3, March, 1982, pp. 591-609

The elastic post-buckling behavior of rectangular plates in combined compression and bending is presented. The theoretical investigation begins with the Marguerre's compatibility equation and proceeds to obtain a solution by the energy method. The computer results are first compared with some available test results. Then the theoretical results are utilized to derive effective width formulas in combined compression and bending. Finally a family of post-buckling strength curves of box sections are presented that are calculated from the derived effective width formulas.

0555 Finite Beams on Elastic Foundation with Restraints

Bing-Yuan Ting, Member, ASCE, (Sr. Consultant, The Marley Cooling Tower Co., 5800 Foxridge Dr., Mission, Kans. 66201)

Journal of the Structural Division, Vol. 108, No. ST3, March, 1982, pp. 611-621

A solution is presented of the differential equation of finite beams with elastic restraints, resting on a Winkler foundation in terms of the initial conditions at one end of the beam. This solution can be used to simulate beams on elastic foundation with different boundary conditions. Numerical examples are given to demonstrate the efficiency of the solution. Results for beams with simulated boundary conditions agree with those for exact boundary conditions.

0556 Strength of Composite Beams at Web Openings

William C. Clawson, Assoc. Member, ASCE, (Research Engr., Exxon Production Research Co., Houston, Tex.) and **David Darwin**, Member, ASCE, (Assoc. Prof. of Civ. Engrg., Univ. of Kansas, Lawrence, Kans. 66045)

Journal of the Structural Division, Vol. 108, No. ST3, March, 1982, pp. 623-641

A model for the strength of steel-concrete composite beams with unreinforced, rectangular web openings is presented. The model includes the contribution of the concrete slab to shear strength, as well as flexural strength. Steel is represented as an elasto-plastic material with no strain hardening. Openings may be concentric or eccentric with respect to the steel section. Comparisons are made with test results. The model provides realistic, and generally conservative, predictions of the strength for the test beams. The contribution of the concrete slab to shear strength at an opening must be included to obtain a realistic model.

0557 Plastic Rotation of Composite Beams

Peter Ansourian, (Sr. lect., University of Sydney, New South Wales, Australia)

Journal of the Structural Division, Vol. 108, No. ST3, March, 1982, pp. 643-659

The sagging rotation capacity of composite beams consisting of a concrete slab attached to a steel beam by a shear connection is studied. Tests are reported on four full-scale beams in the range of ductility parameter χ from 0.65 to 3.0. This parameter is an index of the degree of strain-hardening developed in the steel beam at collapse. Expressions are proposed for the minimum inelastic rotation and deflection available at collapse. Examples are given of the application of these expressions to design problems with continuous composite beams. It is shown that, with a minimum value of χ equal to 1.4, sufficient sagging rotation should be available to develop the conventional plastic design collapse load under the worst combinations of spans and loading.

0558 Small End Restraint Effects on Strength of H-columns

Hironori Sugimoto, (Grad. Asst., School of Civ. Engrg., Purdue Univ., West Lafayette, Ind.) and **Wai F. Chen**, Member, ASCE, (Prof. and Head, Struct. Engrg. School of Civ. Engrg., Purdue Univ., West Lafayette, Ind.)

Journal of the Structural Division, Vol. 108, No. ST3, March, 1982, pp. 661-681

A parametric study of the strength and behaviour of an axially loaded wide-flange steel column is investigated from the viewpoint of a beam column analysis using the approximate deflection method and the bi-linear fitting for test data of moment rotation relationships of beam-to-column connections and taking into consideration the effects of residual stresses, geometric imperfections and small end restraints. A computer program is developed to obtain the numerical results, from which effective length decreasing factor and maximum load increasing factor are proposed. The validity of the computer model is demonstrated by comparison with the results of a test on an actual column.

0559 Interaction of Cable Roofs with Support Structure

Vijaya K. Gupta, (Lect. of Civ. Engrg., Univ. of Roorkee, Roorkee, U.P., India), **Prem Krishna**, Member, ASCE, (Prof. of Civ. Engrg., Univ. of Roorkee, Roorkee, U.P., India) and **Ashok K. Ahuja**, (Research Assoc., Civ. Engrg. Dept., Univ. of Roorkee, Roorkee, U.P., India)

Journal of the Structural Division, Vol. 108, No. ST3, March, 1982, pp. 685-691

Tensioned structures are becoming increasingly popular for covering large column free areas on account of efficient utilization of construction material, speedy construction and economy. Interaction of cable truss and network roofs with the supporting structure modelled as a spring is studied in this paper for a variation of design parameters. Both symmetrical and unsymmetrical loading conditions are considered.

0560 Reliability of Aluminum Beam-Columns

Jacques Chapuis, Assoc. Member, ASCE, (Consulting Engr., Lausanne, Switzerland) and **Theodore V. Galambos**, Fellow, ASCE, (Prof. of Civ. Engrg., Washington Univ., St. Louis, Mo.)

Journal of the Structural Division, Vol. 108, No. ST4, April, 1982, pp. 709-727

A review of reliability computation methods is presented. The strength of aluminum beam-columns is described, taking into account material and cross-sectional properties, initial imperfection and load eccentricity, and comparisons are made to the interaction equation prediction. The reliability of the current aluminum code design criteria for beam-columns is analyzed by different methods. A load and resistance factor design criterion is proposed, and appropriate resistance factors are derived which are a function of the design parameters in order to offer a uniform reliability.

0561 Complete Triaxial Stress-Strain Curves for Concrete

Shuaib H. Ahmad, (Asst. Prof., Dept. of Civ. Engrg., North Carolina State Univ., Raleigh, N.C.) and **Surendra P. Shah**, Member, ASCE, (Prof. of Civ. Engrg., Dept. of Civ. Engrg., Northwestern Univ., Evanston, Ill. 60201)

Journal of the Structural Division, Vol. 108, No. ST4, April, 1982, pp. 728-742

A method to predict the ascending and the descending parts of stress-strain curve of concrete subjected to triaxial compressive stresses is presented. The proposed analytical stress-strain relationship depends on the values of the three principal stresses and strains at the ultimate compressive strength. A criterion to predict the triaxial compressive strength is proposed, which is sensitive to all three stress invariants. To enable predictions of the post-peak stress-strain relation, experiments were conducted on concrete confined by steel tube. Concretes with different compressve strengths were confined by steel tubes of varying thicknesses. The predicted stress-strain curves compared favorably with the available experimental data on uniaxial, biaxial, and triaxial compression.

0562 Effect of Abandoned Roles on Capacity of Wedge Bolts

Edwin G. Burdette, Member, ASCE, (Prof. of Civ. Engrg., Univ. of Tennessee, Knoxville, Tenn.), **Subir Sen**, Member, ASCE, (Engrg. Specialist, Bechtel Power Corp., Gaithersburgh, Md.) and **Ergun Ismen**, (Formerly Engrg. Supervisor, Bechtel Power Corp., Gaithersburgh, Md.)

Journal of the Structural Division, Vol. 108, No. ST4, April, 1982, pp. 743-752

Seventy, 5/8-in. diameter wedge anchors were tested in direct pull-out to determine the effect of various configurations of unused holes and different treatments of these holes on pull-out capacity. The test results indicated that for unused holes as close as approximately 2.0 bolt diameters, there was a measurable reduction in pull-out strength. When the holes were as far away as 3.0 bolt diameters, there was no reduction in capacity. When holes as close as 1.5 bolt diameters were filled with drypack mortar, the anchors behaved as if there were no additional holes present. Thus, no reduction in pull-out load, even the extreme case of holes 1.5 bolt diameters away from an anchor, need be considered when the holes are filled with dry-pack mortar.

0563 Bayesian Approach to Prototype Testing

P. Hauge Madsen, (Research Assoc., Risoe National Lab., 4000 Roskilde, Denmark) and **Niels C. Lind**, (Prof. of Civ. Engrg., Univ. of Waterloo, Ontario, Canada, N2L 3G1)

Journal of the Structural Division, Vol. 108, No. ST4, April, 1982, pp. 753-770

The strength of series-produced structures is uncertain because of modelling error and material parameter uncertainty. By testing of prototypes, information on both uncertainties is obtained. Because of strength correlation in a series, sampling is considered non-random. The Bayesian Method is used to consider the problem of relating prior knowledge and test data to a prescribed characteristic value of the load carrying capacity, i.e., a prescribed fractile of the strength distribution, or a target value of a reliability index. Examples show application to prototype testing structural assemblies or parts. It is concluded that even a slight correlation may have a significant influence on the estimated strength.

0564 Experiments on Closing Reinforced Concrete Corners

Robert L. Yuan, (Prof., Dept. of Civ. Engrg., Univ. of Texas at Arlington Tex. 76019), **Gerald R. McLelland**, (Pres., McLelland Engrg. Co., Dallas, Tex.) and **Wai F. Chen**, Member, ASCE, (Prof. and Head of Struct. Engrg., School of Civ. Engrg., Purdue Univ., W. Lafayette, Ind. 47907)

Journal of the Structural Division, Vol. 108, No. ST4, April, 1982, pp. 771-779

Reinforced concrete corners have inherently more flexural strength across the corner than at other sections when subjected to negative bending moments. The modes of failure and cracking characteristics of negative moment corners, and the reliability of different conventional reinforcement arrangements at corners are studied. The results show that the cracked section moment capacity of the corners should be considered reliable. Channel shaped frame specimens, with and without initial flaws at the corners, are used in the study. Three types of reinforcement arrangements are investigated. Corner efficiency rating is obtained.

0565 Finite Segment Method for Biaxially Loaded RC Columns

Soliman I. Al-Noury, (Asst. Prof. of Civ. Engrg., King Abdulaziz Univ., Jeddah, Saudi Arabia) and **Wai F. Chen**, Fellow, ASCE, (Prof. and Head of Structural Engrg., School of Civ. Engrg., Purdue Univ., West Lafayette, Inc. 47907)

Journal of the Structural Division, Vol. 108, No. ST4, April, 1982, pp. 780-799

The finite segment method is applied here to solve biaxially loaded reinforced concrete columns. The column is treated as a space structure after segmentation. The sections are divided into finite elements of steel and concrete in order to calculate its tangent stiffnesses properties at different levels of strain. The segment stiffness relationship is computed by solving the governing differential equations about the principal axes of the cross section. Direct stiffness method is utilized to construct the whole column stiffness matrix (master stiffness matrix). Nonlinearities due to material plasticity and geometrical change are handled by an iterative procedure based on the modified tangent stiffness approach. Comparisons of failure load and deflection with biaxially loaded reinforced concrete column tests show an excellent correlation between theory and test data.

0566 Cooling Tower Shells Founded on Soil

Osama M. El-Shafee, (Formerly, Post-Doctoral Fellow, Dept. of Civ. Engrg., Washington Univ., St. Louis, Mo.) and **Phillip L. Gould**, Fellow, ASCE, (aH.D. Jolley Prof. and Chmn., Dept. of Civ. Engrg., Washington Univ., St. Louis, Mo.)

Journal of the Structural Division, Vol. 108, No. ST4, April, 1982, pp. 800-813

A dynamic axisymmetric finite element model suitable for shells of revolution founded on ring footings is presented. The model consists of high-precision rotational shell finite elements, representing the axisymmetric shell, supported on an equivalent boundary system, representing the soil medium. The substructure method is used to model the shell and soil components. In addition to the seismic analysis capability of the proposed model, it is also applicable to other dynamic loads such as wind. The dynamic behavior of a hyperboloidal cooling tower shell on discrete supports with a ring footing is studied. Dynamic properties are examined and stress analysis is carried out for a variety of soil conditions.

0567 Load Capacity of Concrete Bridge Decks

David B. Beal, Member, ASCE, (Assoc. Civ. Engr., Engrg. Research and Development Bureau, New York State Dept. of Transportation, Albany, N.Y.)

Journal of the Structural Division, Vol. 108, No. ST4, April, 1982, pp. 814-832

The behavior of two reduced-scale concrete bridge decks subjected to simulated wheel loads was evaluated in a series of tests. One slab was reinforced in accordance with AASHTO requirements; the other had three areas with varying amounts of isotropic reinforcement. Results show that with either reinforcement pattern, service load bending moments are from 40 to 65 percent of those predicted by flexural theory. Failures were by punching shear rather than flexure and occurred at loads at least six times larger than design.

0568 Application of Fracture Mechanics to Concrete Structures

Alberto Carpinteri, (Research Asst., Instituto di Scienza delle Costruzioni, Univ. of Bologna, 2 Viale Risorgimento, 40136 Bologna, Italy)

Journal of the Structural Division, Vol. 108, No. ST4, April, 1982, pp. 833-848

Material properties (heterogeneity and non-linearity) and disturbing phenomena (slow crack growth and microcracking) cause several research-workers to be pessimistic about the applicability of fracture mechanics to concrete structures. A critical review of the preceding works dealing with concrete fracture is presented. During the 1960's and 1970's it became more and more evident that notch sensitivity is not a material property but only a phenomenon dependent on the specimen and crack sizes, and that the degree of heterogeneity is only a matter of scale. A true fracture collapse can occur only with very large structures. On the other hand a real ultimate strength collapse can occur only with very small fracture specimens. Usually, with the sizes of the structures built by man and the lengths of the natural cracks, the two collapses are interacting.

0569 Numerical Bending Analysis of Bridges

Walter J. Austin, Fellow, ASCE, (Prof. of Civ. Engrg., Rice Univ., Houston Tex. 77001), **Timothy J. Ross**, Member, ASCE, (Struct. Engr., Air Force Weapons Lab., Albuquerque, N.M.), **Adel S. Tawfik**, (Struct. Engr., Offshore Structures, Brown and Root, Inc., Houston, Tex.) and **Robert D. Volz**, (Captain, Corps of Engrs., U.S. Army)

Journal of the Structural Division, Vol. 108, No. ST4, April, 1982, pp. 849-868

Related numerical procedures are described for the elastic analysis of arches by large deflection, second order and classical theories. These procedures essentially consist of simple and accurate repeated numerical integrations similar to the Newmark procedure for beams and beam-columns. Iterations are required for the large deflection and second order analyses. Axial strains and temperature effects are considered. The procedure gives high accuracy with few divisions, and computer solutions are inexpensive.

0570 Acrosswind Response of Buildings

Ashan Kareem, Member, ASCE, (Asst. Prof., Dept. of Civ. Engrg., Univ. of Houston, Houston, Tex.)

Journal of the Structural Division, Vol. 108, No. ST4, April, 1982, pp. 869-887

The acrosswind response of buildings is investigated. Expressions for the acrosswind loading are developed through the use of the spatio-temporal measurement of the fluctuating pressure field around a square cross-section building model. Methods of random vibration theory are used to estimate the rms and peak values of displacement, acceleration and jerk components of response. The results are compared with the response of an aeroelastically modeled building. The predicted and measured response agree within the limits of expected experimental errors. The addition of motion induced loads at higher reduced velocities produces still closer agreement between the predicted and measured response. Important features of the measured response characteristics are summarized. A step-by-step procedure and numerical example are presented to illustrate the methodology outlined for predicting the response to be estimated at the preliminary design stages. This would allow early assessment of the requirement to resist the mean deflection and to reduce the acceleration level at the top floors, and to ensure structural safety and occupancy comfort.

0571 Optimal Design Based on Approximate Scaling

Uri Kirsch, (Assoc. Prof., Dept. of Civ. Engrg., Israel Inst. of Tech., Haifa, Israel)

Journal of the Structural Division, Vol. 108, No. ST4, April, 1982, pp. 888-909

A new optimal design procedure, intended to reduce the number of exact analyses during the solution process, is proposed. High quality explicit approximations of the structural behaviour and a scaling procedure are combined to achieve this goal. Explicit behavior models are introduced for a set of lines (directios of move) in the design space. Some algorithms for selecting these lines are presented. For each checked design along a given line the structural behavior is evaluated. The design is then scaled. This procedure, which involves small computational effort, is repeated the various step sizes until the best constrained design is obtained. A new direction of move is selected and a modified approximate model is introduced. Each selected direction requires only a single exact analysis of the structure. The result is an efficient path to the optimum with a relatively small number of exact analyses. Simple examples illustrate the solution process, the quality and the efficiency of the explicit behavior model, and the potential of the proposed procedure.

0572 Concrete Members Under Combined Torsion and Shear

Ahmed A. Ewida, (Frontier Systems Development, Petro-Canada, Calgary, Alberta, Canada) and **Arthur E. McMullen**, Member, ASCE, (Prof. of Civ. Engrg., The Univ. of Calgary, Calgary, Alberta, Canada)

Journal of the Structural Division, Vol. 108, No. ST4, April, 1982, pp. 911-928

The behaviour of reinforced concrete members subjected to combined torsion and shear is studied in this investigation. An analysis is developed for predicting the deformations at all levels of load and for predicting the strength of the members under combined torsion and shear. Twelve specimens were tested in the experimental portion and the test results are presented and compared with the theoretical predictions.

0573 Ductility of Square-Confined Concrete Columns

Robert Park, Member, ASCE, (Prof. and Head of Dept. of Civ. Engrg., Univ. of Canterbury, Christchurch, New Zealand), **M. J. Negel Priestley**, (Reader in Civ. Engrg., Univ. of Canterbury, Christchurch, New Zealand) and **Wayne D. Gill**, (Asst. Engr., Ministry of Works and Development, Huntly, New Zealand)

Journal of the Structural Division, Vol. 108, No. ST4, April, 1982, pp. 929-950

A brief review showed considerable differences in the quantity of rectangular hoops required by various seismic design codes in the potential plastic hinge regions of rectangular reinforced concrete columns. Tests conducted on four nearly full size reinforced concrete columns with a 550 mm (21.7 in) square cross section, various axial load levels, and various quantities of rectangular hoops, are described. The hoops were designed according to the recommendations of the draft New Zealand Concrete Design Code. The columns demonstrated very ductile behaviour when loaded by simulated seismic loading up to displacement ductility factors of at least 6, confirming the adequacy of the proposed design provisions. Considerable enhancement of flexural strength was observed over that calculated by normal code design procedures, but the enhancement could be accounted for a refined flexural strength analysis which included the effect of strain hardening of the longitudinal steel and the increase in strength and ductility of the confined concrete. Maximum compressive concrete strains in the range 0.016 to 0.026 were measured on the surface of the confined concrete core at the peak displacements.

0574 Probability Based Load Criteria: Assessment of Current Design Practice

Theodore V. Galambos, Fellow, ASCE, (Prof. of Civ. Engrg., Univ. of Minnesota, Minneapolis, Minn.), **Bruce Ellingwood**, Assoc. Member, ASCE, (Research Struct. Engr., Center for Building Tech., National Bureau of Standards, Washington, D.C.), **James G. MacGregor**, Fellow, ASCE, (Prof. of Civ. Engrg., Univ. of Alberta, Edmonton, Alberta, Canada) and **C. Allin Cornell**, Member, ASCE, (Prof. of Civ. Engrg., Stanford Univ., Palo Alto, Calif.)

Journal of the Structural Division, Vol. 108, No. ST5, May, 1982, pp. 959-977

This is the first of two papers that describe a study conducted to develop probability-based load factors and load combinations suitable for use with the loads specified in American National Standard A58 on design loads and with all common construction materials and technologies. The first part of the study, described in this paper, involved the selection of a probabilistic methodology for performing the necessary reliability analyses and the collection and examination of statistical data on structural resistance and loads. Levels of reliability implied by the use of current design standards and specifications for common design situations in which performance generally is felt to be satisfactory were then estimated.

0575 Probability Based Load Criteria: Load Factors and Load Combinations

Bruce Ellingwood, Assoc. Member, ASCE, (Research Struct. Engr., Center for Building Tech., National Bureau of Standards, Washington, D.C.), **James G. MacGregor**, Fellow, ASCE, (Prof. of Civ. Engrg., Univ. of Alberta, Edmonton, Alberta, Canada), **Theodore V. Galambos**, Fellow, ASCE, (Prof. of Civ. Engrg., Washington Univ., St. Louis, Mo.) and **C. Allin Cornell**, Member, ASCE, (Prof. of Civ. Engrg., Massachusetts Inst. of Tech., Cambridge, Mass.)

Journal of the Structural Division, Vol. 108, No. ST5, May, 1982, pp. 978-997

This is the second of two papers that describe a study conducted to develop probability-based load factors and load combinations suitable for use with common construction materials and technologies. In the second phase of the study, described in this paper, the checking equation format for the proposed load criteria was selected, and the load factors and load combinations were computed using a constrained optimization procedure. Comparisons of reliabilities obtained using the proposed procedure with existing criteria are made. Guidance is provided for material specification-writing groups to enable them to select compatible resistance criteria without further numerical computations.

0576 Minimax Finite Element Method

Il-Bahng Park, (Sr. Engr., Nuclear Tech. Div., Westinghouse Electric Corp., Pittsburgh, Pa.) and **Walter D. Pilkey**, Member, ASCE, (Prof., Applied Mech. Div., Univ. of Virginia, Charlottesville, Va. 22901)

Journal of the Structural Division, Vol. 108, No. ST5, May, 1982, pp. 998-1011

The minimax method is applied to boundary value problems of structural mechanics. This is a weighted residual method in which the absolute value of a residual is minimized. It is similar to collocation-based finite elements. Computational implementation is achieved using linear programming. Since inequality constraints are acceptable to linear programming, the minimax method can bm used to solve problems formulated using inequality relationships. For example, the minimax finite element method can perform limit analyses or can be applied to such contact problems as offset supports. The proposed approach is formulated in some detail. It is then applied to some elasticity stress analysis problems and to the limit analysis of a beam. The method appears to be relatively easy to set up and gives satisfactory results.

0577 Bridge Loading: Research Needed

Committee on Loads and Forces on Bridges of the Committee on Bridges of the Structural Division, Peter G. Buckland, chmn.

Journal of the Structural Division, Vol. 108, No. ST5, May, 1982, pp. 1012-1020

The need for research in areas of bridge loading is examined. It is argued that one reason research is incomplete is poor definition of the requirements. The main priorities for research are defined. Subjects for research include traffic loading for both long and short span bridges, braking, or longitudinal, forces, fatigue, and environmental effects. Research problems are given.

0578 Design Life Simulation of a PCRV

Christian Meyer, Member, ASCE, (Assoc. Prof., Dept. of Civ. Engrg. and Engrg. Mech., Columbia Univ., New York, N.Y.)

Journal of the Structural Division, Vol. 108, No. ST5, May, 1982, pp. 1021-1033

A nonlinear three-dimensional finite element program was used to analyze the prestressed concrete reactor vessel (PCRV) of a high-temperature gas-cooled reactor. The various loading conditions were selected to simulate the 40-year design life of the vessel. The computer program was developed to consider the nonlinear behavior of concrete, including cracking, crushing, and creep; the elasto-plastic behavior of steel liner plates; prestressing effects due to unbonded tendons and circumferential prestress. The program theory has been described elsewhere. This paper deals with the application of the program to the study of a PCRV. The vessel's response to various normal, abnormal, and extreme loading conditions are discussed together with the approaches taken to assess the safety against failure.

0579 Estimates of Design Wind from Short Records

Mircea Grigoriu, Member, ASCE, (Assoc. Prof. of Civ. Engrg., Cornell Univ., Ithaca, N.Y.)

Journal of the Structural Division, Vol. 108, No. ST5, May, 1982, pp. 1034-1048

Wind speeds of various average return periods are estimated from the series of hourly mean wind speeds assumed independent. This hypothesis of independence is suggested by characteristics of the largest value distribution of Gaussian and non-Gaussian series. The confidence in this estimator of wind speed is subsequently examined when the shape of the instantaneous distribution of the hourly mean wind speed series is available but its parameters are unknown. It is found that the estimators are generally accurate even when the available record is short, e.g. 1 to 5 years of hourly readings. This conclusion is supported by test cases involving records of hourly mean wind speeds in Sydney and Lisbon. On the other hand, the conventional method based on Gumbel lines fitted to yearly maximum wind speeds is inaccurate when based on records of similar sizm.

0580 Investigation of Concrete: PMC Composite

Edward G. Nawy, Fellow, ASCE, (Prof. of Civ. Engrg., Rutgers Univ., Piscataway, N.J. 08854), **Maurice M. Ukadike**, (Lect. of Civ. Engrg., Univ. of Benin, Nigeria) and **Perumalsamy N. Balaguru**, Member, ASCE, (Asst. Prof. of Civ. Engrg., Rutgers Univ., Piscataway, N.J. 08854)

Journal of the Structural Division, Vol. 108, No. ST5, May, 1982, pp. 1049-1063

This paper presents the results of an experimental investigation on the behavior of concrete - Polymer Modified Concrete two-layered composite beams (PMC). Sixteen simply supported beam specimens were tested to failure. While the principal parameter of this investigation was the variation of the top PMC layer thickness; the major aspects studied were the load-deflection and cracking behavior, the mode of failure of the beams, the contribution of PMC to the strength of the composite beam, the initiation and progress of slip in the interface of the two layers, frictional shear resistance of the un-reinforced concrete - PMC interface, and the effect of using shear reinforcement to prevent any slip and shear failure. It is concluded that the two layered system can be effectively used in practice. Based on the experimental results recommendation is made for the allowable interface shear stress. A method with a numerical example is presented for the design of composite beams.

0581 Ultimate Strength of Arches with Bracing Systems

Tatsuro Sakimoto, (Assoc. Prof. of Civ. Engrg., Kumamoto Univ., Kumamoto, Japan) and **Sadao Komatsu**, (Prof. of Civ. Engrg., The Osaka Univ., Osaka, Japan)

Journal of the Structural Division, Vol. 108, No. ST5, May, 1982, pp. 1064-1076

The ultimate load carrying capacity of bridge arches is investigated for the case where they fail by inelastic lateral instability associated with the flexural-torsional deformation of the arch rib. The arch bridge is composed of twin ribs connected to one another by a lateral bracing system. The ribs are subjected to a uniformly distributed vertical load. A parametric study is carried out for several representative cases by using a finite element method. The effects of the stiffness, location and type of the lateral bracing system on the inelastic behavior of twin arches are evaluated. Ultimate strength versus slenderness ratio curves for various twin arches are obtained. The possibility of introducing an effective length concept for lateral buckling is also examined and a comparison is made with the recommendations of existing codes for arch bridges

---

0582 Hysteresis Models for Reinforced Concrete

Mehdi Saiidi, Assoc. Member, ASCE, (Asst. Prof. of Civ. Engrg., Univ. of Nevada, Reno, Nev. 89557)

Journal of the Structural Division, Vol. 108, No. ST5, May, 1982, pp. 1077-1087

Nonlinear dynamic response of a simple reinforced concrete specimen is studied using five different hysteresis models: the Takeda, elasto-plastic, bilinear, Clough, and the Q-hyst model. These models range from simple but unrealistic (the elasto-plastic model) to complicated but realistic (the Takeda model). The Takeda model is used as the basis to evaluate the other models, and to explain the influence of different features used in each model. It is shown that stiffness degradation effects associated with unloading and load-reversal stages had a substantial effect on the amplitudes, waveform, and frequency content of displacement response history of the specimen studied. It is further shown that the Q-hyst model, which takes degrading effects into account while it is simple, resulted in a response history which was in excellent agreement with that obtained using the Takeda model.

0583 Loadings for Electrical Transmission Structures

Committee on Electrical Transmission Structures of the Committee on Analysis and Design of Structures of the Structural Division

Journal of the Structural Division, Vol. 108, No. ST5, May, 1982, pp. 1088-1105

As the costs of high-voltage electrical transmission lines have increased, greater attention is being given to the weather element loads that influence the structure cost. Minimizing the cost of structures can only be accomplished by increased knowledge of loading conditions which influence the possible risks involved. To obtain experience data and utility practice relative to weather element loads an extensive questionnaire was sent to every major utility in the United States. This report has been compiled using the responses from 130 major utilities that own and operate approximately 290,000 miles of high-voltage transmission lines rated 69-kV and above. The report provides data on the practices followed in the design, fabrication, testing, and procurement of transmission structures. Questionnaire responses cover presently applicable legislative codes. Wind, ice, and longitudinal loadings are shown. Utilities evaluate factors such as wind gusts, density of ice formations, wind height factors, and span reduction factors. These factors are normally incorporated into the structure loadings by comparing the special load to a standard load case which controls the design.

0584 Cyclic Shear and Dowel Action Models in R/C

Rafael Jimenez, Member, ASCE, (Asst. Prof. of Civ. Engrg., Univ. of Puerto Rico, Mayaguez, P.R., 00708), **Richard N. White**, Fellow, ASCE, (Prof. of Struct. Engrg. and Dir., School of Civ. and Environmental Engrg., Cornell Univ., Ithaca, NY. 14853) and **Peter Gergely**, Member, ASCE, (Prof. of Struct. Engrg., Cornell Univ., Ithaca, N.Y. 14853)

Journal of the Structural Division, Vol. 108, No. ST5, May, 1982, pp. 1106-1123

The transfer of reversed cyclic shear forces across cracks in reinforced concrete by interface shear transfer and by dowell action is treated in this paper. Experimental cyclic shear load-shear displacement results are given for the interface shear transfer mechanism alone, the dowel action mechanism alone, and the combined mechanism. Empirical prediction expressions are presented for the two independent mechanisms; these mechanisms are then combined to provide a model for predicting the load-displacement relationship for combined interface shear transfer and dowel action, including the effects of load cycling. Model predictions compare reasonably well with experimental results.

0585 Buckling of Pretwisted Columns

Negussie Tebedge, (Assoc. Prof. of Civ. Engrg., Addis Ababa Univ., Addis Ababa, Ethiopia)

Journal of the Structural Division, Vol. 108, No. ST5, May, 1982, pp. 1124-1139

The effect of pretwist on the buckling load of column is presanted. A finite element formulation is developed for the stability analysis of pretwisted columns. Numerical examples are used on pretwisted columns and other related types of columns to demonstrate the finite element model. In all cases, good convergence is observed and satisfactory accuracy is obtained by using few elements, and the required computational effort is found to be minimal. Experimental results are presented for the buckling loads and the associated buckling modes by varying the angles of pretwist. Good agreement between the predicted values and the experimental results is observed. Finite element solutions, of critical loads of pretwisted columns are presented in tabular form and also graphically for various angles of pretwist and different ratios of principal moment of inertia.

0586 Simplified Analysis of Bundled-Type Structures

Alexander Coull, Fellow, ASCE, (Prof. of Civ. Engrg., Univ. of Glasgow, Glasgow, Scotland), **Bishwanath Bose**, (Research Engr., British Steel Corp., Motherwell, Scotland) and **Abdulla K. Ahmed**, (Lect., Civ. Engrg. Dept., Univ. of Khartoum, Sudan)

Journal of the Structural Division, Vol. 108, No. ST5, May, 1982, pp. 1140-1153

A simplified method is presented for the analysis of bundled tube structures subjected to lateral loads. The rigidly-jointed perimeter and interior web frame panels are replaced by equivalent orthotropic plates, whose properties are chosen to represent both the axial and shearing deformation characteristics of the frames. The force and stress distributions in the substitute panels are assumed to be represented with sufficient accuracy by polynomial series in the horizontal coordinates, the coefficients of the series being functions of the height only. The unknown functions are determined from the principle of least work. The influence of stiffer corner columns is included in the analysis. By incorporating simplifying assumptions regarding the form of stress distribution in the frame panels, the structural behaviour can be reduced to the solution of a single second-order linear differential equation, enabling closed-form solutions to be obtained for the standard load cases of a uniformly distributed load, a triangularly distributed load, and a point load at the top. Solutions may then be obtained from design curves.

0587 Beam Idealization

Joachim A. Rodrigues, (Prof. of Civ. Engrg., Univ. of Angola, Luanda, People's Republic of Angola, Africa)

Journal of the Structural Division, Vol. 108, No. ST5, May, 1982, pp. 1154-1173

For any given bemm with varying cross-section and material properties, it is easy to evaluate a number of parameters Q_i and D_i, by numerical integration, accounting for all generalized forces and displacements. Those parameters can be used directly to set up the flexibility and stiffness matrices and also to determine the fixed-end reactions due to any loading. This method provides great accuracy in short computing time and seems to be much simpler than those in use today. Another method to set up the flexibility and stiffness matrices consists in determining an equivalent idealized beam. For any asymmetric beam, it is possible to determine an idealized beam of three inertias, with the same length and constant elastic properties, that is equivalent to the real beam for the purpose of structural analysis and is defined in terms of the parameters Q_i by a simple method (extended to symmetric beams). A program has been developed and tested on an example problem. It makes any standard program capable of analysing structures including members with varying cross-section and material properties.

0588 Buckling in Steel U-Shaped Beams

Yuhshi Fukumoto, Member, ASCE, (Prof., Dept. of Civ. Engrg., Nagoya Univ., Chikusa-ku, Nagoya, Japan) and **Masahiro Kubo**, (Asst. Prof., Dept. of Civ. Engrg., Meijo Univ., Tenpaku-ku, Nagoya, Japan)

Journal of the Structural Division, Vol. 108, No. ST5, May, 1982, pp. 1174-1190

Coupled behavior of overall and local buckling of U-shaped beams subjected to uniform bending moment is presented. A finite strip method is applied to the coupled buckling analysis and the inelastic buckling is analyzed including the residual stresses of the cross section. The buckling tests are carried out with seven welded steel U-beams. Load-deformation behavior, failure modes, the ultimate bending strength and the stiffening effects by the intermediate knee braces and diaphragms are investigated. Interaction of the lateral buckling of compression flanges and the overall buckling of U-beams are made clear both theoretically and experimentally. The lateral buckling strength of laterally unbraced compression flanges can be estimated by tbe approximate method based on the equivalent T-column. The overall buckling strength is primarily affected by the ratio of the span length to spacing of webs.

0589 Dynamic Characteristics of Frame-Wall Systems

Amiya K. Basu, Fellow, ASCE, (Prof. of Civ. Engrg., Indian Inst. of Tech., New Delhi, India), **Ashok K. Nagpal**, (Asst. Prof., Indian Inst. of Tech., New Delhi, India) and **Ashwani K. Nagar**, (Executive Engrg., Post & Telegraph Dept. , (Civil Wing), Ministry of Communication, Government of India, New Delhi, India)

Journal of the Structural Division, Vol. 108, No. ST6, June, 1982, pp. 1201-1218

The first three natural frequencies and the corresponding mode shapes for fixed-base frame-wall systems are presented. The system is modelled as a continuum of uniform properties incorporating the shear deformation in the wall. The resulting fourth-order homogeneous differential equation in terms of lateral displacement is solved exactly using an appropriate numerical technique. The results are presented for various combinations of the two nondimensional parameters which together incorporate all the geometric and material properties of the frame-wall system. The mode shapes are presented in terms of the first three normal modes of a uniform slender cantilever. The nondimensional base shears appropriate for the response spectrum analysis of the system under seismic loading are also given.

0590 Optimum Design of Reinforced Concrete Water Tanks

S. R. Adidam, (Asst. Prof., Dept. of Civ. Engrg., Indian Inst. of Tech., Kanpur, India) and **A. V. Subramanyam**, (Asst. Prof., Dept. of Applied Mechanics, Karnataka Regl. Engrg. Coll., Surathkal, India)

Journal of the Structural Division, Vol. 108, No. ST6, June, 1982, pp. 1219-1231

The conventional method of designing water tanks is irrational and leads to relatively thick walls with a substantial amount of reinforcement. Application of the limit state method results in considerable economy. The limit analysis used to find the strength of the structure is explained. Cost of the tank consisting of the tank wall and floor, which includes cost of concrete, reinforcement and formwork, is chosen as the objective function in the optimization problem formulation. The constrained nonlinear programming problem is solved using DFP algorithm. An example, solved in a textbook by the conventional method, is chosen to get an idea of the saving in the cost that can be obtained. It has been found that optimization coupled with limit state design results in significant reduction in the cost of the tank.

0591 Seismic Hazard Analysis of Lifelines

Jamshid Mohammadi, Assoc. Member, ASCE, (Asst. Prof. of Civ. Engrg., Illinois Inst. of Tech., Chicago, Ill.) and **Alfredo H-S. Ang**, Fellow, ASCE, (Prof. of Civ. Engrg., Univ. of Ill. at Urbana-Champaign, Urbana, Ill.)

Journal of the Structural Division, Vol. 108, No. ST6, June, 1982, pp. 1232-1249

Methods for assessing the seismic safety of a lifeline system are developed; with part of the study being devoted to the development of a relation between the earthquake intensity and the distance for near-source regions. The attenuation of maximum ground motion intensity with distance is examined. Specific attenuation relations are developed and compared with the existing empirical attenuation equations. On this basis, the importance of the source parameters are examined. With regard to the safety analysis of lifeline systems, it is assumed that earthquakes originate as ruptures along geologic faults. Two types of hazards from earthquakes are considered; namely, the hazard of a fault rupture strike on the links of a lifeline system, and the hazard of damage caused by severe ground shaking. The application of the methods presented herein are demonstrated for a specific lifeline system.

0592 Cracking at Openings in Prestressed Beams at Transfer

John B. Kennedy, Member, ASCE, (Prof., Dept. of Civ. Engrg., Univ. of Windsor, Windsor, Ontario, Canada N9B 3P4) and **A. M. El-Laithy**, (Research Asst., Dept. of Civ. Engrg., Ohio State Univ., Columbus, Ohio 43210)

Journal of the Structural Division, Vol. 108, No. ST6, June, 1982, pp. 1250-1265

The behavior of prestressed concrete beams with rectangular openings is investigated at transfer. The theoretical analysis for the potentially-splitting stresses around the opening is performed by the finite element method. The results are verified and substantiated by tests on eighteen post-tensioned concrete beams. Based on the results a simple method is developed to estimate and design for the vertical tension force at the opening. Effects of lateral prestressing around the opening as well as the presence of more than one opening are examined.

0593 Behavior and Design of Reinforced and Composite Concrete Structures

Soliman I. Al-Noury, (Asst. Prof. of Civ. Engrg., King Abdulaziz Univ., Jeddah, Saudi Arabia) and **Wai F. Chen**, Member, ASCE, (Prof. and Head of Struct. Engrg., Purdue Univ., W. Lafayette, Ind. 47906)

Journal of the Structural Division, Vol. 108, No. ST6, June, 1982, pp. 1266-1284

The force-deformation equilibrium equations of reinforced and composite concrete columns are formulated, from which moment-curvature-thrust relationships are developed. The cross section is partitioned into a number of small elemental areas of steel and concrete, and the moment-curvature-thrust relationships are obtained by step-by-step application of the force-deformation equilibrium equations subjected to different loading paths. Contour interaction equation is extended and verified for its applicability to composite sections.

0594 Waffle Slab Concrete Bridges: Ultimate Behavior

John B. Kennedy, Member, ASCE, (Prof., Dept. of Civ. Engrg., Univ. of Windsor, Windsor, Ontario, Canada) and **Ibrahim S. El-Sebakhy**, (Research Asst., Dept. of Civ. Engrg., Univ. of Windsor, Windsor, Ontario, Canada)

Journal of the Structural Division, Vol. 108, No. ST6, June, 1982, pp. 1285-1301

The structural response of reinforced and prestressed concrete waffle slab bridges has been examined under working loads as well as under ultimate collapse loads. Bridges of rectangular and skew planforms are studied. The elastic analysis is based on a Fourier series solution, while the ultimate load analysis is formulated using yield-line theory. The results from the theoretical analyses are compared to those from experiments on five 1/8-scale models of reinforced and prestressed waffle slab bridges. The serviceability and load-carrying capacity of reinforced versus prestressed waffle slab bridges are compared.

0595 The Earthquake Problem in the Eastern United States

Otto W. Nuttli, (Prof. of Geophysics, Dept. of Earth and Atmospheric Sciences, St. Louis Univ., St. Louis, Mo. 63103)

Journal of the Structural Division, Vol. 108, No. ST6, June, 1982, pp. 1302-1312

Earthquakes in the eastern United States have longer recurrence times, less attenuation of damaging ground motion, and occur in more diffuse source zones than their western counterparts. As a consequence, hazard specification, either by deterministic or probabilistic techniques, is more uncertain in the East. Many of the eastern source zones in historic time have not had the largest earthquake of which they are capable. Two eastern source zones, the St. Lawrence valley of Canada and the New Madrid fault of the central Mississippi valley, produced great earthquakes (magnitude larger than 8). They probably are the only zones in the East which can generate such great earthquakes. A number of other zones, however, can produce major earthquakes which will affect a large number of people, because of the high density of population and the low attenuation of damaging ground motion in the East. All critical facilities should be designed to withstand the maximum expected ground shaking, even though it may have a long recurrence time, of the order of 1000 years.

0596 Seismic Response of Friction Damped Braced Frames

Avtar S. Pall, Member, ASCE, (Project Engr., The SNC Group, 1 Complex Desjardins, Montreal, Quebec, Canada) and **Cedric Marsh**, Member, ASCE, (Prof., Centre for Building Studies, Concordia Univ., Montreal, Quebec, Canada)

Journal of the Structural Division, Vol. 108, No. ST6, June, 1982, pp. 1313-1323

A new concept of aseismic design for steel framed buildings is proposed. By providing sliding friction devices in the bracing system of the framed buildings, their earthquake resistance and damage control potential can be considerably enhanced. During severe earthquake excitations, the friction device slips and a large portion of the vibrational energy is dissipated mechanically in friction rather than inelastic yielding of the main structural components. Results of inelastic time-history dynamic analysis show superior performance of the friction damped braced steel frames when compared to computed responses of other structural framing systems. The proposed friction devices act, in affect, both as safety valves and structural dampers. The device may also be conveniently incorporated in existing framed buildings to upgrade their earthquake resistance.

0597 Ultimate Load Testing of Space Trusses

Lewis C. Schmidt, Member, ASCE, (Reader, Dept. of Civ. Engrg., Univ. of Melbourne, Parkville, Victoria, Australia), **Peter R. Morgan**, Member, ASCE, (Sr. Lect., Dept. of Civ. Engrg., Univ. of Melbourne, Parkville, Victoria, Australia) and **Ariel Hanaor**, (Research Asst., Dept. of Civ. Engrg., Imperial College, London, United Kingdom)

Journal of the Structural Division, Vol. 108, No. ST6, June, 1982, pp. 1324-1335

The results of tests to destruction of twelve space trusses are reviewed. The results cover model tests of 3' to 8' span and larger scale tests of 24' and 31' span. The trusses are all double-layer plate-like space structures. Both aluminum and steel members, and welded, bolted or commercial (proprietary) jointing systems are used. Attention is drawn to the significance of factors not normally considered in the analysis and design of such structures. The effects of joint slip, initial force systems caused by lack of fit or welding, combined with the inherent scatter of peak load capacities of the brittle-type compressive elements, furnish reductions of the experimentally derived ultimate loads when compared with theoretical predictions. It appears that the greater the degree of indeterminacy the greater the theoretical overestimate of the experimental value. Load capacities between 3 and 37 percent below those expected were observed.

0598 Reinforced Gypsum for Egyptian Housing

Mahmoud A. Reda Youssef, (Assoc. Prof., Struct. Engrg. Dept., Cairo Univ., Cairo, Egypt) and **Albert G. H. Dietz**, Honorary Member, ASCE, (Prof. Emeritus of Building Engrg., Dept. of Architecture, Massachusetts Inst. of Tech., Cambridge, Mass. 02139)

Journal of the Structural Division, Vol. 108, No. ST6, June, 1982, pp. 1336-1356

Reinforcement of gypsum for wall and roof panels is investigated using Egyptian gypsum combined with fibers and Nile reeds. Panels reinforced with jute, glass wool, wire, polypropylene fabrics, and reeds are studied. The panels were tested mainly under flexural and impact loading to determine the effect of the type and amount of reinforcement and the panel thickness on the strength and deformation characteristics. Roof panels reinforced with Nile reeds are proposed. A 20-year old house of stabilized earth with reed-reinforced gypsum floor and roof panels is described.

0599 Buckling of Stiffened Webs by Local Edge Loads

Tom R. Graves Smith, (Lect., Civ. Engrg. Dept., Southampton Univ., Southampton, U.K., SO9 5NH) and **Jacet T. Gierlinski**, (Resident Fellow, Civ. Engrg. Dept., Southampton Univ., Southampton, U.K., SO9 5NH)

Journal of the Structural Division, Vol. 108, No. ST6, June, 1982, pp. 1357-1366

A finite strip method is used for analysing the buckling of longitudinally stiffened web plates subjected to transverse local edge loading. The probLem is of particular importance for box girders, which have relatively weak flanges for resisting this type of buckling. Critical loads are obtained for square webs stiffened by single asymmetric longitudinal stiffeners positioned at one fifth of the depth from the loaded edge. A sufficient range of stiffener geometries is considered to enable the separate influences of the bending and torsional rigidities to be determined and two simple design rules are proposed. The first gives the lower limit on the bending rigidity for which the stiffener remains straight in the buckled configuration. The second gives the buckling strength of the web as a function of the torsional rigidty of the stiffener.

0600 Collapse Loads of Laced Columns

Sven Eilif Svensson, (Chf. Engr., Ramboll & Hannemann, Inc., Consulting Engrs., Teknikerbyen 38, DK 2830, Virum, Denmark) and **Jan Kragerup**, (Engr., Ramboll & Hannemann, Inc., Consulting Engrs., Teknikerbyen 38, DK 2830, Virum, Denmark)

Journal of the Structural Division, Vol. 108, No. ST6, June, 1982, pp. 1367-1384

The imperfection sensitivity of laced columns is investigated, the imperfection consisting of both a local and an overall geometrical imperfections. A simple numerical method is given and used for solving the linear elastic as well as the linear elastic-ideally plastic problem. The collapse load in the latter case is approximated with the load, initiating yielding in the structure. It is shown that the linear-elastic column, after reaching the collapse load, behaves almost like a shell and must, therefore, be expected to be sensitive to imperfections. It is shown that the relative drop in capacity of the linear elastic-ideal plastic column is largest when the two critical Euler stresses and the yield stress all coincide.

0601 Cyclic Analysis of Fix-Ended Steel Beam-Columns

Shouji Toma, (Asst. Mgr., Office of Welding Research Lab., Kawasaki Heavy Industries Ltd., Chiba-Prefecture, Japan) and **Wai F. Chen**, Member, ASCE, (Prof. and Head of Struct. Engrg., School of Civ. Engrg., Purdue Univ., West Lafayette, Ind. 47907)

Journal of the Structural Division, Vol. 108, No. ST6, June, 1982, pp. 1385-1399

The paper describes an analytical model for theoretically predicting the inelastic cyclic behavior of fix-ended steel beam-columns subjected to one cycle of axial loading. The analysis is based on the hinge-by-hinge method used widely in the study of elastic-plastic behavior of steel structures. In this analysis, the solution of an elastic-plastic beam-column under cyclic and reversed loading is obtained as a sequence of elastic column solutions which follow the history of loading. Using this analytical procedure, closed form expressions are derived for axially loaded beam-columns with constant lateral loads. With these expressions, extensive numerical solutions are presented. Some of the results are compared with available tests on actual tubular beam-columns, providing the needed confirmation of the validity of the simple analytical model for practical use.

0602 Slow Crack Growth in Cement Composites

Methi Wecharatana, (Research Asst., Dept. of Materials Engrg., Univ. of Illinois at Chicago Circle, Chicago, Ill.) and **Surendra P. Shah**, Member, ASCE, (Prof. of Civ. Engrg., Northwestern Univ., Evanston, Ill. 60201)

Journal of the Structural Division, Vol. 108, No. ST6, June, 1982, pp. 1400-1413

The resistance to crack growth in mortar, concrete and fiber reinforced concrete was studied by testing double torsion and double cantilever speciments. R-curves (plots of strain energy release rate vs. crack extension) were obtained from the test data. To include the inelastic strain energy absorbed during the crack extension, the definition of the strain energy release rate was modified. It was found that R-curves for mortar and concrete were similar for both double torsion and double cantilever speciments. R-curves were sensitive to the volume and size of aggregates and to the presence of fibers. It appears that R-curves can be a useful specimen-geometry-independent parameter for chracterizing the fracture behavior of cementitious composites. However, the effects of differing crack-opening displacement on the strain energy release rate must be considered; especially for fiber reinforced concrete.

0603 Composite Masonry: Research Needs

Joseph A. Wintz, Member, ASCE, (Asst. Chf. Engr., Engrg. and Research, Brick Inst. of America, 1750 Old Meadow Rd., McLean, Va. 22102) and **Alan H. Yorkdale**, Member, ASCE, (Vice Pres., Engrg. and Research, Brick Inst. of America, 1750 Old Meadow Rd., McLean, Va. 22102)

Journal of the Structural Division, Vol. 108, No. ST6, June, 1982, pp. 1414-1423

Composite masonry walls (walls built of brick and concrete masonry units bonded so as to exert common action under load) have been used for decades. Until recently, the only method of sizing composite masonry walls was an empirical one using ANSI A41.1 Now, with the completion of ACI-531, there is a rational method available. This standard is actually a concrete masonry standard, but it includes a chapter on composite masonry. The ASCE/ACI 530 Joint Committee on Masonry is currently developing a new design standard for masonry of all products, including composite masonry. This paper presents a short history of past research and gives suggestions for needed research in the area of composite masonry.

0604 Nonlinear Traveling-Wave Effects on Cooling Tower

John P. Wolf, Member, ASCE, (Struct. Engr., Structural Dynamics Div., Electrowatt Engrg. Services Ltd., Bellerivestrasse 36, Zurich, Switzerland) and **Kurt M. Bucher**, (Struct. Engr., Structural Dynamics Div., Electrowatt Engrg. Services Ltd., Bellerivestrasse 36, Zurich, Switzerland)

Journal of the Structural Division, Vol. 108, No. ST6, June, 1982, pp. 1424-1439

The dynamic response of a tall reinforced-concrete hyperbolic cooling tower which has been designed for vertically incident seismic waves, based on a linear analysis, is examined for horizontally propagating waves of the same amplitude. The nonlinear response is also calculated, taking the material law of the reinforcing steel and of the concrete in the columns as well as lift-off and slipping of the separate foundations into account. The global response is increased for retrograde waves and decreased for prograde waves. It is decreased in a nonlinear analysis for both types of waves. The local response is increased for retrograde and prograde waves and is in all cases larger in a nonlinear analysis than in the corresponding linear one. When checking the seismic response of the retrograde wave in a nonlinear analysis with the design determined for vertical incidence, the following applies: The foundations have to be strengthened, the reinforcement of the columns is unaffected, but its area would possibly have to be increased, and the reinforcement in the lower part of the shell is sufficient.

0605 Analysis of Behavior of Steel Liner Anchorages

Daryl R. Armentrout, Member, ASCE, (Asst. Mgr., Office of Engrg. Design & Construction, Tennessee Valley Authority, 400 Commerce Ave., W12A12, Knoxville, Tenn. 37902) and **Edwin G. Burdette**, Member, ASCE, (Prof. of Civ. Engrg., The Univ. of Tennessee, Knoxville, Tenn.)

Journal of the Structural Division, Vol. 108, No. ST7, July, 1982, pp. 1451-1463

A method is presented to analyze the load-deflection behavior of liner anchorage systems used in the design and construction of steel-lined concrete containment structures in nuclear power plants. The PAFEC 75 computer program with its bilinear stress-strain capability produced analytically similar results to those measured in anchorage tests conducted at The University of Tennessee, Knoxville. Three different analyses were made. Two of these were variations in the angle anchorage system and the third was an analysis of a structural tee. In large measure, crushing of the concrete occurs as the load approaches the maximum "peak." Rough approximations of the results of concrete crushing were made by successive analyses where selected elements of the concrete mesh were eliminated. The eliminated elements were those that first reached strains where crushing of the concrete could occur. This analytical approach gave conservative results compared to the data.

0606 Experimental Study of Thin-Web Welded H Columns

R. Richard Avent, Member, ASCE, (Prof., Dept. of Civ. Engrg., Mississippi State Univ., Mississippi State, Miss. 39762) and **Samuel Wells**, (Field Engr., ATEC, Birmingham, Ala.)

Journal of the Structural Division, Vol. 108, No. ST7, July, 1982, pp. 1464-1480

Little attention has heretofore been paid to the strength of welded thin-web columns. The purpose of this research was to determine what factors affected the strength of this type of column. An initial literature search defined the range of testing necessary to provide pertinent data on column strength. Seven columns ranging from 9° (22.9 cm) to 24° (61 cm) deep were used in the testing program. Prior to testing, all columns were measured for residual stress distributions. Stub-columns and long column tests were performed on the 9 inch (22.9 cm) and 12 inch (30.5 cm) deep columns. It was found that residual stresses were larger than those found in either thick-wall welded or rolled sections. The columns were generally stronger in axial compression than would have been predicted by various widely used column curves. This is accounted for, in part, by the straightness of the columns tested and by partial end rotation restraint. However, proposed column design curves may unduly penalize columns of this type. Weld amperage and preparation of edges had relatively small effect on the residual stress pattern and column strength.

0607 Safety Checking Formats for Limit States Design

Bruce Ellingwood, Member, ASCE, (Research Struct. Engr., Center for Building Tech., National Bureau of Standards, Washington, D.C. 20234)

Journal of the Structural Division, Vol. 108, No. ST7, July, 1982, pp. 1481-1493

The growing need for a unified approach to structural design for different construction materials and technologies is met by the limit states design approach. Several standards-writing committees currently are considering how best to implement practical limit states design criteria. This paper examines the relative advantages of several approaches proposed for this purpose, with regard to their ability to provide desired levels of reliability for all probable design situations. The choice of an appropriate approach is found to depend on fundamental characteristics of structural behavior, and thus might not be the same for different construction materials.

0608 Load-Duration Effects in Westerm Hemlock Lumber

Ricardo O. Foschi, (Research Sci., Forintek Canada Corp., Western Lab., 6620 N.W. Marine Drive, Vancouver, British Columbia, Canada) and **J. David Barrett**, (Research Sci., Forintek Canada Corp., Western Lab., 6620 N.W. Marine Drive, Vancouver, British Columbia, Canada)

Journal of the Structural Division, Vol. 108, No. ST7, July, 1982, pp. 1494-1510

Results from a duration of load experiment on western hemlock 2 x 6-inch lumber, visually graded No. 2 and better, are presented. Constant and ramp loadings arm used and, to predict load duration behavior for other load histories, a damage accumulation model is developed and calibrated to the test results. The implications of the test and the damage model for the assessment of reliability of wood structures are discussed. The model's predictions are verified using additional test results from a rate-of-loading experiment with media times-to-failure in the range 0.04 seconds to 11.3 hours.

0609 Probabilistic Models for Prototype Testing

Mircea Grigoriu, Member, ASCE, (Assoc. Prof. of Civ. Engrg., Cornell Univ., Ithaca, N.Y. 14853) and **Niels C. Lind**, (Prof. of Civ. Engrg., Univ. of Waterloo, Ontario, Canada N2L3G1)

Journal of the Structural Division, Vol. 108, No. ST7, July, 1982, pp. 1511-1525

The strength of series-produced or mass produced structures is often correlated because of workmanship or material strength correlation (e.g., steel from the same heat or lumber from similar species grown in the same region). A model of correlation is used to select load factors in design based on prototype testing so as to yield structures with a prescribed reliability. Correlation is usually disregarded in practice. It is found, however, that the correlation in structural resistance is significant and can produce important savings if taken into account. This is illustrated in the design example of an industrial building. In this example the best strategy is to test only one prototype, in agreement with common practice.

0610 Fatigue Crack Propagation from Fillet Weld Toes

Kentaro Yamada, (Asst. Prof., Dept. of Civ. Engrg., Nagoya Univ., Nagoya, Japan) and **Manfred A. Hirt**, Member, ASCE, (Prof., Dept. of Civ. Engrg., Inst. of Steel Construction, Swiss Federal Inst. of Tech., Lausanne, Switzerland)

Journal of the Structural Division, Vol. 108, No. ST7, July, 1982, pp. 1526-1540

The fatigue crack propagation behavior of two types of fillet welded specimens is investigated and the application of fracture mechanics for the prediction of fatigue life is discussed. Small scale fillet welded stiffener and gusset specimens were tested under constant amplitude stress cycles. The crack propagation behavior is estimated from dye-markings on the fatigue fracture surface. The dye penetrant technique permits a crack depth of as small as 0.1 to

0.2 mm to be visibly marked. Along the fillet weld toes of the stiffener specimens, single or multiple cracks, 0.47 to 0.6 mm deep, could be dye-marked, after about 30% of the fatigue life. Cracks could be marked even earlier in the gusset specimens due to a higher stress concentration. Applying the method of fracture mechanics, the fatigue behavior of these dye-marked specimens are analysed. Measured crack shapes and fillet weld toe geometries are taken into account. The computer fatigue life is in good agreement with the test results.

0611 Behavior of Type 2 Steel Frames

Michael H. Ackroyd, (Asst. Prof. of Civ. Engrg., Rensselaer Polytechnic. Inst., Troy, N.Y.) and **Kurt H. Gerstle**, Fellow, ASCE, (Prof. of Civ. Engrg., Univ. of Colorado, Boulder, Colo.)

Journal of the Structural Division, Vol. 108, No. ST7, July, 1982, pp. 1541-1556

In the design of steel building frames, various simplifying assumptions are usually introduced when determining member design forces. Typically, the beams are assumed to be either rigidly connected to the columns or simply supported. However, neither of these conditions is ever realized in actual frames. This paper describes the effects of connection flexibility on beam behavior; a parametric study is described for a family of 32 typical multistory frames representing several design philosophies. The study evaluates the adequacy of preliminary design methods for predicting both member forces and building drifts by comparing these parameters with those obtained from a linear elastic computer analysis which accounts for actual connection flexibility. Example results are shown indicating potential savings of steel from use of refined analytical methods.

0612 Structural Analysis of Wood Floor Systems

Ricardo O. Foschi, (Research Sci., Forintek Canada Corp., Western Lab., 6620 N.W. Marine Dr., Vancouver, B.C., Canada V6T 1X2)

Journal of the Structural Division, Vol. 108, No. ST7, July, 1982, pp. 1557-1574

A combined Fourier series and finite element analysis of wood floors is presented, and its implementation in a computer with small core memory is discussed. Predictions using the model are compared with test results and good agreement is shown. The model includes the lateral and torsional deformation of the joists as degrees of freedom, and this permits the consideration of the effect that joist bridging has on the maximum floor deflection and the maximum bending stressed. In another application, the response of floors under uniformly distributed ad is simulated for different levels of within-fLoor variability in the elastic properties for the joists.

0613 Prediction of Seismic Response of Long-Period Structures

Yozo Fujino, Assoc. Member, ASCE, (Assoc. Prof. of Civ. Engrg., Univ. of Tokyo, Bunkyo-ku, Tokyo, Japan 113) and **Alfredo H-S. Ang**, Fellow, ASCE, (Prof., Dept. of Civ. Engrg., Univ. of Illinois at Urbana-Champaign, Urbana, Ill. 61801)

Journal of the Structural Division, Vol. 108, No. ST7, July, 1982, pp. 1575-1588

This paper describes the application of an analytical method for estimating structures at an intermediate distance from a fault; the method is based on fault dislocation and surface wave theory. For illustration, a potential large earthquake which may occur along the southern portion of the San Andreas fault in California is considered. The prototype event assumed is the 1857 Fort Tejon earthquake. As a site, Long Beach is chosen. Long-period ground motions due to Love waves and their response spectra are calculated for various fault rupture and crustal models. Sloshing wave heights in a large tank during the earthquake are also estimated. The results obtained indicate that: (1) the average velocity response spectra of ground motions with 5% critical damping are nearly uniform (30 to 40 cm/sec) over the period range from 6 to 14 sec; (2) sloshing wave height in a large tank is about 70 cm and may reach 2 m under certain circumstances; (3) the freeboard height specified in current aseismic design appears to be inadequate for sloshing waves produced by a large earthquake. The example discussed herein is just one of a number of potentially important applications of the theory.

0614 Bending Strength of Webs of Cold-Formed Steel Beams

Roger A. LaBoube, Member, ASCE, (Research Engr., Butler Manufacturing Co., Research Center, Grandview, Mo. 64030) and **Wei-Wen Yu**, Fellow, ASCE, (Prof. of Civ. Engrg., Univ. of Missouri at Rolla, Rolla, Mo. 65401)

Journal of the Structural Division, Vol. 108, No. ST7, July, 1982, pp. 1589-1604

The postbuckling strength of web elements in bending and the interaction of beam web and flange have been studied to determine their influence on the moment capacity of beams having either stiffened or unstiffened compression flanges. Based on the test results of 68 beam specimens having stiffened compression flanges and 34 beam specimens having unstiffened compression flanges, four methods are developed for predicting the ultimate moment capacity of cold-formed steel beams with due consideration given to the actual bending strength of web elements.

0615 Nonlinear Analysis of R/C Structures in Practice

Christian Meyer, Member, ASCE, (Assoc. Prof., Dept. of Civ. Engrg. and Engrg. Mech., Columbia Univ., New York, N.Y.) and **Klaus-Jurgen Bathe**, Member, ASCE, (Assoc. Prof., Dept. of Mech. Engrg., Massachusetts Inst. of Tech., Cambridge, Mass.)

Journal of the Structural Division, Vol. 108, No. ST7, July, 1982, pp. 1605-1622

This paper addresses some of the questions facing the engineer who is charged with assessing the safety of concrete structures for which conventional linear analysis is deemed inadequate. The currently available material models are reviewed and discussed under the viewpoint of applicability in engineering practice. Guidelines are given to aid the engineer in selecting the material model most appropriate for this purpose. Questions related to finite element analysis are discussed, including numerical solution techniques, their accuracy, efficiency and applicability. Also considered is the practical significance of solution sensitivity, reliability and cost. An example of a prestressed concrete reactor vessel illustrates the application of nonlinear finite element analysis to concrete structures in engineering practice.

0616 Stayed Column with Initial Imperfection

Kevin C. Wong, (Civ. Engr. Asst., Stone and Webster, Canada Ltd., Toronto, Ontario, Canada) and **Murray C. Temple**, Member, ASCE, (Prof., Dept. of Civ. Engrg., Univ. of Windsor, Windsor, Ontario, Canada)

Journal of the Structural Division, Vol. 108, No. ST7, July, 1982, pp. 1623-1640

The effect of an initial out-of-straightness on the buckling strength of a single-crossarm stayed column is investigated. A geometrical study of the stayed column is used to determine the relationship between the applied load, lateral deflection and the tension in the stays. The geometrical nonlinear behaviour is included in the analysis by using the finite element method and a mixed incremental and iterative procedure. Numerical examples are used to demonstrate the influence of varying the stayed column parameters on the buckling load, deflection rate, as well as the minimum effective, optimum and maximum possible pretensions. Experimental results obtained from a single-crossarm stayed colum with various initial pretensions in the stays are presented and compared with theoretical results, and good agreement was obtained. The results indicate that the presence of an initial out-of-straightness changes the deflection rate and can significantly reduce the buckling load.

0617 Probabilistic Description of Hurricane Wind Speeds

Martin E. Batts, Member, ASCE, (Principal, ParaTech Inc., Gaithersburg, Md. 10879)

Journal of the Structural Division, Vol. 108, No. ST7, July, 1982, pp. 1643-1647

Information is presented on the parameters of the best fitting Weibull distribution estimated for simulated hurricane windspeeds. The effect of the assumption that an Extreme Type I distribution, rather than a Weibull distribution, is the appropriate distribution of the extreme windspeeds, is shown.

0618 Shearing Stress due to Prestress in Variable Depth Beam

Shih Toh Chang, (Assoc. Prof. of Bridge Engrg., Tongji Univ., Shanghai, The People's Republic of China)

Journal of the Structural Division, Vol. 108, No. ST7, July, 1982, pp. 1648-1653

General formulas to estimate shearing stress in prestressed concrete beams due to prestress force are given. These equations also can be used to calculate dead load and live load moment by substituting the corresponding values of shearing force and moment. Finally, it is demonstrated how tension or compression can be calculated.

0619 A Shear Lag Anomaly

Douglas A. Foutch, Assoc. Member, ASCE, (Asst. Prof. of Civ. Engrg., Univ. of Illinois at Urbana-Champaign, 3108 NCEL, 208 N. Romine, Urbana, Ill. 61801) and **P. C. Chang**, (Research Asst., Univ. of Illinois at Urbana-Champaign, 3108 NCEL, 208 N. Romine, Urbana, Ill. 61801)

Journal of the Structural Division, Vol. 108, No. ST7, July, 1982, pp. 1653-1658

This technical note reports an interesting phenomenon associated with shear lag in the flanges of box girders that is quite contrary to well established ideas concerning this subject. A brief description of a beam model and solution technique are given. It is shown that if a cantilever tube is loaded laterally under non-uniform shear, a reversal of the shear lag distribution may occur at some point in the beam; the centerline stress may exceed the edge stress. Although the practical significance is not clear, further study is encouraged to explain the phenomenon and explore its significance.

0620 Limit Analysis without Mechanism Generation

Harry L. Jones, Member, ASCE, (Assoc. Prof., Dept. of Civ. Engrg., Texas A&M Univ., College Station, Tex.) and **Irvin Boaz, Jr.**, (Grad. Student, Civ. Engrg. Dept., Texas A&M Univ., College Station, Tex.)

Journal of the Structural Division, Vol. 108, No. ST7, July, 1982, pp. 1658-1662

It is demonstrated that the static approach to collapse load analysis for frames can be implemented without mechanism generation and virtual work. In addition, the amount of programming effort necessary to automate the generation of the governing linear programming is comparable to that for the elastic analysis of frames.

0621 Vibrations of Inclined Slack Cables

Steven E. Ramberg, (Research Engr., Marine Tech. Div., Naval Research Lab., Washington, D.C.) and **Charles L. Bartholomew**, (Engr., Marine Tech. Div., Naval Research Lab., Washington, D.C.)

Journal of the Structural Division, Vol. 108, No. ST7, July, 1982, pp. 1662-1664

Three previously reported experiments on vibrations of inclined slack cables are examined. It is stated that a linear theory is a valuable tool in the analysis of marine cable vibrations for cables with supports at the same elevation. A method to extend the theory to cables with inclined chords is to view the cable in a coordinate system inclined with the chord. A modified linear analysis for inclined slack cables is suggested.

0622 Multiple Column Curves by Modifying Factors

J. Michael Rotter, (Lect. in Civ. Engrg., Univ. of Sydney, New South Wales 2006, Australia)

Journal of the Structural Division, Vol. 108, No. ST7, July, 1982, pp. 1665-1669

There are two principal disadvantages in the use of multiple column curves as they are currently formulated. First, they are represented either by equations which are cumbersome to use or by tables with a new table for each column curve. Second, the change from one column curve to another is a large step change so that it may be difficult to decide which curve should represent some sections and it is possible for anomalies to arise. These two disadvantages are removed in the proposed method, in which the three SSRC curves are represented by a single curve and a single linear modifying function. The modifying function has been chosen to allow smooth transitions between the existing SSRC curves and to make step changes between the curves unnecessary.

0623 Method of Imaginary Reactions for Taut Cables

Richard A. Skop, (Head, Applied Mechanics Branch, Naval Research Lab., Washington, D.C.) and **F. Rosenthal**, (Head, Ocean Structures Section, Applied Mechanics Branch, Naval Research Lab., Washington, D.C.)

Journal of the Structural Division, Vol. 108, No. ST7, July, 1982, pp. 1669-1670

The problem of determining the three dimensional equilibrium shape of an arbitrarily loaded cable is a classical one in nonlinear structural mechanics. This note presents a new technique, based on an exact representation of cable compliance combined with a Newton-Raphson interation, for solving this problem. As shown by example, this new technique provides convergence to equilibrium far more rapidly than other commonly used methods.

0624 Thermal Convection and Design Wind Speeds

Emil Simiu, Member, ASCE, (Struct. Research Engr., Center for Building Tech., National Bureau of Standards, Washington, D.C. 20234)

Journal of the Structural Division, Vol. 108, No. ST7, July, 1982, pp. 1671-1675

This tmchnical note presents estimates of the extent to which thermal convection effects cause mean wind profiles to deviate from the logarithmic law. It is shown that the deviations from the logarithmic law are significant for wind speeds corresponding to hourly means of the order of 5 m/s above ground in open terrain. However, for strong wind speeds of interest in structural engineering calculations ($\geq$70 mph), these deviations are of the order of 4% or less at 200 m above ground, and 1% at 50 m above ground, and thus may be neglected in practice. For strong wind speeds of interest in extreme wind climatological estimates ($\geq$30 mph), the deviations are of the order of 4% or less at 15 m above ground, and therefore are also negligible for practical purposes.

0625 Meridional Imperfection in Cooling Tower Design: Update

Ajaya K. Gupta, Member, ASCE, (Assoc. Prof. of Civ. Engrg., North Carolina State Univ., Raleigh, N.C.) and **Adam Al-Dabbagh**, Member, ASCE, (Sr. Engr. Analyst, Struct. Analytical Div., Sargent & Lundy, Chicago, Ill.)

Journal of the Structural Division, Vol. 108, No. ST8, August, 1982, pp. 1697-1708

The effect of meridional imperfection in hyperbolic cooling towers is studied in a continuation of an earlier study by the writers; the present paper takes into account the parallel work performed by Croll, et al. In the previous study the writers assumed the meridional shape of imperfection as the combination of two straight lines. Croll, et al considered the shape to be a combination of circular arcs. In the present study, a general shape is assumed, which becomes the straight line or the circular shape as two extremes depending on the value of a parameter 'C'. Simple expressions for the maximum possible imperfection hoop force and meridional moments are obtained. It is proposed that imperfection stresses be considered explicitly in the design of cooling towers. These stresses can be calculated on the basis of the maximum allowable construction tolerances. The writers suggest further professional discussion to decide whether the designer can lower the apparent factor of safety when the imperfection stresses are explicitly accounted for.

0626 Seismic Design of Steel Buildings

Ben Kato, Member, ASCE, (Prof. of Struct. Engrg., Dept. of Architecture, Univ. of Tokyo, Tokyo, 7-3-1 Hongo, Bunkyo-ku, Tokyo 113, Japan) and **Hiroshi Akiyama**, (Assoc. Prof., Dept. of Architecture, Univ. of Tokyo, Tokyo, 7-3-1 Hongo, Bunkyo-ku, Tokyo 113, Japan)

Journal of the Structural Division, Vol. 108, No. ST8, August, 1982, pp. 1709-1721

An approach for the ultimate limit state design of steel buildings against the credible severest earthquake is presented on the basis of energy concept. The safety of a structure is judged by comparing structure's energy dissipating capacity with earthquake input energy to the structure. A general safety criterion is first developed for a simple elasto-plastic shear-type system, and then a procedure is discussed in detail to relate the strength and deformation capacity of steel members to those of the dynamically equivalent elasto-plastic system.

0627 Optimum Design of Prestressed Concrete Tension Members

Antoine E. Naaman, Member, ASCE, (Assoc. Prof. of Struct. Design, Dept. of Materials Engrg., Univ. of Illinois at Chicago Circle, Chicago, Ill.)

Journal of the Structural Division, Vol. 108, No. ST8, August, 1982, pp. 1722-1738

A systematic procedure is proposed for the design of prestressed concrete tensile members. The design criteria considered include maximum allowable compression, ultimate strength, safety against cracking, safety against decompression, minimum reinforcement and maximum instantaneous and long-term deformations. Each design criterion is expressed in function of two unknown variables that are the area of the prestressing steel and the cross sectional area of concrete. It is shown that the relationships representing the various criteria form a feasibility domain bound mostly by linear functions. Using an acceptable design approximation the problem is linearized and reduced to a linear programming problem in which the objective function is to minimize the cost and/or the weight of the member. The geometric solution is explained and recommended for the purpose of design. An example is developed to illustrate the procedure.

0628 Lower Bound Limit Design of Rectangular Slabs

Kanakapura S. Subba Rao, (Assoc. Prof., Civ. Engrg. Dept., Indian Inst. of Science, Bangalore, India), **Nagari Rangaiah**, (Asst. Prof., Civ. Engrg. Dept., College of Engrg., Ananthapur, India) and **B. V. Ranganatham**, (Prof., Civ. Engrg. Dept., Indian Inst. of Science, Bangalore, India)

Journal of the Structural Division, Vol. 108, No. ST8, August, 1982, pp. 1739-1750

A plastic method of design of rectangular slabs based on lower bound solutions of collapse has been developed. The slab is divided into middle and end zones so as to reinforce for beam action in the middle zone and for slab action in the end zones. Earlier developed lower bound limit analysis solutions for rectangular slabs are made compatible with the assumed mode of collapse. Expressions are presented for the plastic moment capacities to be provided in the x and y-directions in the end and middle zones of all the nine possible slab types with combinations of simply supported and fixed edge conditions. Two design illustrations demonstrate the rationale in the application of the method.

0629 Pedestrian Level Winds: A Brief Review

Frank H. Durgin, Member, ASCE, (Assoc. Dir., Wright Brothers Memorial Wind Tunnel, Massachusetts Inst. of Tech., Cambridge, Mass. 02139) and **Alfred W. Chock**, (Engr., Cygna Energy Services, Boston, Mass.)

Journal of the Structural Division, Vol. 108, No. ST8, August, 1982, pp. 1751-1761

The use of analytical and wind tunnel techniques and their limitation for making estimates of pedestrian level winds on or around a building site is reviewed. Some examples of solutions to wind problems both with and without wind tunnel studies are given.

0630 Lateral Stability of I-Beams with Partial Support

Charles W. Roeder, Member, ASCE, (Assoc. Prof., Dept. of Civ. Engrg., Univ. of Washington, Seattle, Wash.) and **Mahyar Assadi**, (Grad. Student, Dept. of Civ. Engrg., Univ. of Washington, Seattle, Wash.)

Journal of the Structural Division, Vol. 108, No. ST8, August, 1982, pp. 1768-1780

The lateral torsional buckling behavior of I beams with tension flange restraint is described. Tension flange restraint occurs frequently in continuous beams and indeterminate structures. A theoretical derivation of the buckling capacity of these beams is described, and the results are compared to other well known solutions. The theory predicts a significant increase in buckling capacity with tension flange restraint when the St. Venant torsional stiffness is large. An experimental setup is designed and tests are performed to verify the theory and to further substantiate the increased moment capacity. Finally, equations which can be used in the design of beams with tension flange restraint are presented. The results of this work show the importance of tension flange restraint in limiting or controlling buckling.

0631 Prestressed Concrete Flexural Members: Design

Shan Somayaji, Member, ASCE, (Asst. Prof., Dept. of Architectural Engrg., California Polytechnic State Univ., San Luis Obispo, Calif. 93407)

Journal of the Structural Division, Vol. 108, No. ST8, August, 1982, pp. 1781-1796

An alternate method for the design of prestressed concrete flexural members using the service load criteria is presented. The method differs from other available methods, in the sense that no preliminary estimation of the prestressing force or sectional dimensions is necessary, and the range of prestress is determined prior to determining the cross-sectional dimensions or section modulus. However, the use of the design equations requires a preliminary assumption of eccentricity ratio. The method can be adopted for design of both fully stressed and non-fully stressed sections, and is applicable for pretensioned as well as post-tensioned sections. The method of construction of interaction diagrams is shown which can be used to determine the range of eccentricity in non-fully stressed sections. Different equations are derived for different types of cross sections for estimation of their thickness or section modulus.

0632 Simulation of a Buried Power Transmission System

William R. Spillers, Member, ASCE, (Prof. of Civ. Engrg., Rensselaer Polytechnic Inst., Troy, N.Y. 12181), **Henry Chu**, (Principal Cable Engr., Transmission and Distribution Dept., Consolidated Edison Co., New York, N.Y.), **Allan N. Greenwood**, (Dir., Center for Electric Power Engrg., Rensselaer Polytechnic Inst., Troy, N.Y. 12181) and **Russell Eaton**, (Power Transmission Specialist, Dept. of Energy, Washington, p.C.)

Journal of the Structural Division, Vol. 108, No. ST8, August, 1982, pp. 1797-1805

This paper examines the simulation of the motion of three power cables (three-dimensional nonlinear beams) which are constrained to move within a pipe under thermal loads. The cables are represented by beam segments which are initially straight between modes; a bilinear constitutive equation is used and changing geometry is accomodated. An example is included. Applications to the pheneomenon of thermo-mechanical bending are covered.

0633 Rigid-Body Response of Base-Isolated Structures

Iradj G. Tadjbakhsh, Member, ASCE, (Prof., Dept. of Civ. Engrg., Rensselaer Polytechnic Inst., Troy, N.Y.) and **Jonathan J. Ma**, (Research Asst., Dept. of Civ. Engrg., Rensselaer Polytechnic Inst., Troy, N.Y.)

Journal of the Structural Division, Vol. 108, No. ST8, August, 1982, pp. 1806-1814

The problem of control of rigid body response of structural systems such as nuclear islands, which are mounted on a gird of bearings with three dimensional damping and stiffness properties, is analyzed. A parametric study is carried out to show the effects of the dimensionless parameters of the foundation on the transmissibilities of the structure due to harmonic ground motion. The characteristics of an optimal base isolation are discussed and results are presented in a form which may be of use in preliminary designs.

0634 Natural Oscillations of Suspension Bridges

Dr. Frank Van der Woude, Member, ASCE, (Sr. Lect. in Civ. Mech. Engrg., Univ. of Tasmania, Hobart, Australia)

Journal of the Structural Division, Vol. 108, No. ST8, August, 1982, pp. 1815-1830

Natural modes and frequencies of a simple span suspension bridge with straight backstays are investigated, theoretically and experimentally. Fourier series representations of cable displacements are used in conjunction with the Lagrange method to represent the system with a small number of degrees-of-freedom. Bridge system matrices are derived which take account of cable, gravity, bending, and torsional stiffnesses. Natural modes and frequencies are obtained as the eigensolutions of small order system matrices. A numerical example is given from which approximate frequency formulas are derived. Experimental results in the form of frequency response curves, are presented for a laboratory model. Agreement between theory and test is within 5 percent or better.

0635 Tensegric Shell Behavior

Oren Vilnay, (Prof., Dept. of Civ. and Struct. Engrg., Univ. College, Cardiff, England) and **Siong S. Soh**, (Grad. Student, Dept. of Civ. and Struct. Engrg., Univ. College, Cardiff, England)

Journal of the Structural Division, Vol. 108, No. ST8, August, 1982, pp. 1831-1845

Tensegric shells are composed of cable or membrane envelopes prestressed by lengthening properly arranged bars connected at both of their ends to the envelope and not touching each other. The prestressing stabilizes the shell and induces tensile forces in the envelope and compression in the bars. The tensegric shells do not comprise a continuous rigid supporting element. The advantage of using tensegric shells for temporary and permanent structure is due to their easy assembly and erection. The nature of the bar arrangement in tensegric shells is shown. The stability criteria of a tensegric shell are formulated, and the effect of prestressing the envelope by lengthening some of the bars is examined. The analysis of the shell geometrical distortion due to prestressing is formulated. The effect of external load applied to the structure is investigated. It is shown that the effect of external load on the geometrical configuration of the shell depends on the type of the shell. External load applied to some types of shell causes too large geometrical distortion, whereas the distortion of other types is due to the elasticity of its members only. Different types of tensegric shells are analyzed as numerical examples.

0636 Dynamic Response of Coupled Wall Systems

J. D. Aristizabal-Ochoa, Member, ASCE, (Asst. Prof. of Civ. Engrg., Vanderbilt Univ., Nashville, Tenn.)

Journal of the Structural Division, Vol. 108, No. ST8, August, 1982, pp. 1846-1857

Results of an experimental investigation to determine the inelastic dynamic behavior of reinforced concrete coupled wall systems under severe earthquake motions are reported. These tests show that the natural frequencies of reinforced concrete structures are reduced to as much as 50% of the initial "uncracked" values during the first few seconds of a strong earthquake, additional reductions in the natural frequencies are not as significant as those caused by the first shock even if the aftershocks are significantly stronger than the first one, the size, strength, and ductility of the coupling beams are the controlling factors in the mode of failure of coupled wall systems, and although the test structures were subjected to severe inelastic behavior, some characteristics, such as modal shapes, nodal points, and amplification (filtering) along the height of the structure were similar to those of linear elastic systems.

0637 Circularly Curved Beams under Plane Loads

David J. Just, (Lect., Dept. of Civ. Engrg., Univ. of Aston, Birmingham, England)

Journal of the Structural Division, Vol. 108, No. ST8, August, 1982, pp. 1858-1873

The exact 6 x 6 stiffness matrix for a circularly curved beam subjected to loading in its own plane, derived from the governing differential equations and from the finite element procedure, is presented, the elements of the stiffness matrix being obtained explicitly, thus eliminating the need for any numerical matrix inversion procedures in the formulation of the matrix. The displacement transformation matrix, which allows the beam element to be incorporated into a general frameworks program is then formulated, after which the treatment of distributed loading is described. The accuracy and correctness of the stiffness and transformation matrices are verified by comparison with a range of convergent solutions obtained from subdivision of the curved members into a number of straight segments.

0638 2-D Finite Element Analysis of Massive RC Structures

Niels Saabye Ottosen, (Research Engr., Riso National Lab., Engrg. Dept., DK-400 Roskilde, Denmark)

Journal of the Structural Division, Vol. 108, No. ST8, August, 1982, pp. 1874-1893

Nonlinear analysis of concrete structures using finite elements is discussed. The applications include a thick-walled top-closure for a pressure vessel as well as the delicate problems of beams failing in shear. The top-closure analysis evaluates the effect of two different failure criteria and modeling of a realistic post-failure behavior is demonstrated to be mandatory for accurate structural predictions. For shear beams it is shown that the primary cause of failure is strain softening adjacent to the load point. This softening causes a strain localization which in turn results in a tendency toward diagonal cracking. Stirrups prevent this tendency and a shear-compression failure follows. Otherwise, a diagonal tension failure results. This again highlights modeling of strain softening. The influence of the shear retention factor is found to be relatively moderate. Variation of the uniaxial tensile strength influences the results insignificantly. Modeling of secondary cracks is found to be essential. Finally, it is shown that dowel action must be treated through the bending of the bars and not through their shear deformation.

0639 Epoxy-Repaired Concrete Walls Under Fire Exposure

Joseph M. Plecnik, Assoc. Member, ASCE, (Assoc. Prof., Dept. of Civ. Engrg., North Carolina State Univ., Raleigh, N.C. 27650), **Boris Bresler**, Fellow, ASCE, (Principal, Wiss, Janney, Elstner and Assoc., Emeryville, Calif.), **Hon M. Chan**, (Grad. Research Assoc., California State Univ., Long Beach, Calif.), **M. Pham**, (Assoc. Prof., Dept. of Civ. Engrg., California State Univ., Long Beach, Calif.) and **James Chao**, (Assoc. Prof., Dept. of Civ. Engrg., California State Univ., Long Beach, Calif.)

Journal of the Structural Division, Vol. 108, No. ST8, August, 1982, pp. 1894-1908

The strength properties and other important parameters regarding the behavior of epoxy repaired concrete walls during fire exposure are experimentally determined. Both ASTM E-119 and the SDHI time-temperature "pseudo" fire exposures are considered in evaluating the fire ratings of epoxy-repaired concrete walls. Conclusions are provided regarding the inter-relationship between fire performance and wall thickness, crack width, duration and intensity of fire exposure and the type of epoxy adhesives.

0640 Load Redistribution of Cracked Girders

Conrad P. Heins, Member, ASCE, (Prof., Inst. of Physical Science and Tech. and Civ. Engrg. Dept., Univ. of Maryland, College Park, Md. 20742) and **Hisato Kato**, (Grad. Research Asst., Civ. Engrg. Dept., Univ. of Maryland, College Park, Md. 20742)

Journal of the Structural Division, Vol. 108, No. ST8, August, 1982, pp. 1909-1915

Design of many bridges is controlled by the fatigue requirements. Such control can be minimized if the classification from a non-redundant to redundant structure is selected. Such a classification may be selected, if secondary members are considered in the design. In order to illustrate such effects, a two girder system is examined, with and without bottom lateral bracing. One girder is considered in a cracked and uncracked flange condition. The preliminary results show the effect bracing has on the load redistribution.

0641 Loads due to Drifted Snow

J. T. Templin, (Research Officer, Building Structures Section, Div. of Building Research, National Research Council of Canada K1A 0R6) and **W. R. Schriever**, Member, ASCE, (Former Head, Building Structures Section, Div. of Building Research, National Research Council of Canada, Ottawa, Canada K1A0R6)

Journal of the Structural Division, Vol. 108, No. ST8, August, 1982, pp. 1916-1925

Roof failures due to snow loads are often the result of heavy drifts rather than uniform snow loads. Codes and standards such as Standard A58.1 of the American National Standards Institute (ANSI) and the National Building Code of Canada contain procedures for estimating snow loads on basic structures only. An understanding of elementary fluid mechanics and snow accumulation phenomena allows a designer to predict regions of potentially heavy snow loading in situations not covered by the codes. Several basic principles are presented and illustrated by examples of air flow around commonly shaped structures and the regions of snow accumulation. Modeling of snow drifting supplies qualitative information on snow accumulation in complex situations; advantages and limitations are examined.

0642 Probabilistic Evaluation of Welded Structures

Achintya Haldar, Member, ASCE, (Asst. Prof., School of Civ. Engrg., Georgia Inst. of Tech., Atlanta, Ga.)

Journal of the Structural Division, Vol. 108, No. ST9, September, 1982, pp. 1943-1955

Some deficiencies in structual welding, particularly when thousands of welds are installed, are expected. This deficiency may or may not compromise the strength of the joint or the structural element. A deterministic evaluation of this deficiency may not be possible. A probabilistic procedure is suggested. Limited sampling information from the site is used to develop this methodology. Sampling information obtained from a project should be checked before the information is used for any statistical prediction. The χ^2 test of homogeneity could be used to check the homogeneity of the samples. The hyperbinomial distribution may be the most appropriate distribution to study the characteristics of a group of welds. It considers the effect of sample size on the prediction. A large sample size is always beneficial, but the cost-benefit ratio for the collection of samples should be studied before any decisions are made. The method suggested here would help to study this area. Repairing the welds, whenever necessary, is the final step. This method is a safe, economical and conservative solution to this complex problem.

0643 Spectra of Road Surface Roughness on Bridges

Hideyuki Honda, (Asst. Prof., Dept. of Civ. Engrg., Kanazawa Inst. of Tech., 7-1 Ogigaoka, Nonoichi-Machi, Ishikawa 920, Japan), **Yasuo Kajikawa**, (Assoc. Prof., Dept. of Civ. Engrg., Kanazawa Univ., 2-40-20 Kodatsuno, Kanazawa 920, Japan) and **Tameo Kobori**, (Prof., Dept. of Construction and Environmental Engrg., Kanazawa Univ., 2-40-20 Kodatsuno, Kanazawa 920, Japan)

Journal of the Structural Division, Vol. 108, No. ST9, September, 1982, pp. 1956-1966

The power spectral density (PSD) of road surface roughness on highway bridges is presented. Eighty-four lines of 56 bridges were measured and the road surface roughness was measured by surveyor's level. The PSD is assumed by a stationary normal probability process with a zero mean value, and is calculated by a maximum entropy method (MEM) to a measured value of road surface roughness. Numerical examples are presented to obtain the information of road surface roughness on bridges in spectral characteristics, and to give the appropriate PSD using the dynamic response analysis of bridges under moving vehicles.

0644 Buckling of Angles: State of the Art

John B. Kennedy, Member, ASCE, (Prof., Dept. of Civ. Engrg., Univ. of Windsor, Windsor, Ontario, Canada) and **Murty K. S. Madugula**, (Assoc. Prof., Dept. of Civ. Engrg., Univ. of Windsor, Windsor, Ontario, Canada)

Journal of the Structural Division, Vol. 108, No. ST9, September, 1982, pp. 1967-1980

A summary of the theoretical and experimental studies conducted to investigate the buckling of angle columns and beams is presented. The theoretical analyses for flexural, torsional-flexural and plate buckling, and experimental investigations on single and built-up angle sections are expounded. Approximate methods of analysis and interaction diagrams for flexural buckling and equivalent radius of gyration method for torsional-flexural buckling are reviewed. North American and European design practices are compared. Recommendations for further research are given.

0645 Reliability Analysis of Wood Transmission Poles

Alain H. Peyrot, Member, ASCE, (Prof., Dept. of Civ. and Environmental Engrg., Univ. of Wisconsin, Madison, Wisc. 53706), **Marvin E. Criswell**, Member, ASCE, (Assoc. Prof., Dept. of Civ. Engrg., Colorado State Univ., Fort Collins, Colo.), **Mike D. Folse**, (Grad. Student, Dept. of Civ. Engrg., Colorado State Univ., Fort Collins, Colo.) and **Jean-Pierre Aznavour**, (Grad. Student, Dept. of Civ. Engrg., Univ. of Wisconsin, Madison, Wisc.)

Journal of the Structural Division, Vol. 108, No. ST9, September, 1982, pp. 1981-1994

The feasibility of probability-based design procedures for electrical transmission lines is discussed. Single pole structures have been selected in this initial study because of their simplicity as well as the lack of consistency in their design procedures. For the reliability analysis of a line, the paper emphasizes the "design point," or advanced mean value second order method. All equations needed for a "design point" analysis are presented. These equations were implemented in the reliability program PAPOLE. The computer program can perform a reliability or "design point" analysis for a line consisting of poles with random properties and the line can be subjected to random wind and ice loads. Any probability distribution can be prescribed for the variables. Sample reliability analyses are presented. It is shown how reliability analyses of existing lines can be used to calibrate new proposed design procedures. A design procedure based on the (LRFD) Load Resistance Factor Design format is recommended.

0646 Simplified P-Delta Analyses for Asymmetric Structures

Avigdor Rutenberg, Member, ASCE, (Assoc. Prof. of Civ. Engrg. and Head of Structures, Technion-Israel Inst. of Tech., Haifa, Israel)

Journal of the Structural Division, Vol. 108, No. ST9, September, 1982, pp. 1995-2013

Computer oriented and manual procedures are presented that enable a direct second order analysis of asymmetric building structures to be performed by means of first order formulation. This is achieved by modeling the geometric-stiffness of the structure as a fictitious column with negative lateral and torsional stiffness properties. Simplified methods are proposed for structures amenable to some form of uncoupling. The approach is also applied to elastic-plastic analysis of asymmetric structures by combining center-of-rigidity and sway-subassemblage procedures. A numerical example is provided.

0647 Lateral-Distribution Factors for Fatigue Design

Charles G. Schilling, Member, ASCE, (Assoc. Research Consultant, U.S. Steel Corp., Research Lab., Monroeville, Pa. 15146)

Journal of the Structural Division, Vol. 108, No. ST9, September, 1982, pp. 2015-2033

An extensive finite-element parametric study developed a chart giving lateral-distribution factors for the fatigue design of steel highway bridges. The chart gives conservative upper-limit factors for interior and exterior beams. The factor for interior beams is 0.5. The factor for exterior beams varies with the distance from the beam to the outer traffic lane and reaches a value of 0.9 when the center of the lane is directly over the beam. These upper-limit values can be used to make an initial fatigue check. If this check is satisfactory, as it would be in many cases, no further check is required. Otherwise, a second check can be made by using a more precise criterion that depends on the beam span and moment of inertia. The factors from the chart are generally much less than the American Association of State Highway and Transportation Officials (AASHTO) factors. Lateral-distribution factors were calculated from field measurements on 21 steel bridges. The factors ranged from 0.20–0.52 and were in good agreement with factors from the lateral-distribution chart.

0648 Impact Factors for Fatigue Design

Charles G. Shilling, Member, ASCE, (Assoc. Research Consultant, U.S. Steel Corp., Research Lab., Monroeville, Pa. 15146)

Journal of the Structural Division, Vol. 108, No. ST9, September, 1982, pp. 2034-2044

Theoretical and experimental information is presented on impact factors for the fatigue design of steel highway bridges. Theoretical studies suggest that the impact factors for individual trucks in actual traffic should generally be less than 0.25, but that much higher factors would occur occasionally due to unusual conditions, such as a bump at a critical location. Measurement of impact factors in actual steel bridges confirms these theoretical conclusions. Impact factors approaching 100 percent have been reported in studies aimed at determining peak impact factors for nonfatigue design. However, impact factors determined for individual trucks in traffic, or for test trucks, are generally much lower. This suggests that the impact factors given by the present American Association of State Highway and Transportation Officials (AASHTO) formula are representative of actual bridge loading and, therefore, are generally appropriate for fatigue design, which should be based on typical, rather than extreme, conditions.

0649 Prediction and Control of Pedestrian-Induced Vibration in Footbridges

John E. Wheeler, (Sr. Engr., Bridge Design Main Roads Dept., Waterloo Crescent, East Perth, Western Australia, Australia 6000)

Journal of the Structural Division, Vol. 108, No. ST9, September, 1982, pp. 2045-2065

Footbridges vibrate because the forces imparted by the user are applied with the frequency of the pace and movement, however small, is therefore a forced vibration. In instances where the pace coincides with a structure resonant frequency dynamic movement can be quite large and the user may be disturbed or even alarmed. Vibration is recognized as a serviceability limit state and a process that can be employed as a design tool permitting an early assessment of behavior is proposed. The subject is investigated in terms of the pedestrian excitation source, the structure response, and acceptance criteria. Calculated response compares favorably with test results. Remedial or preventative options in the form of various damping devices are reviewed.

0650 Eccentric Bracing in Tall Buildings

Andrew T. Merovich, Member, ASCE, (Project Engr., URS/John A. Blume & Assocs., Engrs., 130 Jessie St. , (at New Montgomery), San Francisco, Calif. 94105), **Joseph P. Nicoletti**, Fellow, ASCE, (Sr. Vice Pres., URS/John A. Blume & Assocs., Engrs., 130 Jessie St. , (at New Montgomery), San Francisco, Calif. 94105) and **Earl Hartle**, Member, ASCE, (Chf. Engr., John Portman & Assocs., Atlanta, Ga.)

Journal of the Structural Division, Vol. 108, No. ST9, September, 1982, pp. 2066-2080

Primary concerns in the design of high-rise buildings, particularly those located in seismically active regions, are the control of interstory drift and the ability of the structure to withstand inelastic deformations. The design of the 47 story 4 Embarcadero Center building in San Francisco, California, utilizes a recently developed structural system that provides economical control of interstory drift, as well as sufficient ductility to enable the structure to perform satisfactorily in the event of a major earthquake. The use of eccentric bracing elements is the key to this structural system. This paper describes how an eccentric bracing scheme was used to satisfy the requirements for both drift control and ductility in the design of a high-rise structure located in a zone of high seismic exposure.

0651 Finite Element Analysis of RHS T-Joints

Robert M. Korol, Member, ASCE, (Assoc. Prof., Dept. of Civ. Engrg. and Engrg. Mechanics, McMaster Univ., Hamilton, Canada, L8S 4L7) and **Farooque A. Mirza**, (Asst. Prof., Dept. of Civ. Engrg. and Engrg. Mechanics, McMaster Univ., Hamilton, Canada L8S 4L7)

Journal of the Structural Division, Vol. 108, No. ST9, September, 1982, pp. 2081-2098

The finite element method (FEM) is used to model the behavior of rectangular hollow section (RHS) T-joints beyond the elastic limit, and is applied to determine their ultimate and working strengths. Punching shear and rotational stiffnesses under branch axial force and bending moment have also been calculated. The unreinforced and haunch type joints were analyzed in the study. In both cases the chord top flange was modelled as a thin plate supported by coupled edge springs. Loading was incorporated through line loads along the perimeter of a rigid inclusion (the branch). In the elastic range, joint stiffness values were found to be in good agreement with the available test results. The sensitivity of joints to different geometric parameters, i.e., width ratio λ, haunch size λ_1, and chord thickness t was studied using the FEM model in the elastic-plastic range. Ultimate branch moments or punching shear forces were found to be about five times the corresponding first yield values, thus indicating the need to base design loads on an ultimate strength or maximum deflection criterion.

0652 Elastic Buckling of Laterally Continuous I-Beams

Peter F. Dux, (Lect. in Civ. Engrg., Capriconia Inst. of Advanced Education, Rockhampton, Qld., Australia) and **Sritawat Kitipornchai**, Member, ASCE, (Sr. Lect. in Civ. Engrg., Univ. of Queensland, Qld., Australia)

Journal of the Structural Division, Vol. 108, No. ST9, September, 1982, pp. 2099-2116

An approximate method for determining the elastic flexural-torsional buckling loads of laterally continuous structures is developed. The method can be applied to structures loaded at braced points. The braces and supports are assumed to prevent lateral deflection and twisting. The procedure locates a critical segment and adjacent restraining segments which together form a subassemblage. The elastic critical load is obtained by determining the effective length of the critical segment. Charts of effective length factors are presented and the factors are shown to depend on the amount of end restraint and on the critical segment beam parameter and end moment ratio. The proposed method is applied to a worked example. The result, along with other results presented in the paper, is in close agreement with accurate numerical solutions.

0653 Bolted Connections Dynamically Loaded in Tension

Leendert P. Bouwman, (Research Engr., Dept. of Civ. Engrg., Delft Univ. of Tech., Delft, The Netherlands)

Journal of the Structural Division, Vol. 108, No. ST9, September, 1982, pp. 2117-2129

A report of fatigue tests on tensile loaded bolted connections is presented. The tests clearly demonstrate that the structural design is of major importance with regard to the fatigue strength of such connections. By designing connections which are structurally well-detailed, it is possible to entirely eliminate bolt fatigue in tension, while in that case an elaborate analysis of the fatigue strength of the bolts can be eliminated.

0654 Full-Scale Tests on Double-Tapered Glulam Beams

Richard M. Gutkowski, Member, ASCE, (Assoc. Prof. of Civ. Engrg., Colorado State Univ., Fort Collins, Colo.), **George R. Dewey**, (Instr. of Engrg. Sci., Univ. of Idaho, Moscow, Idaho) and **James R. Goodman**, Member, ASCE, (Prof. of Civ. Engrg., Colorado State Univ., Fort Collins, Colo.)

Journal of the Structural Division, Vol. 108, No. ST10, October, 1982, pp. 2131-2148

Analytical and experimental studies were performed on straight-soffited, double-tapered glulam beams. Full-scale tests were conducted under working loads to evaluate the applicabiLity of the finite element method for theoretical analysis. Two Douglas-fir and two Southern pine beams, each with a span of 30 ft (9.1 m), were studied. Each was instrumented for strain and deflection measurements. Experimental flexural and perpendicular to grain tensile stresses were compared with theoretical predictions. Load tests were also conducted to failure. Both Southern pine members failed in tension due to bending, whereas the Douglas-fir members failed in tension perpendicular to grain. The interaction of stresses concept of failure is evaluated in light of the test results observed in this study.

0655 Full-Scale Tests on Single-Tapered Glulam Members

Richard M. Gutkowski, Member, ASCE, (Assoc. Prof. of Civ. Engrg., Colorado State Univ., Fort Collins, Colo. 80523), **George R. Dewey**, (Instr. of Engrg. Sci., Univ. of Idaho, Moscow, Idaho) and **James R. Goodman**, Member, ASCE, (Prof. of Civ. Engrg., Colorado State Univ., Fort Collins, Colo. 80523)

Journal of the Structural Division, Vol. 108, No. ST10, October, 1982, pp. 2149-2161

Analytical and experimental studies were conducted on Glulam members with tapered configuration. Full-scale flexural tests were conducted under working loads. Three different geometries were investigated; namely a beam, an arch leg and an arch rafter. Each geometry was replicated in Douglas-fir and Southern pine species. Strain and deflection were measured and compared with theoretical predictions obtained by a finite element modeling. Load tests were also performed to failure. Stress results indicate that current design computations yield highly conservative stress levels. A commonly used interaction equation is evaluated in light of the ultimate strength results observed in the tests to destruction.

0656 Buckling of Cooling Towers

John F. Abel, Member, ASCE, (Assoc. Prof. of Struct. Engrg., Cornell Univ., Ithaca, N.Y.), **David P. Billington**, Fellow, ASCE, (Prof. of Civ. Engrg., Princeton Univ., Princeton, N.J.), **Dennis A. Nagy**, Member, ASCE, (Prof. Leader, SDRC-Engrg. Gmbh, Wiesbaden, West Germany) and **Christine Wiita-Dworkin**

Journal of the Structural Division, Vol. 108, No. ST10, October, 1982, pp. 2162-2174

The main goal of this study is to provide a clearer basis for judging the succeding safety of wind-loaded cantilever hyperboloidal shells. New predictions for cooling tower buckling are presented and compared with newly available experimental information. Various analysis techniques are evaluated, and the equivalent axisymmetric approach (including suction) is identified as a useful compromise between the accuracy and economy of calculation.

0657 Wave and Earthquake Response of Offshore Structures: Evaluation of Modal Solutions

Stavros A. Anagnostopoulos, Member, ASCE, (Acting Head, Inst. of Engrg. Seismology and Earthquake Engrg., Hapsa 1, Thessaloniki, Greece)

Journal of the Structural Division, Vol. 108, No. ST10, October, 1982, pp. 2175-2191

Direct numerical integration of the equations of motion is used to evaluate the adequacy of modal solutions, as applied to the analysis and design of large offshore structures under wave and earthquake loadings. Two such solutions are considered: mode acceleration and simple modal superposition. Both these methods are widely used by the offshore industry but recently their adequacy in predicting member forces under wave loadings has been questioned. After reviewing the underlying assumptions and the differences between the two solutions in relation to the two types of loading, results from analyses of a template platform, modeled with 1,428 degrees of freedom, are compared to those obtained by a direct solution in geometric coordinates. It is found that for wave loadings the mode acceleration method gives excellent results, while simple modal superposition can lead to erroneous member forces and stresses even if a large number of modes is used. For earthquake loadings, on the other hand, mode acceleration is not any better than simple modal superposition, which gives answers that may be considered acceptable for design applications.

0658 Beam Element Stiffness Matrix for X-Braced Truss

R. Richard Avent, Member, ASCE, (Prof., Dept. of Civ. Engrg., Mississippi State Univ., Mississippi State, Miss.) and **Raja R. A. Issa**, (Grad. Research Asst., Dept. of Civ. Engrg., Mississippi State Univ., Mississippi State, Miss.)

Journal of the Structural Division, Vol. 108, No. ST10, October, 1982, pp. 2192-2210

The techniques of discrete field analysis are utilized to investigate the response of X-braced trusses to various static loading conditions. Deflection formulas for arbitrary static loads are presented in finite Fourier series form. These formulas are used to derive stiffness coefficients which comprise the equivalent beam element stiffness matrix for a given X-braced truss. Fixed end moments for arbitrary static loads are also derived from the deflection formulas. All stiffness formulas are exact in the context of linear, elastic truss analysis and are valid for an arbitrary number of panels. These stiffness coefficients when compared to approximate estimates are found to vary by as much as 100% within a given range of span lengths. The formulas can be used to analyze a variety of structural systems which have X-braced trusses as the basic component. Examples include two-way truss systems, trussed frames, and planar trusses continuous over several supports. The analysis of such systems would be identical to that in which the individual components were beams except that here modified stiffness coefficients would be used.

0659 Creep of Concrete in Variable Environments

William L. Gamble, Member, ASCE, (Prof. of Civ. Engrg., Univ. of Illinois at Urbana-Champaign, Urbana, Ill. 61801)

Journal of the Structural Division, Vol. 108, No. ST10, October, 1982, pp. 2211-2222

Creep and shrinkage measurements are reported on three steam-cured and one moist-cured control concrete. Half the specimens were stored in a constant environment of 70° F and 50% RH, and half were stored outdoors in central Illinois. The shrinkage of the outdoor specimens was much less than in the laboratory, as expected, but the creep strains were comparable in spite of the large difference between the average relative humidities of the two environments. This high creep in a relatively moist environment must be taken into account if the loss of prestress in prestressed concrete structure and the time-dependent deformations of all concrete structures exposed to variable environments are to be properly predicted. Annual cyclic expansions and contractions occurred in all outdoor specimens.

0660 Internal Pressure and Building Safety

Henry Liu, Member, ASCE, (Prof. of Civ. Engrg., Univ. of Missouri, Columbia, Mo. 65211) and **Patrick J. Saathoff**, (Research Assoc., Center for Environmental Assessment, National Oceanic and Atmospheric Administration, Columbia, Mo. 65201)

Journal of the Structural Division, Vol. 108, No. ST10, October, 1982, pp. 2223-2234

The wind-induced pressure inside buildings has a profound effect on building safety. Yet, little attention has been given to it in the past. The effect is either ignored or improperly considered in contemporary building codes or standards. This paper seeks to clarify the matter and improve the way internal pressure is considered in the design of buildings. Both static and dynamic effects of internal pressure are considered herein. The dynamic effects include the overshooting of internal pressure following the breakage of a windward window during high wind, and the Helmholtz resonance created by any disturbance. Combined effects of internal and external pressure variations are also reviewed.

0661 Mutual Interference Drag on Signs and Luminaires

James F. Marchman, III., (Wind Tunnel Dir. and Assoc. Prof. of Aerospace and Ocean Engrg., Virginia Polytechnic Inst. and State Univ., Blacksburg, Va. 24061) and **Todd D. Werme**, (Undergraduate Asst., Virginia Polytechnic Inst. and State Univ., Blacksburg, Va. 24061)

Journal of the Structural Division, Vol. 108, No. ST10, October, 1982, pp. 2235-2244

Wind tunnel tests were conducted to determine the increase in drag on flat plate and luminaire models due to the proximity of nearby models of like shape. Test results showed drag increases of up to twenty percent due to interference between the flows around the two bodies. The maximum interference drag generally occurred at a spacing of about 20% of a single model width. Several models exhibited a pronounced dip in the data where interference drag first decreased and then increased again. On the luminaire model this resulted in an optimum spacing where the interference drag went to zero.

0662 Lateral Buckling Tests on Welded Continuous Beams

Yuhshi Fukumoto, Member, ASCE, (Prof., Dept. of Civ. Engrg., Nagoya Univ., Chikasa-ku, Nagoya, Japan), **Yoshito Itoh**, (Research Asst., Dept. of Civ. Engrg., Nagoya Univ., Chikusa-ku, Nagoya, Japan) and **Ryoji Hattori**, (Grad. Student, Dept. of Civ. Engrg., Nagoya Univ., Chikusa-ku, Nagoya, Jmpan)

Journal of the Structural Division, Vol. 108, No. ST10, October, 1982, pp. 2245-2262

Twenty-one two-span continuous beams with nominally identical cross sections of welded I-250x100x6x8 mm are tested in seven groups of three, each group having different span length. These continuous beams are tested under a concentrated load on the midspan of one side span. The continuous test beams are on the simple and clamped supports at the both ends against in and out-of plane displacements with an intermediate support. The test results are compared directly with the buckling behavior of simple beams and also with the buckling results based on the effective length concept. The ultimate strengths of the present continuous beams are higher than those of simple beams and scatter between the elastic buckling strength curves of simple beams and of propped cantilever beams which are considered to have a nonloaded span with infinite rigidity. However, in the plot of (P_u/P_y) vs ($\lambda = P_y/P_E$) the test results of continuous beams scatter on the lower part of the results of simple beams and the coefficients of variation are close to the values of simple beams.

0663 Rigidities of Nonorthogonally Shaped Waffle Slabs

John B. Kennedy, Member, ASCE, (Prof., Dept. of Civ. Engrg., Univ. of Windsor, Ontario, Canada) and **K. J. Iyengar**, (Research Officer, Atomic Energy of Canada Ltd., Mississauga, Ontario, Canada)

Journal of the Structural Division, Vol. 108, No. ST10, October, 1982, pp. 2263-2278

Theoretical formulae are presented for the various rigidities of non-orthogonally shaped waffle concrete slabs for both the precracking and postcracking stages. The deduced expressions for the flexural, coupling and torsional rigidites are applicable to both reinforced and prestressed concrete constructions. Experimental results derived from tests on twelve non-orthogonally shaped waffle concrete slabs verify and substantiate the theoretical derivations. The effect of creep on the rigidites is discussed and a method to account for it is suggested. The use of reliable estimates of the rigidities should lead to a more economical and better design of such waffle-type structures.

0664 Inelastic Cyclic Analysis of Pinned-Ended Tubes

Shouji Toma, (Assoc. Prof., Faculty of Engrg., Hokkaigakuen Univ., W-11, S-26, Chunouku, Sapporo, Japan) and **Wai F. Chen**, Member, ASCE, (Prof. and Head of Struct. Engrg., School of Civ. Engrg., Purdue Univ., W. Lafayette, Ind. 47907)

Journal of the Structural Division, Vol. 108, No. ST10, October, 1982, pp. 2279-2294

An analytical study of the inelastic cyclic load-deflection behavior and load-shortening behavior of axially loaded steel tubular bracing members subject to cyclic loading is presented. Expressions are first derived for the moment and axial strain expressed explicitly in terms of curvature and thrust for tubular sections with geometric imperfections and residual stresses. For the case of reversed loading, several approximations of the moment-thrust-curvature curves and moment-thrust-axial strain curves are made. Using these relations, cyclic solutions of pin-ended columns subjected to one cycle of axial loading are obtained and compared with some available experimental tests. Good agreement is generally observed.

0665 Dynamic Tests and System Identification of Bridges

Bruce M. Douglas, Member, ASCE, (Prof., Dept. of Civ. Engrg., Univ. of Nevada, Reno, Nev. 89557) and **Wayne H. Reid**, Assoc. Member, ASCE, (Grad. Student, Dept. of Civ. Engrg., Univ. of Nevada, Reno, Nev. 89557)

Journal of the Structural Division, Vol. 108, No. ST10, October, 1982, pp. 2295-2312

A five-span, four-hundred-foot long, reinforced concrete box girder birdge supported on single column piers that are pile-founded was subjected to extensive dynamic tests. The pullback and quick-release method of excitation was used. Two D-8 crawler tractors were used to apply total release loads that had a magnitude of about one-quarter of the earthquke design loads. The transverse mode shapes and natural frequencies obtained from these tests were used in conjunction with the system identification procedure outlined herein. By using these techniques, it is possible to identify the in situ dynamic stiffnesses of the pier foundations as well as the abutment structures. In addition, the dynamic bending and shear stiffness factors for the superstructure deck were also obtained. Thus, carefully conducted dynamic tests of full-scale highway bridges provide detailed information about the structure. At the same time, they shed much experimental light on the soil-structure interaction aspects of the problem, which in many cases greatly affects the distribution of seismic loads in bridges.

0666 Ultimate Strength of Biaxially Loaded Composite Sections

Luc Lachance, (Assoc. Prof., Dept. of Civ. Engrg., Laval Univ., Quebec City, Canada G1K 7P4)

Journal of the Structural Division, Vol. 108, No. ST10, October, 1982, pp. 2313-2329

A theory is developed for the non-linear or linear analysis of arbitrary composite sections subjected to biaxial loading. The accuracy of the theory is verified by comparison with known uniaxial non-linear analyses of steel wide-flange and channel sections encased in reinforced concrete, with classical linear theory based on principal moments of inertia, and with test results from various authors. Three basic shapes of concrete stress-strain diagram are considered, for two allowable compressive strains and two ultimate strengths. This gives twelve different types of concrete stress distribution which are used to analyse four different composite sections. A comparative study of the results indicates that the factor which most influences ultimate biaxial bending strength and curvature is the allowable concrete compressive strain. The shape of concrete stress distribution has little effect on the section behavior.

0667 Sum of Dependent Gamma Loads

K. Breitung, (Dipl.-Math., Research Asst., Institut fur Massivbau, Technische Universitat Munchen, Arcisstr. 21, D-8000 Munchen, West Germany)

Journal of the Structural Division, Vol. 108, No. ST10, October, 1982, pp. 2333-2338

A model for correlated sustained live loads is outlined. The live loads are assumed to be gamma distributed random variables. It is shown that already small correlations between loads can substantially increase the probability that the sum of the loads exceeds a specified level. The asymptotic behavior of the cumulative distribution function is derived by Laplace transform methods.

0668 Balancing of Trunnion-Type Bascule Bridges

Sung Y. Lu, (Assoc. Prof., Dept. of Engrg. Sci., Univ. of Florida, Gainesville, Fla. 32611), **Lawrence E. Malvern**, (Prof., Dept. of Engrg. Sci., Univ. of Florida, Gainesville, Fla. 32611), **David A. Jenkins**, (Assoc. Engr., Dept. of Engrg. Sci., Univ. of Florida, Gainesville, Fla. 32611), **S. Frederick Allred**, (Prof. Engr., Civ. Engr., Dept. of Transportation, Tallahassee, Fla. 32301) and **Lynn W. Biwer**, (Prof. Engr., Mech. Engr., Dept. of Transportation, Tallahassee, Fla. 32301)

Journal of the Structural Division, Vol. 108, No. ST10, October, 1982, pp. 2338-2343

This paper reports an improved balancing technique using continuous strain-gage recording of drive-shaft torque during both opening and closing of bascule bridges. Explicit formulas for correction weight changes are developed. Static measurements are compared with the continuous recording. It is shown that satisfactory measurement can be made by either method, provided that a damped oscillation occurs in the observed torque when the opening is stopped so that the static friction effect is reduced.

0669 Artificial Intelligence in Simple Beam Design

Martin F. Rooney, Assoc. Member, ASCE, (Sr. Systems Analyst, Stone and Webster Engrg. Corp., Boston, Mass. 02107) and **Steven E. Smith**, (Research Asst., Dept. of Civ. Engrg., Rensselaer Polytechnic Inst., Troy, N.Y. 12181)

Journal of the Structural Division, Vol. 108, No. ST10, October, 1982, pp. 2344-2348

The concept of an intelligent computer program is introduced using the design of a simply-supported wide-flange beam as a model. The key to the approach is component selection through a database organized by frequency of past usage. The programming implications are divided into acquisition of the database and application of the database. For the truly intelligent program, a dynamic feedback loop connects acquisition to application. The intelligent approach is compared to a conventional approach, binary search, with favorable results.

0670 Simultaneous Airblast and Ground Motion Response

Charles H. Dowding, (Assoc. Prof. of Civ. Engrg., Northwestern Univ., Evanston, Ill. 60201), **Craig S. Fulthorpe**, (Grad. Student, Northwestern Univ., Evanston, Ill. 60201) and **Robert T. Langan**, (Research Sci., Gulf Science and Tech. Co., Dept. of Seismology, Pittsburgh, Penn.)

Journal of the Structural Division, Vol. 108, No. ST11, November, 1982, pp. 2363-2378

A single-degree-of-freedom approach is used to model structural response to superimposed airblast and ground vibration from conventional explosions. A method is used whereby the force applied to the structure by the airblast is combined with the force due to the ground motion into a single exciting force. The use of field data permits comparison of model response to actual measured structural response. Natural frequencies and damping values used in this study are typical values obtained for residential structures in other studies. The model response is similar to the actual response in frequency content, and the peak velocities are predicted within a factor of two.

0671 Circular Voided Concrete Slab Stiffnesses

G. Elliott, (Lect., Dept. of Engrg., Univ. of Lancaster, Lancaster, England) and **L. A. Clark**, (Lect., Dept. of Civ. Engrg., Univ. of Birmingham, England)

Journal of the Structural Division, Vol. 108, No. ST11, November, 1982, pp. 2379-2393

Finite element procedures for determining the plate bending stiffnesses of a circular voided slab are presented and the results are compared with those obtained from tests on model elastic voided plates. Preliminary investigations showed that a simplified plane-strain formulation, rather than a more complex general plane-strain formulation, could be used to determine the transverse flexural stiffness. Torsional stiffnesses were obtained from an analysis using the Prandtl stress-function formulation of the torsion problems. Tests were carried out on four epoxy-resin model plates having different void sizes. Good agreement was obtained between the finite element and experimental results. For design purposes, charts are presented which enable a designer of a concrete voided slab, having a Poisson's ratio of 0.2, to determine the values of the plate bending stiffnesses required for a thin plate analysis of such a slab. It is also shown that simple calculations based upon replacing a circular void with an equivalent square void are often adequate for design purposes.

0672 Prestressed Decks in Continuous Composite Bridges

John B. Kennedy, Member, ASCE, (Prof., Dept. of Civ. Engrg., Univ. of Windsor, Windsor, Ontario, Canada) and **Nabil F. Grace**, (Research Asst., Dept. of Civ. Engrg., Univ. of Windsor, Windsor, Ontario, Canada)

Journal of the Structural Division, Vol. 108, No. ST11, November, 1982, pp. 2394-2410

A convergent series solution, using the concept of equivalent orthotropy, is employed to analyze continuous composite bridges under static loads. The effect of the interaction between prestressing a portion of the deck slab and shear connectors in the region of the intermediate supports is examined. It is shown that prestressing the deck slab increases substantially the cracking load as well as the stiffness of such bridges. Results from tests on two bridge models verify and substantiate the theoretical analysis.

0673 Tornado Technology and Professional Practice

Joseph E. Minor, Fellow, ASCE, (Dir., Inst. for Disaster Research, Texas Tech Univ., Lubbock, Tex.)

Journal of the Structural Division, Vol. 108, No. ST11, November, 1982, pp. 2411-2422

Advancements in tornado technology make it possible to plan and design for the tornado. Tornado proof designs are being employed in the nuclear industry and in the design of shelters. Understandings of the tornado and design experiences can be utilized to plan facilities and to design conventional buildings to resist the tornado. Literature which represents modern tornado technology is summarized, examples of designs for tornadoes are presented, and potential uses for tornado technology in engineering practice are outlined. Barriers to the effective use of tornado technology are discussed.

0674 Longitudinally Reinforced Cold-Formed Steel Beam Webs

Richard P. Nguyen, Assoc. Member, ASCE, (Sr. Struct. Engr., Black & Veatch Consulting Engrs., Kansas City, Mo.) and **Wei-Wen Yu**, Fellow, ASCE, (Prof. of Civ. Engrg., Univ. of Missouri-Rolla, Rolla, Mo.)

Journal of the Structural Division, Vol. 108, No. ST11, November, 1982, pp. 2432-2442

In cold-formed steel structures, the bending strength of beams can be improved by adding longitudinal stiffeners to web elements. This study deals with the required minimum rigidity of the longitudinal stiffeners and the ultimate bending capacity of cold-formed steel beams having adequately stiffened webs.

0675 Reinforcement Concrete Plate: Frictional Limit Design

Zdenek P. Bazant, Fellow, ASCE, (Prof. of Civ. Engrg. and Dir., Center for Concrete and Geomaterials, Technological Inst., Northwestern Univ., Evanston, Ill. 60201) and **Chuan Lin**, (Engr., Dynamic Load Evaluation Unit, Nuclear Energy Business Operation, General Electric Co., San Jose, Calif. 95125)

Journal of the Structural Division, Vol. 108, No. ST11, November, 1982, pp. 2443-2459

The limit design of a regular rectangular reinforcing net in a concrete plate subjected to arbitrary normal and shear forces and bending and twisting moments is studied. The assumptions are: The reinforcement is plastic; concrete resists no tension; a normal crack of any direction may exist in concrete; the normal stress in concrete is constant; in-plane shear stresses can be transmitted in concrete, are of constant magnitude but possibly of alternating sign, and their ratio to the normal compressive stresses may not exceed the friction coefficient. The previously developed frictional (no-slip) limit design for membrane loading is the special case when the compression zone covers the whole thickness of the plate. Design charts for the case of no membrane forces are computed and plotted. They differ up to about 20% from the classical solution. In contrast to pure membrane loading, the frictional design for failure may yield a lighter reinforcement than the classical design. Mostly, however, it yields a heavier reinforcement. The difference vanishes when the bars are laid in the direction of principal bending moments. In contrast to the classical design, the angle between the crack and the bars does not have to be 45° for the optimum design.

0676 Formulas for Wood Shear Walls

John T. Easley, Member, ASCE, (Prof., Dept. of Civ. Engrg., Univ. of Kansas, Lawrence, Kans. 66044), **Mehdi Foomani**, (Grad. Student, Dept. of Civ. Engrg., Univ. of Kansas, Lawrence, Kans. 66044) and **Robert H. Dodds**, Assoc. Member, ASCE, (Asst. Prof., Dept. of Civ. Engrg., Univ. of Kansas, Lawrence, Kans. 66044)

Journal of the Structural Division, Vol. 108, No. ST11, November, 1982, pp. 2460-2478

Formulas are presented for the analysis of typical wood-frame shear walls with sheathing attached by nails or other types of discrete fasteners. Formulas are derived for the sheathing fastener forces, for the linear shear stiffness of a wall, and for the nonlinear shear load-strain behavior of a wall. These formulas are shown to be in agreement with the results of load tests and of finite element, linear, and nonlinear analyses. Details of the load tests and finite element analyses are given. The formulas are relatively simple and should be applicable to the design of wood frame shear walls with plywood or other types of sheathing.

0677 Flexural Strength of Flat Slabs at Exterior Columns

Harinatha B. Goli, (Reader in Civ. Engrg., Andhara Univ., Waltair, Andhra Pradesh, India) and **Hans Gesund**, Fellow, ASCE, (Prof. of Struct. Engrg., Univ. of Kentucky, Lexington, Ky. 40506)

Journal of the Structural Division, Vol. 108, No. ST11, November, 1982, pp. 2479-2495

The flexural strength of exterior panels of flat slab structures is investigated by use of Yield Line Theory. Six controlling, geometrically admissible, collapse mechanisms are identified, three of which involve local flexural failures around the columns. The loading is uniformly distributed, the columns are rectangular, and the reinforcement is arranged orthotropically. The coefficient of orthotropy and the ratio of negative to positive yield moment may vary arbitrarily. The governing equations for the six mechanisms are derived and presented in dimensionless form. Based on numerical solutions of four of the equations, a set of graphs is provided which covers practical ranges of the important physical and geometrical parameters. These graphs are useful for the analysis or design of the exterior slab panels. In addition, an equation is derived which gives an upper bound on the bending moment which the slab can transfer into the exterior column at collapse load through flexural action.

0678 Planar Strength of Restrained Beam Columns

Sriramulu Vinnakota, Member, ASCE, (Assoc. Prof. of Civ. Engrg., Marquette, Univ., Milwaukee, Wisc.)

Journal of the Structural Division, Vol. 108, No. ST11, November, 1982, pp. 2496-2516

A theoretical method that describes the planar, elastic-plastic behavior of a beam-column loaded at the ends and along the length is described. The theory considers the three important parameters: residual stresses, geometrical imperfections and nonlinear end-restraints. A computer program using finite difference technique has been developed to obtain numerical results. Several numerical results are presented to show the influence of the three parameters mentioned on the maximum strength of I-section columns and beam-columns.

0679 Reliability Procedure for Fixed Offshore Platforms

William D. Anderson, (Supervising Engineer, Exxon Company U.S.A., New Orleans, La.), **Mark N. Silbert**, (Proj. Engr., Exxon Production Malaysia Inc., Kuala Lumpur, Malaysia) and **James R. Lloyd**, Member, ASCE, (Research Advisor, Exxon Production Research Co., Houston, Tex.)

Journal of the Structural Division, Vol. 108, No. ST11, November, 1982, pp. 2517-2538

This paper presents a method for calculating the "notional" risk of overload associated with fixed offshore platforms subjected to random wave loading. The lifetime exposure of the structure to potentially damaging storms is treated within a rigorous mathematical framework. The procedure used to establish the long term extreme wave load distribution is emphasized. This procedure utilizes presently available oceanographic information in a statistical model of the wave environment. The composite probability of exceedance associated with all significant loading events is computed for a lifetime exposure to the ocean environment. For the purposes of demonstrating this procedure, only the resistance of pile foundations to overturning forces is investigated. A simplified technique for estimating the ultimate foundation capacity is presented and compared to a more rigorous method. Failure mechanisms in the jacket are not investigated, i.e., the structural jacket is assumed to be adequate to transfer loads into the pile foundation.

Finally, the resistance and load effects are combined to evaluate the reliability of the pile foundation in a wave environment. Results are presented for a recently installed Gulf of Mexico platform. The sensitivity of the results to various input parameters is discussed.

0680 Anton Tedesko: Thin Shells and Esthetics

David P. Billington, Fellow, ASCE, (Prof. of Civ. Engrg., Princeton Univ., Princeton, N.J. 08544)

Journal of the Structural Division, Vol. 108, No. ST11, November, 1982, pp. 2539-2554

Rarely can historians attribute to one person the introduction into a society of a new and widely useful engineering idea. We have no difficulty, however, in attributing to one structural engineer, Anton Tedesko, the introduction of thin-shell concrete roof structures into the United States. This achievement merits some reflection not only on the events themselves, but also on the background and personality of the individual engineer. Anton Tedesko's work and ideas in thin shell concrete studies demonstrate: First, his educational background broadened his vision, emphasized design at least as much as analysis, and included aesthetics as a serious part of structural design. Second, his experience in European practice gave him not only a solid grounding in thin shell construction, but also a first-hand view of the operation of designer-builders. Third, the history of thin shell structures shows how new ideas in structural engineering came from designers and especially those with construction experience. Fourth, Tedesko's long career has focused on the design of full-scale works in which behavior, cost, and appearance have been the criteria for success.

0681 Laterally Loaded Cylinders in Half Space

Isamu A. Okumura, (Assoc. Prof. of Civ. Engrg., Kitami Inst. of Tech., Kitami, Japan) and **Jin Yoshimura**, Member, ASCE, (Prof. of Civ. Engrg., Univ. of Hakkaido, Sapporo, Japan)

Journal of the Structural Division, Vol. 108, No. ST11, November, 1982, pp. 2555-2570

The extended Neuber solution which has the curl of a harmonic vector added to Papkovich-Neuber solution is proposed as a useful solution to asymmetric problems referring to cylindrical coordinates. A three-dimensional analysis for a laterally loaded semi-infinite cylinder embedded in an elastic half-space is presented on the basis of that solution. The potentials of displacement for a semi-infinite cylinder and an elastic half-space are given. The components of displacement and stress are expressed in those potentials of displacement. The Fourier, Bessel, and Dini expansion are used in order to satisfy the boundary and continuity conditions. Numerical calculations are made for various values of Poisson's ratios and the shear moduli of a semi-infinite cylinder and an elastic half space. Some available data for the behavior of such a structure as single piles in a foundation, anchor bolts, and reinforcements subjected to a shearing force in a concrete with cracks are presented.

0682 Short-Term Records and Extreme Wind Speeds

Emil Simiu, Member, ASCE, (Struct. Research Engr., Center for Building Tech., National Bureau of Standards, Washington, D.C. 20234), **James J. Filliben**, (Mathematician Statistician, Center for Applied Mathematics, National Bureau of Standards, Washington, D.C. 20234) and **James R. Shaver**, (Civ. Engr., Center for Manufacturing Engrg., National Bureau of Standards, Washington, D.C. 20234)

Journal of the Structural Division, Vol. 108, No. ST11, November, 1982, pp. 2571-2577

An empirical study was conducted to determine whether design wind speeds can be estimated confidently from short records, i.e., records extending over periods of a few years. The study was based upon the analysis of records of daily, weekly, and monthly largest wind speeds taken at 36 United States weather stations over periods ranging from three to 10 years and upon largest yearly wind speeds recorded at these stations over periods ranging from approximately 25-45 years. Investigations were presented into the type of probability distributions that best fit sets of largest daily, weekly, and monthly wind speeds and into the relationship between winds estimated from short records of largest daily and weekly wind speeds on the one hand, and from 25-year to 45-year records of largest yearly wind speeds on the other hand. The main effort in the paper was focused on the relationship between estimates of 50-year wind speeds based on three-year records of largest monthly speeds and on 25-year to 45-year records of largest yearly speeds. It was found that 50-year speeds can be estimated confidently from three-year records of the largest monthly speeds.

0683 Preliminary Planning of Steel Roof Space Trusses

Jerome S. B. Iffland, Fellow, ASCE, (Pres., Iffland, Kavanagh, and Waterbury, New York, N.Y.)

Journal of the Structural Division, Vol. 108, No. ST11, November, 1982, pp. 2578-2590

During the preliminary plan phase of designing a steel roof space truss, a number of parameters must be considered and evaluated. These include the geometry of the framing system, the possible use of a proprietary system, the type of nodal joint, grid spacing, method of support, and depth of truss. Twelve types of truss framing systems that have been used for actual structures are presented. These systems are developed from basic elements of either rectangular or triangular prisms or from pyramids with either rectangular or triangular bases. Eleven different proprietary systems are briefly described. These proprietary systems are adaptable to closely spaced grids and shallow trusses and result in low weights per square area of structural material. Alternatively, a space truss can be developed utilizing a fabricated joint. Two examples are presented. These types of structures are usually deeper in depth with a larger module size. The weight of the structure is generally higher than that obtained by using a proprietary system. The actual weight of the truss is a secondary cost consideration. Three methods of analyses for preliminary designs are referenced. The review highlights some of the important factors that must be considered during the planning state of a steel roof space truss design.

0684 Minimization of Composite Dead Load Effects

David R. Schelling, Member, ASCE, (Assoc. Prof., Dept. of Civ. Engrg., Univ. of Maryland, College Park, Md. 20742) and **Norman Clerman**, Assoc. Member, ASCE, (Grad. Asst., Dept. of Civ. Engrg., Univ. of Maryland, College Park, Md. 20742)

Journal of the Structural Division, Vol. 108, No. ST11, November, 1982, pp. 2591-2608

A method is presented whereby the effects of the sequencing of dead load effects on continuous composite highway bridges may be determined. All relevant time-dependent concrete properties are accounted for and codified into a comprehensive computer program. The effects of staging for 12 highway bridges is presented along with a schedule for reducing the cracking in the concrete through the use of an optimization technique.

0685 Composite Cold-Formed Steel-Concrete Beams

George Abdel-Sayed, Member, ASCE, (Prof. of Civ. Engrg., Univ. of Windsor, Windsor, Ontario, Canada, N9B 3P4)

Journal of the Structural Division, Vol. 108, No. ST11, November, 1982, pp. 2609-2622

Experimental study was conducted at the University of Windsor and a composite beam system using soffits made of cold-formed steel channels was developed. The test results show good characteristics of these beams, especially with regard to their ultimate load carrying capacity and bond strength. The study shows that by replacing the standard reinforcing bars into a cold-formed steel section of equal area, the structural performance of the beam could be almost unchanged, while saving is achieved in the cost of forms and shorings. Soffits made in the form of stiffened channels with embossments performed very well as integral parts of the composite beam. The combined action of the embossments and the channel's lips lead to excellent bond characteristics between the soffit and the concrete. This can be explained by the fact that the concrete has to be lifted up in order to slide over the embossments, while this movement is restrained by the lips of the channel.

0686 Finite Element Model for Composite Masonry

Subhash C. Anand, Member, ASCE, (Prof. of Civ. Engrg., Clemson Univ., Clemson, S.C. 29631) and **David T. Young**, (Design Engr., Piedmont Engrs. and Architects, Greenville, S.C. 29631)

Journal of the Structural Division, Vol. 108, No. ST12, December, 1982, pp. 2637-2651

A two-dimensional composite element is developed in this study that is capable of predicting the out-of-plane interlaminar shearing stresses between the brick and block wythes without using a three-dimensional finite element model. This composite element is utilized in two-dimensional analyses in a plane-strain condition. The results of these analyses are compared with those using the plane-strain finite element models. These comparisons indicate that the two-dimensional composite element models predict interlaminar shearing stresses in the collar joint that are in good agreement with those obtained from the plane-strain models. Although the composite element is initially developed for only elastic conditions in an unreinforced collar joint without any consideration to fracture and failure, these criteria can be easily incorporated in the proposed model.

0687 Composite Beams with Web Openings: Design

C. Michael Donoghue, Assoc. Member, ASCE, (Principal, Porter Donoghue Consulting Engrs., Austin, Tex.)

Journal of the Structural Division, Vol. 108, No. ST12, December, 1982, pp. 2652-2667

A general method for the analysis and design of composite steel-concrete beams with openings centered in the web of the steel section is developed. The approach is based on ultimate strength concepts and allows for the presence of variable amounts of reinforcements in the web and partial shear connection. Design equations and a worked example are presented.

0688 Serviceability Limit States: Connection Slip

Theodore V. Galambos, Fellow, ASCE, (Prof. of Civ. Engrg., Univ. of Minnesota, Minneapolis, Minn. 55455), **T. A. Reinhold**, Member, ASCE, (Research Struct. Engr., Center for Building Tech., National Bureau of Standards, Washington, D.C.) and **Bruce Ellingwood**, Member, ASCE, (Research Struct. Engr., Center for Building Tech., National Bureau of Standards, Washington, D.C.)

Journal of the Structural Division, Vol. 108, No. ST12, December, 1982, pp. 2668-2680

The serviceability limit state for slip of bolted steel joints is the slip-resistance. The statistical properties of the parameters which define this resistance are presented and analyzed. Based on these properties, and using First-Order Second-Moment probabilistic analysis, limit-states design criteria are developed for friction-grip bolted joints.

0689 Tension Stiffening Concept Based on Bond Slip

Helmut Floegl, (Research Asst., Dept. of Civ. Engrg., Technical Univ. of Vienna, Vienna, Austria) and **Herbert A. Mang**, Member, ASCE, (Assoc. Prof., Dept. of Civ. Engrg., Technical Univ. of Vienna, Vienna, Austria)

Journal of the Structural Division, Vol. 108, No. ST12, December, 1982, pp. 2681-2701

Consideration of tension stiffening in reinforced concrete panels by means of so-called "tension stiffening factors", appearing in total and incremental equilibrium equations, has proved to be very effective. These factors result from bond slip between the reinforcement and the surrounding concrete. They are directly proportional to the average bond stress between adjacent cracks, the crack spacing, the cosine of the angle enclosed by the normal to the crack, and the reinforcement and inversely proportional to the average steel stress and the diameter of the reinforcement. The proposed concept does not require introduction of relative displacements. The method is applied to predict the short-term behavior of panels. The Finite Element Method is used as an analysis tool. Results agree well with experimental values obtained by other researchers. If, however, tension stiffening is disregarded, failure loads are underestimated whereas steel stresses are overestimated.

0690 Analytical Model for Concrete Confinement in Tied Columns

Shamim A. Sheikh, (Asst. Prof., Dept. of Civ. Engrg., Univ. of Houston, Houston, Tex. 77004) and **S. M. Uzumeri**, Member, ASCE, (Prof., Dept. of Civ. Engrg., Univ. of Toronto, Toronto, Canada)

Journal of the Structural Division, Vol. 108, No. ST12, December, 1982, pp. 2703-2722

An analytical model for the confinement mechanisms in tied columns has been developed. The model is based on the test results reported previously. A complete stress-strain curve for confined concrete is proposed. The strength of the confined concrete is calculated by using the concept of the effectively confined concrete area within the nominal concrete core. The area of the effectively confined concrete is determined by the tie spacing, the distribution of longitudinal steel around the core perimeter and the resulting tie configuration. According to the model, better distribution of longitudinal steel and smaller tie spacing result in higher strength and ductility of concrete. Other variables examined include the volumetric ratio of lateral reinforcement and the characteristics of steel. The results from the model agree well with the test results. Practical examples demonstrating the application of the model have been presented. Building code requirements for confinement are examined and it has been concluded that in certain circumstances the Codes provisions may be unsafe, while in others they may be unnecessarily conservative.

0691 Wind Pressures on Low Buildings with Parapets

Theodore Stathopoulos, Member, ASCE, (Asst. Prof. of Engrg., Centre for Building Studies, Concordia Univ., Montreal, Quebec, Canada)

Journal of the Structural Division, Vol. 108, No. ST12, December, 1982, pp. 2723-2736

The effect of parapets on the wind loading of low-rise buildings has been examined in the past, primarily under conditions of uniform flow not representing the variation of the mean speed with height and the turbulence characteristics of the natural wind. This paper presents the results of a recent experimental study carried out in a boundary layer wind tunnel in order to determine the wind loads on low-rise buildings of different configurations with and without parapets. Data show that the addition of parapets reduces the magnitude of roof peak suction coefficients with values greater than about 1.75 for open country and about 3.0 for suburban exposures, whereas lower suction values remain unaffected or slightly increase. On the other hand, parapets tend to increase the mean suction locally close to roof corners. The implication of the data presented for current Standards and Codes of Practice is analyzed and suggestions are made.

0692 Forest Service Timber Bridge Specifications

Lawrence D. Bruesch, Member, ASCE, (Asst. Staff Dir., Engrg., Forest Service Northern Region, Missoula, Mont.)

Journal of the Structural Division, Vol. 108, No. ST12, December, 1982, pp. 2737-2746

Approximately one half of the 12,000 Forest Service road and trail bridges are constructed of wood or have wood as a major component. The number constructed each year fluctuates with the cost of timber and the complexity of the individual bridges. The typical Forest

Service road bridge is a pressure treated, single span, single lane, 40 ft. long structure with timber abutments, glu-lam or solid sawn stringers, laminated deck, timber curbs and rail posts, and a steel deep beam railing. Design, with minor exceptions such as truck loadings and railing, is in accordance with the AASHTO Standard Specifications for Highway Bridges. Installation is controlled by Forest Service Standard Specifications for Construction of Roads and Bridges. The specifications are quite similar to the FP-79 specifications developed by the Federal Highway Administration. There are regional differences depending on availability of construction materials and on the environmental conditions.

0693 Computer Simulated Flexural Tests of Partially Prestressed Concrete Sections

M. Z. Cohn, Fellow, ASCE, (Prof., Dept. of Civ. Engrg., Solid Mechanics Div., Univ. of Waterloo, Waterloo, Ontario, Canada N2L 3G1) and **M. Bartlett**, (Research Asst., Dept. of Civ. Engrg., Univ. of Waterloo, Waterloo, Ontario, Canada N2L 3G1)

Journal of the Structural Division, Vol. 108, No. ST12, December, 1982, pp. 2747-2765

Partially prestressed concrete is redefined in terms of three variables related to the total amount of steel reinforcement, the proportion of high strength to total amount of steel reinforcement, and the degree of prestressing of the high grade steel. This definition enables any structural concrete section design to be identified by a point in the three dimensional space of the chosen variables. Limit states for all types of structural concrete sections are also defined. A computer program for the analysis of structural concrete sections is developed and used to trace the loading history of 264 computer-simulated tests. Two series of tests explore the extreme cases of maximum permissible prestressing and maximum mixed reinforcement index. The effects of the three primary variables above inelastic behavior and ductility of partially prestressed concrete sections are investigated.

0694 Strengthening of Single-Span Steel-Beam Bridges

Wayne F. Klaiber, Member, ASCE, (Prof. of Civ. Engrg., Iowa State Univ., Ames, Iowa), **Kenneth F. Dunker**, Member, ASCE, (Asst. Prof. of Civ. Engrg., Iowa State Univ., Ames, Iowa) and **Wallace W. Sanders, Jr.**, Fellow, ASCE, (Prof. of Civ. Engrg. and Asst. Dir. of the Engrg. Research Inst., Iowa State Univ., Ames, Iowa)

Journal of the Structural Division, Vol. 108, No. ST12, December, 1982, pp. 2766-2780

The use of post-tensioning to increase the strength of steel beams in beam and concrete-slab bridges is shown to be feasible. The purpose of the study is to determine a technique for increasing the capacity of the bridges to meet today's loadings. The results of testing a half-scale model of an existing bridge indicate that post-tensioning can be used with only minor deck cracking. Comparisons of the data from the model with results using orthotropic plate theory indicate that this theory can be used as a preliminary design methodology for determining the required prestress force and its location for strengthening existing bridges.

0695 Uniform Snow Loads on Structures

Michael J. O'Rourke, Member, ASCE, (Assoc. Prof., Civil Engrg., Rensselaer Polytechnic Inst., Troy, N.Y. 12181), **Robert Redfield**, Member, ASCE, (Research Civ. Engr., U.S. Army Cold Regions Research and Engrg. Lab., Hanover, N.H.) and **Peter von Bradsky**, Assoc. Member, ASCE, (Former Grad. Student, Rensselaer Polytechnic Inst., Troy, N.Y. 12181)

Journal of the Structural Division, Vol. 108, No. ST12, December, 1982, pp. 2781-2798

Data on ground and roof snow loads for 199 structures are analyzed. Relationships between ground-to-roof conversion factor for uniform roof loads and parameters such as roof slope, exposure and thermal characteristics are investigated. The conversion factor was found to be most strongly influenced by exposure. On the average, there was about a 35% difference in conversion factors for windswept and sheltered roofs. The conversion factor was influenced to a lesser degree by thermal characteristics. On the average, there was about a 20% difference in conversion factors for heated and unheated structures. Roof slopes between 0° and 35° did not have an observable effect upon the conversion factors. An expected value relationship for the conversion factor was established. The variation of observed values about the expected value was modeled by a log-normal error term with a standard deviation for $\ln(\epsilon) = 0.42$. Comparisons with existing codes and standards indicate that their recommended conversion factors are, on average, larger than measured values.

0696 Creep Buckling of Reinforced Concrete Columns

John E. Behan, (Lect., Dept. of Civ. Engrg., Univ. of Queensland, St. Lucia, Queensland, Australia 4067) and **Colin O'Connor**, Member, ASCE, (Prof., Dept. of Civ. Engrg., Univ. of Queensland, St. Lucia, Queensland, Australia 4067)

Journal of the Structural Division, Vol. 108, No. ST12, December, 1982, pp. 2799-2818

An analysis is described for the long-term buckling behavior of reinforced concrete columns using the superposition model for creep analysis. The analysis considers the complete strain history and allows for cracking and age-hardening of concrete. The analysis is nonlinear, providing for yielding of steel and the inelastic response of the concrete at all levels of stress. The shape of the deflected column is determined in the analysis, and is not constrained to any particular geometry. Allowance is made for initial imperfection and eccentricity of loading. The only assumptions made are that full composite action exists between steel and concrete, and a linear strain profile exists for combined axial force and bending. The results of 51 short-term and 16 long-term tests are presented. The analysis gives excellent prediction of short-term load capacities and deflected shapes. A reasonable estimate for life under long-term loads is obtained. A method is given for the accurate estimation of long-term load capacities, which agrees well with the results of tests.

0697 Assessment of Relaxation Loss Estimates for Strands

Arcady V. Koretsky, (Sr. Lecturer in Civ. Engrg., Univ. of Queensland, St. Lucia, Queensland, Australia) and **Ross W. Pritchard**, (Civ. Engr., Main Roads Dept., Brisbane, Queensland, Australia)

Journal of the Structural Division, Vol. 108, No. ST12, December, 1982, pp. 2819-2836

Significant variations in the techniques of measuring relaxation losses in prestressing strands and in the relaxation values recommended by the Australian and International Standards and Design Codes are discussed. A specially designed test frame is reported and its capabilities are demonstrated on a series of twenty-five relaxation tests. Areas of particular emphasis are test method, relaxation losses during first minute after completion of stressing, isothermal relationships between the relaxation values at 1,000 hours and at ultimate, and the time dependent relaxation relationships when steam curing is employed. It is concluded on the basis of the test data obtained that test equipment should be more rigidly specified and standardized and that modifications are required to the current methods of estimating the magnitude of the strand relaxation. Modifications to the Australian Standards are suggested.

0698 Bolted Splice Connections with Undeveloped Fillers

Joseph A. Yura, Member, ASCE, (Prof., Dept. of Civ. Engrg., The Univ. of Texas, Austin, Tex. 78712), **Mikkel A. Hansen**, Assoc. Member, ASCE, (Civ. Engr. II, Central Offshore Engrg. Dept., Texaco, New Orleans, La.) and **Karl H. Frank**, Assoc. Member, ASCE, (Assoc. Prof., Dept. of Civ. Engrg., The Univ. Of Texas, Austin, Tex.)

Journal of the Structural Division, Vol. 108, No. ST12, December, 1982, pp. 2837-2849

The use of undeveloped filler plates has a definite influence on the loading and deformation behavior of bolted splice joints. Two replicate tests were completed on each of five different filler plate assemblages. Bolted splice joints were tested with fillers ranging from 0 - 0.75 in. (0 - 19.05 mm) in thickness. Single-ply and multi-ply fillers were used. Deformations, slip loads, and ultimate loads were studied for the filler combinations. Increasing the thickness of undeveloped fillers causes a significant increase in the deformation at maximum load but only a slight decrease in ultimate load. Comparison of test results to the AASHTO and AISC specifications are made along with recommendations for design.

0699 Multipurpose Land Information Systems: A Canadian Perspective

John McLaughlin, (Assoc. Prof., Dept. of Surveying Engineering, University of New Brunswick, Fredericton, Canada) and **Salem Masry**, (Prof., Dept. of Surveying Engrg., Univ. of New Brunswick, Fredericton, Canada)

Journal of the Surveying and Mapping Division, Vol. 108, No. SU1, April, 1982, pp. 1-9

The growing involvement of the state in the land management process has been accompanied by the increasing presence and influence of professional managers and formal planning processes. It has also lead to demands for more formal and systematic information facilities. One result has been the gradual evolution of computerized, multifunctional land information systems and concomitant information tehnologies. The evolving demand for new information products and services from a systems perspective and from an information technology perspective is examined. The impact of this demand on the surveying profession in Canada is assessed.

0700 Research Needs in Marine Surveying

Harold D. Palmer, Affiliate Member, ASCE, (Chf. Sci., Interstate Electronics Corp., 1745 Jefferson Davis Hwy., Arlington, Va. 22202)

Journal of the Surveying and Mapping Division, Vol. 108, No. SU1, April, 1982, pp. 10-17

Research needs in oceanographic and hydrographic surveying and charting are discussed. It is suggested that the two most crucial elements which effect all measurements in hydrography and oceanography are rates and scales. The importance of understanding the effect of coupling two or more environmental factors is also essential in the study of these two disciplines. It is concluded that improved data collection and analysis systems for oceanographic measurements are necessary for better comprehension and prediction of marine processes.

0701 Surveying: The Profession and Its Requirements

Hallas H. Ridgeway, Member, ASCE, (Lect., Civ. Engrg. Dept., Univ. of Connecticut, Storre, Conn.)

Journal of the Surveying and Mapping Division, Vol. 108, No. SU1, April, 1982, pp. 18-22

Two definitions of surveying are discussed. The A.S.C.E. definition which is a broad definition that includes a. land surveying, b. engineering surveyors, c. cartographic surveying, aerial surveying services, and cartography; and a definition typical of those used by boards of registration that is confined to land surveying are presented. The requirements depend on the definition. Whatever definition of surveying is used by a state board of registration, the board is obligated to assure that professional areas of practice permitted by the registered surveyor are reflected in the requirements for education, experience and examination.

0702 Geodetic Monuments for Eccentric Property Corners

James E. Stem, Member, ASCE, (Geodesist, Office of the Dir., National Geodetic Survey, National Oceanic and Atmospheric Administration, Rockville, Md.)

Journal of the Surveying and Mapping Division, Vol. 108, No. SU1, April, 1982, pp. 23-33

Problem areas in land information systems are identified and described and solutions are proposed. The main areas discussed include a method of equitable property taxation, replacement of an archaic land conveyance system and implementation of intelligent land use and resource planning. It is suggested that independent solutions to these problems are inadequate. A multipurpose cadastre, which would be a land conveyance system, a property taxation system, and a database for land use and resource planning, is proposed.

0703 Acoustic Techniques for Offshore Positioning

John B. Ciani, Member, ASCE, (Research Civ. Engr., Naval Civ. Engrg. Lab., U.S. Navy, Port Hueneme, Calif.) and **Harold D. Palmer**, (Assoc., Dames & Moore, Washington, D.C.)

Journal of the Surveying and Mapping Division, Vol. 108, No. SU1, April, 1982, pp. 34-45

Acoustic positioning systems for accurately determining the location of a site, structure, or vessel in the ocean are described. The advantages of acoustic positioning systems over electromagnetic and satellite techniques are identified. It is shown that acoustic positioning systems offer a reliable, portable, and precise means of determining the location of objects, vessels, divers, or all three at any water depth. For specific installations, the use of acoustics and graphic displays provides the tools necessary to accomplish difficult tasks in remote sites.

0704 New Adjustment of North American Datum

Capt. John D. Bossler, (Dir., National Geodetic Survey, National Ocean Survey, NOAA, Rockville, Md. 20852)

Journal of the Surveying and Mapping Division, Vol. 108, No. SU2, August, 1982, pp. 47-52

The new adjustment of the North American Datum will be completed in 1984. The numerical problems associated with the simultaneous solution of 500,000 unknown parameters appear to be minimal. Results of several large test adjustments using the Helmert blocking software system indicate that the adjustment can be achieved in an efficient manner. The new adjustment will be based on a geocentric datum, known as the North American Datum of 1983 (NAD 1983), which will span the entire North American continent. In addition, the National Geodetic Survey will continue to publish and support two plane coordinates in the United States: the Universal Transverse Mercator System, and the State Plane Coordinate.

0705 Minimal Constraints in Two-Dimensional Networks

Alfred Leick, Assoc. Member, ASCE, (Asst. Prof., Dept. of Civ. Engrg., Univ. of Maine, Surveying Engrg. Program, Orono, Me. 04469)

Journal of the Surveying and Mapping Division, Vol. 108, No. SU2, August, 1982, pp. 53-68

The use of coordinates in the adjustment of two-dimensional networks leads to inherently singular solutions. Various minimal constraints are applied to eliminate the singularity. The variant and invariant quantities, with respect to a choice of minimal constraints, are identified. Inner constraints, which constitute a subset of all sets of minimal constraints, are useful in displaying the internal network geometry through error ellipses whose size, shape and orientation are independent of the definition of the coordinate system. Inner constraint solutions are helpful in simulation studies where the optimum station locations and observing program are investigated. Any minimal constraint solution can be used for checking the quality of observations and the detection of blunders.

0706 Community Geographic Data Base

Henry A. Emery, (Div. Mgr., Kellogg Corp., 5601 South Broadway, Suite 400, Littleton, Colo. 80121)

Journal of the Surveying and Mapping Division, Vol. 108, No. SU2, August, 1982, pp. 69-79

The need for a community geographic and utility management system is discussed. Information needs are identified, as are potential user groups such as local government agencies, utility companies, and engineering and surveying companies. The uses of such a management system are also considered. Potential methods to accomplish a community geographic and utility management information system are discussed. The results of several projects in both the public and private sectors are presented.

0707 Serviceability Failures due to Structural Movements

Eric Downey, (Dir., Macdata Unit., Dept. of Civ. Engrg., Paisley Coll. of Tech., Paisley, Scotland) and **Iain A. MacLeod**, Member, ASCE, (Prof. of Struct. Engrg., Univ. of Strathclyde, Glasgow, Scotland)

Journal of the Technical Councils of ASCE, Vol. 108, No. TC1, May, 1982, pp. 1-9

A significant number of serviceability failures of buildings may be attributed to structural movements of one type or another. The reasons for these failures in general appear to relate either to a breakdown of communication between the architect and the structural engineer or to a failure on the part of either of these parties or the contractor to appreciate the implications of action taken during the design or construction of the building. Examples of this type of problem are cracking of nonload bearing walls on suspended floors, relative movements of reinforced concrete frames and claddings, normal structural deflections, thermal movements, and settlement. There is a need for better dissemination of information about serviceability failures so that designers are more aware of problems experienced by others.

0708 Seismic Risk Analysis of Infrastructural Systems

Hector Monzon-Despang, (Doctoral Candidate and Research Asst., John A. Blume Earthquake Engrg. Center, Dept. of Civ. Engrg., Stanford Univ., Stanford, Calif. 94305) and **Haresh C. Shah**, Member, ASCE, (Prof. and Dir., John A. Blume Earthquake Engrg. Center, Dept. of Civ. Engrg., Stanford Univ., Stanford, Calif. 94305)

Journal of the Technical Councils of ASCE, Vol. 108, No. TC1, May, 1982, pp. 10-23

This paper discusses the quantitative evaluation of seismic damage and risk for lifelines and other infrastructural systems. The central subject is the determination of the likelihood of attaining various Levels of system impairment, due to earthquakes, and what to do with that information. Seismic damage is measured in terms of system global response to damaging events. Possible parameters to describe global response are mentioned though not discussed at length (references are given). The emphasis is placed in the philosophy of using such parameters as seismic damage measures. The general structure of the algorithms, the assumptions to be made, and the modeling techniques to implement the analysis are discussed. Risk analysis, using the concepts of reliability and expected performance, is addressed. It is concluded that the required analysis techniques are currently founded on a great deal of uncertain and subjective information or both. Nevertheless, these methods do provide a very valuable insight into the seismic performance of infrastructural systems.

0709 Seismic Design of Seattle Waterfront Facilities

Frank W. Pita, Assoc. Member, ASCE, (Mgr., Seattle Geotechnical Dept., CH2M HILL, Bellevue, Wash.), **Bradley P. Erickson**, Assoc. Member, ASCE, (Mgr., Seattle Geotechnical Dept., CH2M HILL, Bellevue, Wash.) and **Arthur J. O'Brien**, Assoc. Member, ASCE, (Geotechnical Engr., CH2M HILL, Bellevue, Wash.)

Journal of the Technical Councils of ASCE, Vol. 108, No. TC1, May, 1982, pp. 24-33

The waterfront facilities of the Port of Seattle are part of the region's transportation lifelines. The lifeline integrity of the water transportation system depends on maintaining these facilities in operation following a major earthquake. An extensive development program to improve the Port of Seattle's existing waterfront facilities and provide new facilities is being implemented. Current codes do not provide comprehensive earthquake design criteria for port facilities. It is recommended that a national building code be expanded to address the peculiar characteristics of waterfront facilities. It is also recommended that more information regarding lifeline design requirements and recommendations for waterfront facilities is required.

0710 Control of Snow and Ice in Hokkaido, Japan

Seiiti Kinosita, (Prof., Inst. of Low Temperature Sci., Hokkaido Univ., N19 W8, Sapporo, 060, Japan)

Journal of the Technical Councils of ASCE, Vol. 108, No. TC1, May, 1982, pp. 34-43

The Hokkaido Islands are marked by a cold climate despite being low in latitude; as a result, snow, ice, sea ice and freezing of ground have caused disaster in large numbers. Members of the Institute of Low Temperature Science and another group have conducted studies concerning the disaster and developed a variety of devices to prevent it from occurring. This paper presents a short review on the work devoted to prevent the disaster related to frost heaving, avalanche, snow drifting, snow removal, snow on the roof, snow accretion on electric lines, sea ice, and ice accretion on ships.

0711 Computers in Civil Engineering Practice

Steven J. Fenves, Member, ASCE, (Prof., Dept. of Civil Engineering, Carnegie-Mellon Univ., Pittsburgh, Pa. 15213)

Journal of the Technical Councils of ASCE, Vol. 108, No. TC1, May, 1982, pp. 44-52

The current use of computers in civil engineering is discussed. Engineers' expectations from computer use are outlined and realistic considerations are presented. Advantages and shortcomings of computers are compared. Future technical and organizational needs are discussed and trends in computer applications are reviewed.

0712 Computer Applications in Water Resource Planning

Richard M. Males, Assoc. Member, ASCE, (Vice Pres., W. E. Gates and Assoc., 1515 Batavia Pike, Batavia, Ohio 45103)

Journal of the Technical Councils of ASCE, Vol. 108, No. TC1, May, 1982, pp. 53-59

Due to the complexity and inter-disciplinary nature of water resource planning and management, and because many of the problems are site and situation-specific, general-purpose computer programs are less prevalent than in other fields. Much of the work is model oriented, and many tools have been borrowed from related disciplines of hydraulics, hydrology, and environmental engineering. The combination of computerized spatial data management and analytical mathematical modeling appears well-suited to the need for generalized computer tools. Spatial data management involves the computer storage of information such as population, soils, terrain, etc., that is organized and referenced geographically. Analytical mathematical models can operate upon this stored data, as physical or economic/demographic simulations, or as statistical and combinatorial analyses. It is likely that such tools, organized as systems of inter-related models operating upon a data base, will become increasingly popular and important in the near future.

0713 Seismic Reliability of Lifeline Networks

Masoud Moghtaderizadeh, (Grad. Student, Dept. of Civ. Engrg., Univ. of California, Berkeley, Calif.), **R. Kevin Wood**, (Grad. Student, Operations Research Center, Univ. of California, Berkeley, Calif.), **Armen Der Kiureghian**, Assoc. Member, ASCE, (Assoc. Prof., Dept. of Civ. Engrg., Univ. of California, Berkeley, Calif.) and **Richard E. Barlow**, (Prof., Dept. of Industrial Engrg. and Operations Research, Univ. of California, Berkeley, Calif.)

Journal of the Technical Councils of ASCE, Vol. 108, No. TC1, May, 1982, pp. 60-78

An efficient method for seismic reliability assessment of lifeline networks, such as water distribution systems and communication or transportation networks, is developed. Lifeline component failures resulting from ground shaking and fault differential movements are analyzed using structural performance functions given in terms of earthquake variables. An improved fault-rupture model which considers the ruptured area produced by an earthquake on the fault plane is utilized. A new, polynomially bounded method for computer lifeline network reliability is developed. It is shown that for a fixed earthquake magnitude on a fault, the network will take on at most $2n$ stages where n is the number of network components. Computing the seismic reliability of large networks becomes feasible using these new techniques. A water distribution system is analyzed using newly developed computer programs.

0714 Computerized Specification System Saves Resources

Edgar Samuel Neely, Jr., Member, ASCE, (Principal Investigator, U.S. Army Construction Engrg. Research Lab., P.O. Box 4005, Champaign, Ill. 61820)

Journal of the Technical Councils of ASCE, Vol. 108, No. TC1, May, 1982, pp. 79-88

Application of a computer-aided specification preparation system, known as EDITSPEC, results in production of higher-quality project specifications and uses 30 percent fewer resources. Specifications produced by EDITSPEC are more complete, accurate, and consistent. Use of EDITSPEC reduces the elapsed time, specification writer and typist time, and costs required to produce a project specification. Change orders due to specification errors are reduced, thus lowering total construction costs. Preparing Corps of Engineers guide specifications, program-oriented guide specifications, and project specifications with EDITSPEC are discussed. The benefits and operation of EDITSPEC are also described.

0715 Data Acquisition and Process Control

Suresh G. Pinjarkar, Member, ASCE, (Sr. Struct. Consultant, Construction Tech. Lab., Portland Cement Assoc., Stokie, Ill.)

Journal of the Technical Councils of ASCE, Vol. 108, No. TC1, May, 1982, pp. 89-95

Automatic data acquisition and control systems are discussed. Computerized systems are used in laboratory, field, and industrial applications to acquire, process, and control test parameters. The new generation of data acquisition systems are microprocessor controlled and offer direct interfaces with minicomputers. This paper presents an analysis of data acquisition systems, computer interfacing, characteristic of data transmission and other hardware requirements. A complete description of an in-house computer controlled system is provided. An example of the operation of the system is presented to highlight the benefits of the system to the engineers.

0716 Infrequent Earthquake Regions: Seismic Resistance

Anshel J. Schiff, Member, ASCE, (Dir., Center for Earthquake Engrg. and Ground Motion Studies and the School of Mech. Engrg., Purdue Univ., W. Lafayette, Ind. 47907)

Journal of the Technical Councils of ASCE, Vol. 108, No. TC1, May, 1982, pp. 96-105

The earthquake damage experienced by electric power systems is briefly reviewed. Low cost methods for improving the seismic response of five categories of equipment damage are discussed. Securing substation transformers prevent catastrophic losses due to transformers falling from elevated pedestals, and the failure of lighting arrestors, bushings and control cables due to lack of slack in interconnections. Securing substation and power station batteries to their racks and the racks to their foundation prevents the serious consequences of their loss. Securing computers can prevent damage and disruption due to their loss. Properly supporting ceilings in control rooms can prevent damage and disruption caused by restriction in their operation. While securing circuit breakers may reduce damage, improved design, materials or installation methods may be required.

0717 Damping as Measure of Damage in Composites

R. Chandra, (Lect., Dept. of Mech. Engrg., Regional Engrg. Coll., Srinagar, India), **A. K. Mallik**, (Asst. Prof., Dept. of Mech. Engrg., Indian Inst. of Tech., Kanpur, India) and **R. Prabhakaran**, (Assoc. Prof., Dept. of Mech. Engrg. and Mech., Old Dominion Univ., Norfolk, Va. 23508)

Journal of the Technical Councils of ASCE, Vol. 108, No. TC1, May, 1982, pp. 106-111

The increase of damping with damage induced by static and fatigue loading of a glass fiber reinforced composite has been measured and is shown to be greater than the corresponding decrease in stiffness. A simple test rig was designed to measure damping and was calibrated using aluminum beam specimens. The loss factor for undamaged specimens was measured for 0°, 15° ×, 30°, 45° fiber orientations and was found to be independent of the strain amplitude up to 0.04 per cent strain. Damage was induced by static and cyclic preloading. Static preloading was done at 50, 60, 75, 85 percent of the fracture strength while cyclic preloading was carried out at 50 percent of the fracture strength for 1, 5, 10, 25 cycles. Both types of preloading increased the specimen loss factor. For comparable levels of preloading, the loss factor increased in the order of 0°, 15°, 30°, 45° fiber orientation; for a given fiber orientation, the loss factor was more for greater number of cycles and level of prestressing.

0718 Piling in Frozen Ground

Frederick E. Crory, Member, ASCE, (Research Civ. Engr., Geotechnical Research Branch, U.S. Army Cold Regions Research and Engrg. Lab., Hanover, N.H. 03755)

Journal of the Technical Councils of ASCE, Vol. 108, No. TC1, May, 1982, pp. 112-124

Recent construction in northern regions has emphasized the need for economical and reliable foundations, particularly in areas having permanently frozen ground. In many permafrost areas piling can provide cost effective and reliable foundation systems, avoiding the undesirable freeze and thaw cycles at the surface and the potentimlly large settlements which accompany the thawing of permafrost. The design, construction and performance of piles in frozen ground requires a thorough knowledge of the soil and ice conditions at depth and an understanding of the theraml regime. Techniques are available to expedite pile installation and freezeback. Pile testing procedures have been developed for frozen ground, providing design information on short and long term loadings. Field design changes are essential for the installation of pile foundations such as used in elevated pipelines, which are designed on limited soil and permafrost data. Since subsurface conditions in permafrost areas can vary widely over short distances the requirement for close field inspection and ability to redesign is critical to match the actual conditions. Recognizing and avoiding thermal changes created by adjacent foundations, roads, utilities and flowing water will become increasingly important as northern regions develop.

0719 Implementing Large Programs on Microcomputers

Martin F. Rooney, Assoc. Member, ASCE, (Asst. Prof., Dept. of Civ. Engrg., Rensselaer Polytechnic Inst., Troy, N.Y. 12181), **Jonathan D. Howland**, (Doctoral Student, Dept. of Civ. Engrg., Rensselaer Polytechnic Inst., Troy, N.Y. 12181) and **Ronald J. Molz**, (Undergraduate Student, Dept. of Mech. Engrg., Rensselaer Polytechnic Inst., Troy, N.Y. 12181)

Journal of the Technical Councils of ASCE, Vol. 108, No. TC1, May, 1982, pp. 125-137

The capabilities of running large programs on micro-computers and the size limitations imposed on software are examined. Suggestions are provided based on practical experience gained during the conversion of a large program rather than abstract theories. A test case (slope stability analysis) is run on a minimal hardware configuration. The examination is divided into three major topics: storage issues, translation issues, and economic issues. A summary of lessons learned about converting large programs is provided.

0720 Wastewater Treatment for Cold Regions

Daniel W. Smith, Member, ASCE, (Prof. of Civ. Engrg., Univ. of Alberta, Edmonton, Canada)

Journal of the Technical Councils of ASCE, Vol. 108, No. TC1, May, 1982, pp. 138-149

The objectives of wastewater treatment in cold regions have been dictated by guidelines and regulations established for other regions. The impact of this occurrence has been extremely high cost for the protection of the aquatic environment and utilization of complex technology. The imposed and natural constraints that influence the selection, and may influence the performance, of wastewater treatment systems in cold regions are discussed. Factors such as natural dissolved oxygen depletion during ice cover and microorganism survival at low stream temperatures, population patterns and methods of water supply are addressed. Wastewater treatment methods are related to the water delivery and collection systems used. Wastewater strength can be divided into undiluted, moderately dilute, conventional and very dilute. The need for continued engineering research into more economic and reliable methods of treatment within the existing constraints is emphasized. Also, more research is required to increase understanding of natural constraints. The latter will aid in proper adjustment of imposed constraints.

0721 Dade County Rapid Transit: Management Scheduling Techniques

Ronald S. Steiner, Member, ASCE, (Vice Pres., Fogel & Assoc., Inc., New York, N.Y.)

Journal of the Technical Councils of ASCE, Vol. 108, No. TC1, May, 1982, pp. 150-158

In order to reduce the size and maintenance problems of CPM schedules for large public works programs, a system of summarization networking was developed and utilized on the Dade County, Miami, Transit Program. By summarizing contractor provided CPM schedules into a 2,000 activity management network while working directly with the contractor's networks, the control of a large public works program is greatly simplified. Additional summarization by computer is provided for by means of a unique coding structure which allows for reducing the 2,000 activities to higher levels of detail and ultimately to a one activity definition of the entire program. It is believed that by reducing the bulk of CPM scheduling, the respect and usage it deserves will again increase.

0722 Stormwater-Combined Sewage Pollution Abatement

Richard P. Traver, Assoc. Member, ASCE, (Staff Engr., Storm and Combined Sewer Section, Municipal Environmental Research Lab., U.S. Environmental Protection Agency, Edison, N.J. 08837)

Journal of the Technical Councils of ASCE, Vol. 108, No. TC1, May, 1982, pp. 159-171

The mission of the USEPA's Storm and Combined Sewer Pollution Control Research and Development Program is to quantify the urban storm and combined sewer overflow pollution problems and develop countermeasure controls. As a result of the Program's efforts an advanced technology or state-of-the-art has been developed, demonstrated, and disseminated. This paper addresses the problem and characterization of stormwater and combined sewage impacts upon our Nation's receiving waters while presenting various levels of pollution control technology. Efforts of urban runoff management examined deal with source control, collection system control, storage and treatment, and integrated (complex) systems.

0723 Transverse Seismic Waves at Pipeline Junctions

Michael J. O'Rourke, Member, ASCE, (Assoc. Prof., Dept. of Civ. Engrg., Rensselaer Polytechnic Inst., Troy, N.Y.), **Richard R. Pikul**, Member, ASCE, (Assoc., Clough Assocs., Albany, N.Y.) and **Leon R. L. Wang**, Member, ASCE, (Prof. of Civ. Engrg. and Environmental Science, Univ. of Oklahoma, Norman, Okla.)

Journal of the Technical Councils of ASCE, Vol. 108, No. TC1, May, 1982, pp. 173-177

The effects of transverse seismic wave propagation at junctions in a buried pipeline network are studied. Relationships for the bending strain induced in the pipeline segments intersecting at a network junction are presented. It is shown that the bending strain is a function of the peak ground velocity, the propagation velocity of the transverse seismic waves with respect to the ground surface, the flexural rigidity of the pipeline segments, and the lateral stiffness of the soil surrounding the pipeline. An example problem is presented to illustrate the procedures for determining these bending strains.

0724 Engineering Data Analysis

Albert J. Klee, (Sr. Sci. Advisor, Program Operations Office, Industrial Environmental Research Lab., U.S. Environmental Protection Agency, Cincinnati, Ohio) and **Lawrence A. Minton**, (Engrg. Aide, Program Operations Office, Industrial Environmental Research Lab., U.S. Environmental Protection Agency, Cincinnati, Ohio)

Journal of the Technical Councils of ASCE, Vol. 108, No. TC2, November, 1982, pp. 187-199

Solutions to two major data defects (autocorrelation and multicolinearity) commonly encountered in the use of multivariable linear regression analysis when applied to engineering data are described. A computer program, incorporating a new method for the determination of the ridge parameter, is introduced for their implementation. The methods described are then applied to a complex data set involving a demonstration of the use of densified refuse-derived fuel burned in an industrial spreader stoker boiler. The analysis shows that, had only the more conventional regression analysis been applied, the fits to some response variables would have been overvalued while others would have been undervalued, and that the relationships between many of the response and predictor variables would have been determined incorrectly.

0725 Ice Engineering Design of Marine Piling and Piers

C. Allen Wortley, (Assoc. Prof., Dept. of Engrg., Univ. of Wisconsin-Extension, Madison, Wis.)

Journal of the Technical Councils of ASCE, Vol. 108, No. TC2, November, 1982, pp. 200-213

Design criteria and structure designs for pilings, piers, and docks subject to ice are presented. Included are estimates of stationary and moving horizontal and vertical ice forces, types of constructions and materials, facility layouts to mitigate ice problems, and methods of ice suppression. Design recommendations are based on actual design experiences, field investigations in several hundred U. S. and Canadian Great Lakes harbors and ports over a period of years, and laboratory studies.

0726 Hydraulic and Hydrologic Computer Applications

John J. Cassidy, Fellow, ASCE, (Chf. Hydr. Engr., Bechtel Civ. and Minerals, Inc., P.O. Box 3965, San Francisco, Calif. 94119) and **Elias C. Tonias**, Member, ASCE, (Owner, Tonias Engrs., Rochester, N.Y.)

Journal of the Technical Councils of ASCE, Vol. 108, No. TC2, November, 1982, pp. 214-225

This paper has been prepared as a part of the project *Introductory Manual of Computer Services* undertaken by the Committee on Coordination of the Technical Council on Computer Practices of the American Society of Civil Engineers. Several Generally available computer programs are briefly described. The programs, chosen because of wide usage, include both single-purpose programs and large, comprehensive system programs for both hydraulic-engineering and engineering-hydrology applications. An extensive bibliography is provided which lists author and source. Applications to urban hydrology, sewer design, transient flows in closed conduits, pipe network analysis, watershed hydrology, flood-hydrology development and routing, water-surface profiles, and reservoir operations are considered. The intent is to summarize usage and identify program sources for practicing engineers who have a need to use programs but are not intimately familiar with source and availability.

0727 Research and Practice in Urban Runoff in Canada

Donald Weatherbe, (Head, River Systems Unit, Water Resources Branch, Ontario Ministry of the Environment, 135 St. Clair Ave. West, Toronto, Ontario, Canada M4V 1P5), **Jiri Marsalek**, (Hydr. Engr., Hydr. Div., National Water Research Inst., Environment Canada, Burlington, Ontario, Canada) and **George Zukovs**, (Project Mgr., Pollution Control Branch, Ontario Ministry of the Environment, Toronto, Ontario, Canada)

Journal of the Technical Councils of ASCE, Vol. 108, No. TC2, November, 1982, pp. 226-240

Much of the research of the 1970's has entered into the practice of urban runoff control in Canada. Urban runoff quality inputs on the Great Lakes are slight; although additional research into organic trace contaminants is needed. Urban rainfall-runoff modelling has concentrated on design storms and hydraulic aspects mostly for use in design. Quantity control technology includes inlet and storage devices with quality control limited to consideration of ponds as settling devices, sometimes with chemicals added. Comprehensive river basin studies typically consider urban drainage effects. Municipal drainage guidelines generally require consideration of the major and minor system. Continued research needs in urban runoff quality and quantity would be assisted by establishment of research catchments for long-term monitoring.

0728 Application of HEC-2 for Ice-Covered Waterways

D. J. Calkins, (Research Hydr. Engr., U. S. Army Cold Regions Research and Engrg. Lab., Hanover, N.H.), **R. Hayes**, (Hydr. Engr., Hydrologic Engrg. Center, U.S. Army Corps of Engrs., Davis, Calif.), **S. F. Daly**, (Research Hydr. Engr., U.S. Army Cold Regionss Research and Engrg. Lab., Hanover, N.H.) and **A. Montalvo**, (Hydr. Engr., Hydrologic Engrg. Center, U.S. Army Corps of Engrs., Davis, Calif.)

Journal of the Technical Councils of ASCE, Vol. 108, No. TC2, November, 1982, pp. 241-248

HEC-2, the widely known open channel flow water surface profile computer program developed by the U. S. Army Corps of Engineers' Hydrologic Engineering Center, has been recently updated for the U. S. Army Cold Regions Research and Engineering Laboratory to account for the presence of a floating ice cover. It has been shown by many writers that at uniform flow the normal flow depth can be increased by as much as 30% by a floating ice cover. HEC-2 with the ice cover option will allow the Corps of Engineers and other users of the program to evaluate effectively the effect of an ice cover on the flow depth, flow velocity, unit discharge, etc. in a river reach or in an entire river system. This paper presents an overview of the modifications to the uniform flow equation, the required input data, and an analysis. Specific examples are presented for a straight rectangular channel with a downstream control, which exhibits gradually varying flow to uniform flow conditions of an ice cover, grown from the shores and progressively covering more and more of the channel width. Finally, the paper describes the effects on the water surface profile, the flow discharge in the open water section, and the departure from open channel uniform depth.

0729 Prestandard Programs and United States System of Standards and Codes

William G. Kirkland, Fellow, ASCE, (Acting Exec. Dir., Building Seismic Safety Council, Washington, D.C.)

Journal of the Technical Councils of ASCE, Vol. 108, No. TC2, November, 1982, pp. 249-256

A concise description of the United States system of codes and standards is given, with the objective of presenting highlights of the system of interest to the civil engineer and the introduction of the concept of prestandards programs. The voluntary standards programs in the United States involves over 300,000 individuals who volunteer some of their time to writing standards. If all standards and prestandard activities and programs were included, the individual individuals involved would far exceed the 300,000 estimate, and overall expenditures related to standards would amount to hundreds of millions of dollars. A program related to the National Earthquake Hazards Reduction Program, which can be defined as a prestandards program, involves the follow-up work on the "Tentative Provisions for the Development of Seismic Regulations for Buildings" prepared by the Applied Technology Council (ATC) and widely distributed in 1978. This program, under the overflow authority of the Building Seismic Safety Council (BSSC) is a unique undertaking in the building community for which there may be no precedent. It provides the opportunity for technically qualified individuals as well as all other diverse groups with an interest in the subject to participate before code and standards work begins under the voluntary standards procedures.

0730 Imposed Ground Strain and Buried Pipelines

Leon R. L. Wang, Member, ASCE, (Prof. of Civ. Engrg. and Environmental Sci., Univ. of Oklahoma, Norman, Okla. 73018), **Richard R. Pikul**, Member, ASCE, (Assoc., Clough Assoc., Albany, N.Y.) and **Michael J. O'Rourke**, Member, ASCE, (Assoc. Prof. of Civ. Engrg., Rensselaer Polytechnic Inst., Troy, N.Y.)

Journal of the Technical Councils of ASCE, Vol. 108, No. TC2, November, 1982, pp. 259-263

A technique for the analysis of buried pipelines subjected to longitudinal soil strain resulting from earthquakes or manmade underground explosions is presented. The effects of this soil strain on the axial strain in the pipeline as well as relative displacement at the pipeline joints are determined using a beam on elastic foundation approach. The pipeline is assumed to be straight and of prismatic cross section. The pipeline, pipeline joints and surrounding soil are assumed to behave as linear elastic materials. Non-dimensional curves are presented relating pipeline strains and relative joint displacement to the relative stiffness of the soil, joint and pipeline.

0731 Des Moines Skywalk System

C. Todd Heglund, Member, ASCE, (Principal Assoc., Barton-Aschman Assoc., Inc., Minneapolis, Minn.)

Transportation Engineering Journal, Vol. 108, No. TE1, January, 1982, pp. 1-13

A method is proposed for estimating pedestrian traffic on a proposed skywalk system in Des Moines, Iowa. Factors that influence usage are identified. Two mathematical models are presented to predict savings in vehicle delay which would result from the system. A pedestrian accident analysis is discussed. A cost benefit analysis of the system is described.

0732 Intra-CBD Secondary Travel Patterns of Downtown Workers

Murthy V. A. Bondada, Member, ASCE, (Project Mgr., Transportation Planning, Gannett Fleming Corddry and Carpenter, Inc., Engrs. and Planners, P.O. Box 1963, Harrisburg, Pa. 17105)

Transportation Engineering Journal, Vol. 108, No. TE1, January, 1982, pp. 15-26

Previous Downtown People Mover (DPM) feasibility studies have shown that the majority of the average daily DPM trips will be made by downtown workers during their travel to and from the downtown (Primary Travel) and within the downtown (Secondary Travel). An understanding of the travel patterns of downtown workers has become a prerequisite in developing alignments and in maximizing patronage of DPMs. A preliminary engineering study for the Detroit (Michigan) DPM, an alternatives analysis study for a distribution and circulation system in downtown Atlanta (Georgia) and a feasibility study for a Norfolk (Virginia) DPM form the basis for this paper. These projects were performed for the Southeastern Michigan Transportation Authority, Georgia Department of Transportation and the City of Norfolk respectively. During these studies, questionnaire surveys were conducted to understand the existing travel patterns of downtown workers and to calibrate DPM patronage forecasting models. The intra-CBD secondary travel patterns obtained from the surveys of CBD workers in these three cities are discussed.

0733 Full-Depth and Granular Base Course Design for Frost Areas

Robert A. Eaton, (Research Civ. Engr., Geotechnical Research Branch, USACRREL, 72 Lyme Road, Hanover, N.H. 03755) and **James O. Payne**, (USACRREL, 72 Lyme Road, Hanover, N.H. 03755)

Transportation Engineering Journal, Vol. 108, No. TE1, January, 1982, pp. 27-39

When properly designed and constructed, the Asphalt Institute full-depth pavement concept can be a viable design alternative for seasonal frost areas. The Corps of Engineers reduced subgrade strength frost design proved to be an upper bound or conservative design under these test conditions. For each design, two different thicknesses were studied in test sections placed over 12 in. of prepared subgrade and tested under light traffic conditions in Hanover, New Hampshire. After design traffic loading was exceeded, pavement failure occurred as expected in the thinner full-depth section. The thinner reduced subgrade strength section was still in good condition after experiencing twice its design loading. Frost penetrations, pavement n-factors (surface transfer coefficients), Benkelman Beam deflections, and the spring subgrade moisture contents are also compared for the two designs.

0734 Optimization Method for Railroad Curve Alinement

Milan Velebit, Assoc. Member, ASCE, (Chf. Designer, Illinois Central Gulf Railroad, Chicago, Ill.), **Robert B. Schattke**, (Engrg. Analyst, Illinois Central Gulf Railroad, Chicago, Ill.) and **Steve G. Citko**, Assoc. Member, ASCE, (Systems Designer, Illinois Central Gulf Railroad, Chicago, Ill.)

Transportation Engineering Journal, Vol. 108, No. TE1, January, 1982, pp. 41-50

Realignment of existing railroad curves is cast in terms of a quadratic optimization problem and a method of minimizing the objective function is developed. Geometry relating track curvature and movement are presented, and quadratic objective functions which measure curve smoothness are derived. The objective function is minimized using a modified form of Cauchy's method, and integer constraints are applied. Results of the procedure are track movements required to move the existing alignment to a new smooth alignment. Computer results are presented for a typical curve. The method works well in finding smooth alignments and is not very sensitive on choice of objective function. The method presented shows much improvement over the current, hand trial and error method employed by railroads which is called Stringlining.

0735 Economic Evaluation of Development Projects on Transportation ROW

Snehamay Khasnabis, Assoc. Member, ASCE, (Assoc. Prof., Dept. of Civ. Engrg., Wayne State Univ., 667 Merrick, Detroit, MI 48202), **Kenneth S. Opiela**, Assoc. Member, ASCE, (Mgr., Transportation Engrg. and Planning, Goodell, Grivas, Inc., 17320 W. Eight Mile Road, Southfield, MI 48075) and **Ronald G. Arbogast**, Member, ASCE, (Vice-Pres., Refuse Resource Recovery, Inc., 30555 Southfield Rd., Southfield, Mich. 48075)

Transportation Engineering Journal, Vol. 108, No. TE1, January, 1982, pp. 51-70

The development and application of a computer model for testing the feasibility of multiple agency development projects at proposed transit stations is described. Given a development strategy, the model allocates project costs and revenues among a number of participating entities and uses the 'Internal Rate of Return' technique to compute the project returns to each entity. An iterative technique is used to derive the return by converging the numerical value of the 'Net Present Worth' of the cash flow to zero by systematically altering the interest rate. The return for each entity is calculated the model in the form of a three dimensional matrix as a function of the type, location and strategy of the project. The model can also estimate annual net return/cost for each entity for a specified interest rate. The study shows that the model can be used to identify optimal development strategies for a given project, economically feasible development projects at a given site, and desirable sites for development projects from a number of possible alternatives.

0736 Simulation Approach to Assess Highway Performance

Fred L. Mannering, (Doctoral Candidate in Civ. Engrg., Massachusetts Inst. of Tech., Cambridge, Mass. 02139) and **Kumares C. Sinha**, Member, ASCE, (Prof. of Civ. Engrg., Purdue Univ., West Lafayette, Ind. 47907)

Transportation Engineering Journal, Vol. 108, No. TE1, January, 1982, pp. 71-86

A procedure is presented that can be utilized to estimate the impacts of alternative financing policies and other highway related factors on future highway system performance. The approach is to first simulate the present highway performance characteristics, such as pavement condition, volume/capacity ratio, lane width and so on, and then apply a series of equations to forecast the impacts of total vehicle and commercial vehicle travel growth, future vehicle weight distributions, the availability of future highway improvement funds, and inflationary pressures. The proposed procedure provides projections of highway performance with minimal data requirements and is designed with considerable flexibility to facilitate present usage and future modification.

0737 Design of Pedestrian Space in Airport Terminals

Richard de Neufville, Member, ASCE, (Prof. of Civ. Engrg., Massachusetts Inst. of Tech., Cambridge, Mass. 02139) and **Michel Grillot**, (Airport Planner, Direction du Transport, Air France, Orly Cedex, France)

Transportation Engineering Journal, Vol. 108, No. TE1, January, 1982, pp. 87-102

A method to determine the required width of corriders and size of passenger staging areas in airport terminals is presented. It introduces two novel elements into the planning process for terminals. First, it draws attention to the critical period over which loads actually build up, and substitutes the loads which occur during this time for the typical peak hour passengers, thus giving a realistic view of the problem. Secondly, it permits the designer to focus on the particular problem areas which will limit the capacity of the entire terminal. Primarily intended for the design of new facilities, the method can also provide objective measures of the space available in existing terminals for ancillary commercial or first amendment activities. Being an integration of problem techniques, it can be used with confidence, as demonstrated by application to the new terminals at Paris/Charles deGaulle airport.

0738 Residual Stresses and Strains and Fatigue Cracking

W. O. Yandell, (Sr. Lect., School of Civ. Engrg., Box 1, P.O. Kensington, New South Wales, Australia)

Transportation Engineering Journal, Vol. 108, No. TE1, January, 1982, pp. 103-116

A mechano-lattice analysis is used for simulating one million traverses of a wheel load along an asphaltic pavement. It is shown in the example presented that the accumulation of residual stresses and strains may cause severe crack provocation to the asphaltic layer in its early life when the asphaltic layer behaves more elastically than the base and subgrade. This situation should improve in the remaining life when the asphaltic layer behaves more plastically than the base and subgrade. These situations can be controlled by the engineer by considering relative plastic behaviour of pavement components in any design procedure if fatigue life is to be maximized.

0739 Improved Mobility for Elderly and Handicapped Persons

James A. Bautz, (Chf., Paratransit and Special User Group Div., Office of Service and Methods Demonstrations, Urban Mass Transportation Administration, 400 7th St., S.W., Washington, D.C. 20590), **Robert Casey**, (Operations Research Analyst, Transportation Systems Center, Cambridge, Mass.), **Donald Kendall**, (Operations Research Analyst, Transportation Systems Center, Cambridge, Mass.), **Elizabeth Page**, (Economist, Transportation Systems Center, Cambridge, Mass.) and **Bruce Spear**, (Community Planner, Transportation Systems Center, Cambridge, Mass.)

Transportation Engineering Journal, Vol. 108, No. TE1, January, 1982, pp. 117-127

The mobility of elderly and handicapped persons can be improved in an economical manner by using existing resources more efficiently. Techniques such as coordination of social agency transportatiion and user subsidies can provide improved service to these groups at a lower cost than starting new transportation services. These approaches are described and findings from demonstrations and studies are presented. In addition, recent experience with the use of wheelchair accessible transit buses is presented. Low ridership levels have disappointed transit operators but hard data on the reasons for lack of use are not available.

0740 Selection of Pavement Joint Sealant Material

B. Frank McCullough, (Dir., Center for Transportation Research, Univ. of Texas at Austin, Austin, Tex. 78712)

Transportation Engineering Journal, Vol. 108, No. TE2, March, 1982, pp. 137-155

A review of literature shows joint sealant selection should be based on the design, construction and environmental variables. Satisfactory performance for one set of conditions does not imply the same results for another set of conditions. Based on studies of climatological conditions, material properties and expected transverse joint movement, a joint width and sealant depth should be recommended for preformed and liquid type sealants. Guidelines for preparing specifications for specific site conditions to maximize sealant material, the criteria to be considered include skill, experience and difficulty in installing the material, sealant properties such as adhesion, cohesion, and tensile strength, environmental condition, and joint properties such as movement and faulting. Procedures are presented to compute the joint sealant shape factor based on the expected movement and the material properties.

0741 Decision Strategy for Asphalt Pavement Recycling

Telimoye M. Oguara, Assoc. Member, ASCE, (Sr. Lect., Dept. of Civ. Engrg., Univ. of Science and Tech., Port Harcourt, Nigeria)

Transportation Engineering Journal, Vol. 108, No. TE2, March, 1982, pp. 157-167

A decision model that uses utility theory to the decision making process in selecting optimum techniques for asphalt pavement recycling is presented. The model requires the identification of all the quantitative and qualitative decision criteria and the determination of a utility function for each decision criterion. It then considers the uncertainties associated with the estimates of decision criteria and, in conjunction with the utility functions, determines expected decision criteria utilities. Weighting factors or regression coefficients are used to weight the decision criteria utilities and the optimum recycling technique is defined based on the maximum expected utility associated with the technique. The model provides a systematic approach to analyzing the various factors with should be considered before a pavement recycling decision could be made.

0742 New Structural Systems for Zero-Maintenance Pavements

Surendra K. Saxena, Fellow, ASCE, (Prof. and Chmn., Dept. of Civ. Engrg., Illinois Inst. of Tech., Chicago, Ill. 60616)

Transportation Engineering Journal, Vol. 108, No. TE2, March, 1982, pp. 169-182

Unique and infrequently used concepts and systems suitable for zero-maintenance pavements are identified. For this study, a zero-maintenance pavement is defined as one that will provide at least 20 years of maintenance-free operation and at least 10 more years of routine maintenance operation on a high volume road. Based on the criterian and guidelines established in this study, seven innovative concepts and one modified conventional system are presented and discussed. Detailed analytical investigation of one of the seven concepts is presented. The environmental and mechanical load imposed stresses are analyzed for an anchored pavement and a conventional slab and comparisons are presented. The study concludes that since a significant portion of pavement distress mechanism is contributed by subgrade failure, the transmission of loads to a deeper zone within the subgrade would considerably reduce related failure.

0743 Highway Fiscal and Program Management for the 1980s

Kent Rao, (Deputy Dir., Pennsylvania Transportation Inst., Pennsylvania State Univ., University Park, Pa. 16802) and **Thomas D. Larson**, (Secy. of Transportation, Pennsylvania Dept. of Transportation)

Transportation Engineering Journal, Vol. 108, No. TE2, March, 1982, pp. 183-195

The 1980s are a time of rapid change to most transportation agencies in the country—change in a funding base, which has remained essentially stable for the last 60 years, and change in programs at both state and federal levels as efforts are made to recapitalize our vast highway network. Fiscal and program management strategies are described for responding to this change — strategies which have been put into practice in Pennsylvania and have worked well to date in providing a better climate for transportation. Various efficiency improvements and methods of managing both intra- and inter-agency communication are also described.

0744 Automated Accident Location Systems

James D. Decker, Fellow, ASCE, (Vice Pres., Wilbur Smith and Assoc., 194 East Bay St., Charleston, S.C.) and **Robert W. Hawkins**, (Assoc., Wilbur Smith and Assoc., Columbia, S.C.)

Transportation Engineering Journal, Vol. 108, No. TE2, March, 1982, pp. 197-206

State transportation agencies continue to seek better ways to identify road improvement needs. Research and development funds have been allocated to seek efficient methods for identifying accident locations using computer-associated procedures. Two states, North Carolina and Kentucky, have engaged in these efforts on a statewide basis. With the aid of consultants, these agencies have sought ways to increase the coverage and detail of accident location data. The research examined the needs of the states and their present methodology. An array of possible new methodologies were examined and recommendations were made to the states for implementation in subsequent phases. The recommended accident location systems are being implemented in subsequent phases, with a primary objective being the use of the locator system for other highway purposes.

———————

0745 Recreational Travel: Gasoline Shortages and Price Increases

Edward C. Sullivan, Member, ASCE, (Assoc. Research Engr., Inst. of Transportation Studies, Univ. of California, Berkeley, Calif.) and **Kevin C. Picha**, (Transportation Engr., JHK Assoc., Emeryville, Calif.)

Transportation Engineering Journal, Vol. 108, No. TE2, March, 1982, pp. 207-216

An empirical study utilizes visitation data from a number of recreational attractions in California to estimate the reduction of travel due to the gasoline price increases that occurred during 1977-79 and the gas shortages that occurred in 1979. Differences in the price elasticities and the impacts of the gas shortage are determined by comparing private and public recreation sites, located in both urban and rural areas. It is shown that local zoos suffer the greatest long term visitation reduction due to increased gas prices, and that recreation attractions that draw from the national tourism market are least affected by gas prices. However, national attractions are the most affected by a gas shortage. It is recommended that public officials should consider mass transit for providing mobility to low cost local recreation areas and plan for possible mitigation of adverse local economic consequences if major recreational attractions are affected by future fuel shortages.

———————

0746 Expressway Route Optimization by OCP

Jocelyn F. B. Shaw, Fellow, ASCE, (Assoc. Prof., School of Civ. Engrg., Oklahoma State Univ., Stillwater, Okla.) and **Bernard E. Howard**, (Prof., Dept. of Mathematics, Univ. of Miami, Coral Gables, Fla.)

Transportation Engineering Journal, Vol. 108, No. TE3, May, 1982, pp. 227-243

The optimum curvature principle (OCP) is applied to the horizontal alignment optimization of an expressway in flat South Florida between two given points. Construction cost is the criterion optimized; any other criteria which can be quantized (e.g., environmental), can be optimized by the OCP in a similar manner. The OCP is shown to be a practical aid in the location and optimization of highways and other transportation routes by practical application. The global optimum route, not easily available by other methods, is obtained deterministically. An error analysis is used together with an intrinsic equation integration algorithm to establish error bounds when the true answer is unknown. The error bound (less than one hundredth of 1%) is much less than the 5% approximate construction cost saving over the straight line route with a length increase of about 1/2 of 1%.

0747 Porous Pavement: Research, Development, and Demonstration

Richard Field, (Chf., Storm and Combined Sewer Sector, Municipal Environmental Research Lab., Cincinnati, Ohio; U.S. Environmental Protection Agency, Edison, N.J.), **Hugh Masters**, (Staff Engr., Storm and Combined Sewer Sector, Municipal Environmental Research Lab., Cincinnati, Ohio; U.S. Environmental Protection Agency, Edison, N.J.) and **Melvin Singer**, (Staff Engr., Storm and Combined Sewer Sector, Municipal Environmental Research Lab., Cincinnati, Ohio; U.S. Environmental Protection Agency, Edison, N.J.)

Transportation Engineering Journal, Vol. 108, No. TE3, May, 1982, pp. 244-258

This paper discusses the U.S. Environmental Protection Agency's porous pavement research program along with the economies, advantages, potential applications, and status and future research needs of porous pavements. Porous pavements are an available stormwater management technique which can be used on parking lots and low volume roadways to reduce both stormwater runoff volume and pollution. In addition, groundwater recharge is enhanced. Also, cost reductions result due to elimination of curbs, drains, and small sized storm sewers. Porous asphalt pavements consist of a relatively thin course of open graded asphalt mix over a deep base of large size crushed stones. Water can be stored in the crushed stone base until it can percolate into the subbase or drain laterally. Other porous pavement types include concrete lattice blocks and a porous concrete mix.

0748 Barriers in Highway Work Areas

Bradley T. Hargroves, (Asst. Prof., Dept. of Civ. Engrg., Univ. of Virginia, Charlottesville, Va.)

Transportation Engineering Journal, Vol. 108, No. TE3, May, 1982, pp. 259-266

A design process has been developed to determine the necessity for temporary positive barriers systems in highway work zones. The methodology includes two prescreening steps to determine whether site and traffic conditions constitute a probable or significant hazard. In the third step, the expected number of hazardous vehicle encroachments into the work area is estimated based on the duration and length of the hazard, the roadside clearance and cross slope, the roadway curvature and the traffic speed and volume. The procedure draws heavily on the guidelines developed for the installation of permanent barriers and on empirical results from a number of studies.

0749 Buried Pipelines: Providing Flexibility at Structures

John H. Scarino, Fellow, ASCE, (Principal Assoc., Clinton Bogert Assoc., Fort Lee, N.J.)

Transportation Engineering Journal, Vol. 108, No. TE3, May, 1982, pp. 267-281

When differential settlements occur bmtween a structure and the connected buried piping, the pipe is subjected to longitudinal bending, and joints deflect. The length of the pipe section immediately adjacent to the structure must be selected to keep both of these considerations within allowable limits. A method of determining the appropriate length of pipe section is developed. The method can be applied to pipe of differing materials and with different types of joints, thus permitting rational selections to be made for specific site conditions. Representative pipe and joint properties are presented with examples.

0750 Aircraft Evaluation in Air Network Planning

Adib Kanafani, Member, ASCE, (Prof. of Transportation Engrg., Univ. of California, Berkeley, Calif.) and **Atef Ghobrial**, (Research Asst., Univ. of California, Berkeley, Calif.)

Transportation Engineering Journal, Vol. 108, No. TE3, May, 1982, pp. 282-300

The relationship between aircraft technology and the structure of air service networks in short haul systems is investigated. The analysis is based on an optimal frequency assignment model that is integrated with a route choice model. Together these models allow the optimization of aircraft utilization and define an optimal network structure for a given demand pattern and aircraft technology. Large aircraft have a compelling advantage due to the scale economies they enjoy. Small aircraft, on the other hand, can offer level-of-service advantages and an opportunity for reducing the dependence on congested major transfer hubs in the network. A desirable technology development trend would be aimed at eliminating the diseconomies of distance suffered by small aircraft due to their limited payload-range characteristics.

0751 Airport-Passenger Parking Design

Richard de Neufville, Member, ASCE, (Prof. and Chairman, Technology and Policy Program, Massachusetts Institute of Technology, Cambridge, Mass. 02139)

Transportation Engineering Journal, Vol. 108, No. TE3, May, 1982, pp. 302-306

Despite standard procedures for estimating the quantity of parking space for airports, great difficulties persist at many terminals. These principally concern the amount of space available for the short-term parking of vehicles used in the pickup and delivery (PUD) of passengers. The problem is that of determining the right quality or mix of parking facilities. A need is identified for a different kind of facility: space for vehicles remaining for 30 minutes or less during PUD activities. This represents an intermediate kind of facility between curb space, which is most useful for very rapid PUD by taxis, courtesy cars, and the like, and the standard short- and long-term storage facilities. When this kind of space is not available close to a terminal, drivers will be forced to jam the curbside. A way to estimate the amount of PUD space desirable for airports is proposed. The practical examples of Boston and Los Angeles are discussed.

0752 Safety Studies Related to RRR Projects

Charles P. Brinkman, (Physicist, Federal Highway Administration, Washington, D.C.)

Transportation Engineering Journal, Vol. 108, No. TE4, July, 1982, pp. 307-312

The Federal Highway Administration has several research efforts specifically aimed at determining the impact of the Resurfacing, Restoration, and Rehabilitation (RRR) Program and, in general, the improvements often associated with RRR projects on safety. Studies that analyze traffic accident data are presented. The safety implications of resurfacing, skid resistance, lane and shoulder width delineation, and the removal of roadside and fixed objects are discussed.

0753 Bus Stop Location and Design

Michael J. Demetsky, Member, ASCE, (Assoc. Prof., Dept. of Civ. Engrg., Univ. of Virginia and Virginia Highway and Transportation Research Council, Charlottesville, Va.) and **Brian Bin-Mau Lin**, (Transportation Planner, Southern California Assoc. of Governments, Los Angeles, Calif.)

Transportation Engineering Journal, Vol. 108, No. TE4, July, 1982, pp. 313-327

Those elements associated with the location and design of bus stops that affect the efficiency of transit and traffic operations are identified. This evaluation was accomplished using two questionnaires that provided opinions of city traffic engineers and transit operators regarding current practice, criteria appropriate for evaluating the utility of bus stops, and strategies for improving the utility of bus stops. It was found that the criteria used to determine the location and design of bus stops should be in the form of guidelines, not numerical warrants. In locating stops each should be treated individually. A case study is presented to show how the individual location and design measures interact.

0754 Energy Use and Emissions Impact Measurement in TSM

Frank Southworth, (Asst. Prof. of Transportation Planning, Civ. Engrg. Dept., 2114 NCEL, Univ. of Illinois at Urbana-Champaign, Urbana, Ill.) and **Bruce N. Janson**, (Asst. Prof. of Transportation Planning, Civ. Engrg. Dept., Carnegie-Mellon Univ., Pittsburgh, Pa. 15217)

Transportation Engineering Journal, Vol. 108, No. TE4, July, 1982, pp. 328-342

An equilibrium traffic assignment model is linked to a modal choice model to estimate the route and mode shifting implications of a bus-only lane on Chicago's North Lake Shore Drive. These models are in turn linked to equations for calculating the energy usage and emissions implications of such Transportation System Management (TSM) tactics. The energy equations are based upon General Motors Laboratory figures. The hydrocarbon, carbon monoxide and nitrogen oxide impacts are derived using USEPA procedures. The paper emphasises model evaluation, and finds the models very useful for interpreting TSM impacts within heavily congested urban corridors.

0755 Transit Performance Evaluation Model

Richard P. Guenthner, Assoc. Member, ASCE, (Asst. Prof. of Civ. Engrg., Marquette Univ., Milwaukee, Wisc. 53233) and **Kumares C. Sinha**, Member, ASCE, (Prof. and Head, Transportation and Urban Engrg., School of Civ. Engrg., Purdue Univ., West Lafayette, Ind. 47907)

Transportation Engineering Journal, Vol. 108, No. TE4, July, 1982, pp. 343-361

An analytical methodology, known as the Transit Performance Evaluation Model, has been extended and applied. The model is used to evaluate the effect on transit performance from short term policy changes including: fares, service frequencies, route coverage, and route alignment. Long term trends in ridership due to a shift in the demand function caused by factors beyond the control of transit operators were also considered. The model was intended for use by bus operators in small to medium sized cities. Model tests were performed in several small midwestern cities using a variety of policy changes. Analysis of available "after" data indicated that the model results are reasonable and the model can be effectively used to make decisions regarding transit improvement options.

0756 Impact of Gasoline Price on Transit Ridership in Fort Worth, Texas

Gary J. Wolff, Assoc. Member, ASCE, (Transportation Engr., Dept. of Transportation, Ft. Worth, Tex.) and **Dayle M. Clark**, Member, ASCE, (Asst. Prof., Dept. of Civ. Engrg., Univ. of Texas at Arlington, Tex.)

Transportation Engineering Journal, Vol. 108, No. TE4, July, 1982, pp. 362-375

The impact of gasoline price, gasoline supply, automobile ownership costs, and automobile operating costs on transit ridership is examined. Monthly data on transit ridership, gasoline supply, gasoline price, and consumer price indexes for automobile repair and maintenance, automobile insurance rates, motor oil, used cars, and automobile finance charges are all plotted to compare monthly, seasonal and yearly trends. The differences between the two energy shortages of 1974 and 1979 are analyzed, indicating that latent demand from by-choice transit riders for transit service exists in Fort Worth, Texas.

0757 Gravel Road Corrugations

James Stoddart, (Technical Adviser, Inst. of EngrQ., Kathmandu, Nepal), **Robert B. L. Smith**, (Prof. of Civ. Engrg., Univ. of Nairobi, Nairobi, Kenya) and **Richard M. Carson**, (Sr. Lect. in Mech. Engrg., Univ. of Nairobi, Nairobi, Kenya)

Transportation Engineering Journal, Vol. 108, No. TE4, July, 1982, pp. 376-382

The significance and causes of corrugation of gravel roads are discussed together with a brief account of theories advanced over the past 60 years. A test rig to simulate the performance of a vehicle wheel with hard and soft tires on various soil compositions of road surface is described and results analyzed. It is concluded that the main factors in the formation of corrugations are the effective stiffness of the tired wheel and the sand/clay ratio of the road surface soil.

0758 Turnkey Projects for Downtown People Movers

Carl W. Goepfert, Fellow, ASCE, (Mgr., Transportmtion Projects, Bechtel Civ. & Minerals, Inc., San Francisco, Calif.)

Transportation Engineering Journal, Vol. 108, No. TE4, July, 1982, pp. 383-391

Los Angeles, Detroit, and Miami received proposals for construction of downtown people mover systems. The requests for bids contemplated turnkey type construction contracts. The multiplicity of requirements for functional and environmental compatibility lead to a complex proposal preparation process but allows the owner to deal with a single contractual entity in negotiations. The background of the people mover projects, the design, and contract development, and a comparison with more conventional procedures are all briefly presented. Advantages of the alternative procedures are described in the context of three ongoing projects.

0759 Lawsuits Arising Out of Reconstruction

David C. Oliver, (Attorney Advisor, Motor Carrier and Highway Safety Law Div., Office of the Chf. Counsel, Federal Highway Administration, Room 4224, Nassif Building, 400 7th St., SW., Washington, D.C.)

Transportation Engineering Journal, Vol. 108, No. TE4, July, 1982, pp. 392-407

Liability judgements resulting out of negligence in the performance of highway design, construction or maintenance activities have continued to increase in the past decade, notwithstanding improved inspection and signing practices. This can be attributed to increased familiarity with the literature on the subject and with codes and guidelines addressing all aspects of highway work. The increase in judgements evidences impatience with those administrators or contractors who ignore the guidelines or who appear to have made a conscious decision to gamble with the public's safety. Application of the case law which has been developed can result in a decrease in future judgements and a safer system. It is time for contractors and administrators to recognize that the protections relied upon in the past to cover culpability are no longer present. The Federal Government is in an obvious period of retrenchment, State legislators are responding to public pressures to make Government accountable and the courts refuse to accept the English notion of the sovereign.

0760 Guideway Geometry for AGT Systems

David Allen, Member, ASCE, (Project Mgr., Kaiser Engrs., Inc., 35 East Wacker Dr., Chicago, Ill. 60601)

Transportation Engineering Journal, Vol. 108, No. TE4, July, 1982, pp. 408-418

A logical method of defining horizontal and vertical Automated Guideway Transit (AGT) guideway geometry is presented. Only geometric accelerations and jerk are considered, which are defined as those which a passenger feels if the guideway is smooth and rigid and the vehicle fully damped. No account is taken of vibrational effects. The interaction of the roll flexibility of the vehicle with the superelevation on curves is discussed and simplified expressions developed for design applications. A new horizontal transition curve is presented which is a refinement of the Mast equation and ensures continuity of unbalanced lateral acceleration at both ends of the curve for small and large spiral angles. Expressions are developed for transition lengths based on limitation on lateral jerk, vertical acceleration at superelevation buildup, and on rate of twist. These lengths are compared in a general way with those currently in use for highways and railroads. A new vertical transition curve, which ensures a gradual buildup of vertical accelerations is presented. By limiting the value of maximum acceleration, the length of the vertical curve is established.

0761 Traffic Signals at New and Rebuilt Intersections

David L. Guell, Member, ASCE, (Assoc. Prof., Dept. of Civ. Engrg., Univ. of Missouri-Columbia, Columbia, Mo.)

Transportation Engineering Journal, Vol. 108, No. TE4, July, 1982, pp. 419-427

The need for traffic signals at an intersection is generally determined by the application of the Manual on Uniform Traffic Control Devices (MUTCD) warrants. When the volume warrants are not satisfied it may be because the absence of signals is discouraging motorists from using the intersection and that if signalized, the volume warrants would be satisfied. The function of the intersection and the streets involved should be considered when evaluating the need for signals. A case study is given which shows substantial volume increases resulted on the minor street when signals were installed at a previously unsignalized intersection.

0762 Structural Design of Cemented Pavement Layers

Eddie Otte, (Assoc., Van Wyk and Louw, Inc., Consulting Civ. Engrs., P.O. Box 905, Pretoria 0001, South Africa), **Phillip F. Savage**, (Prof., Dept. of Civ. Engrg., Univ. of Pretoria, Pretoria, South Africa) and **Carl L. Monismith**, Fellow, ASCE, (Prof. of Civ. Engrg., Dept. of Civ. Engrg., Univ. of California, Berkeley, Calif.)

Transportation Engineering Journal, Vol. 108, No. TE4, July, 1982, pp. 428-446

A mechanistic structural design procedure for pavements containing cement- and lime-treated layers is outlined. The procedure includes provision for use of a range in traffic wheel loadings, fundamental material properties, and failure criteria, including fatigue. Layered elastic theory is utilized to calculate stresses and strains at critical positions in the pavement components, and these are compared with allowable values to minimize specific distress modes. Use of standard designs resulting from the more detailed analyses are discussed and explained. Reasonable agreement between predicted and actual response and behavior of several trial pavements are noted. Three design examples illustrating the procedure are included.

0763 Highways, Transport, and Energy: A Look Ahead

Herbert S. Levinson, Fellow, ASCE, (Prof. of Civ. Engrg., Univ. of Connecticut, School of Engrg., Storrs, Conn.)

Transportation Engineering Journal, Vol. 108, No. TE5, September, 1982, pp. 447-456

Changes in urban development and travel will affect transportation energy consumption. Methods for reducing energy use in the transportation sector in the decades ahead are examined. Principal possibilities include: reducing the need to travel, improving vehicle fuel economy, maximizing vehicle occupancy, and making each vehicle-mile travelled as energy efficient as possible. The gains from using fuel-efficient vehicles will encompass the entire population. Consequently, they will be far greater than those that would be achieved by better transportation system management, new road construction, transit improvements, or land-use changes. Such measures, while desirable complements to better vehicle fuel efficiency, should be assessed in terms of the broader benefits they confer. Therefore, the promising first step toward reducing transportation energy is to substantially increase the miles per gallon achieved by the motor vehicle fleet. This should be supported by development and pricing policies that foster compactness, thereby reducing the journey to work, concentrating commuter travel patterns and improving public transport potentials.

0764 Equilibrium Concepts in Modal Split Modelling

James Beattie Morison, (Transportation Planner, Transportation Planning Div., City of Calgary, Calgary, Alberta, Canada)

Transportation Engineering Journal, Vol. 108, No. TE5, September, 1982, pp. 457-467

The potential usefulness of extrapolating Wardrop's first principle regarding traffic equilibrium on road systems to modal split modelling is evaluated. A conceptual framework based upon equilibrium theory is presented, a model for peak hour home based work trips is developed and the price eguality prediction of the equilibrium model is tested. The test, which used data from the 1971 City of Winnipeg work trip survey, provides support for the equilibrium theory. It is noted that a logit model can be derived from this model. It is concluded that this model and research based upon it can lead to a better understanding of modal choice behavior and to a reliable and stable modal choice model.

0765 Visual Intrusion Impact of Residential Traffic

Gerald R. Brown, (Assoc. Prof., Dept. of Civ. Engrg., Univ. of British Columbia, Vancouver, British Columbia, Canada) and **Donald F. Park**, (Grad. Student, Community and Regional Planning, Univ. of British Columbia, Vancouver, British Columbia, Canada)

Transportation Engineering Journal, Vol. 108, No. TE5, September, 1982, pp. 468-483

An investigation of the nature of visual intrusion impact of traffic in residential neighbourhoods is presented. The specific purpose of the study is to discover if residents perceive and respond to visual intrusion of traffic in accordance with their situation and socioeconomic characteristics. The method is an attitudinal survey questionnaire of 111 residents of three socially different districts in a medium size city. Results show that residents of different situation and socioeconomic characteristics perceive the visual intrusion problem differently, and that parked vehicles may have a greater impact than anticipated. A second result showed that, based on resident responses, the tolerable street traffic in the districts studied for visual intrusion impact was between 650 and 990 vehicles per hour. It is concluded that better management of parked cars in residential districts will reduce visual intrusion impact, and that traffic flow for visually-related environmental capacity of residential streets is the same order of magnitude as noise-related environmental capacity. A design procedure for application is suggested.

0766 Structural Materials in Aerospace Applications

Task Committee on Materials of the Committee on Aerospace Structures and Materials of the Aerospace Division

Transportation Engineering Journal, Vol. 108, No. TE5, September, 1982, pp. 484-494

In the selection of materials for aerospace applications, consideration is given to factors, such as strength and weight, as well as to economics and energy requirements for both manufacture and in-service operation of the completed structure. The proper combination of all factors must be considered before new applications can be undertaken. Aerospace structural materials cover a wide range of both metal alloys and nonmetallics. Because coverage of all materials is not feasible, for the civil engineer who is probably not familiar with many of these materials, three major metal alloying systems—steel, aluminum, and titanium—are examined. Heat treatments, fabrication, product forms and applications are reviewed to give the reader a broad introduction to some materials that could satisfactorily meet future structural requirements.

0767 Vehicle Lateral Placements on Urban Roads

Eric J. Miller, (Research Assoc., Univ. of Toronto-York Univ., Joint Program in Transportation, 42 St. George St., Toronto, Ontario, Canada M5S 2E4) and **Gerald N. Steuart**, (Assoc. Prof., Dept. of Civ. Engrg., Univ. of Toronto, Toronto, Ontario, Canada)

Transportation Engineering Journal, Vol. 108, No. TE5, September, 1982, pp. 495-508

Time-lapse photographs were taken of traffic flows on five streets of varying lane width in the Toronto region. From these photographs, time histories of lateral placements and forward speeds were obtained for each vehicle. The analysis of these data indicated that lateral placement is a function of vehicle size, lane width and lane type. It was further found that, in general, the smaller the expected variance in driver response for a given roadway, the closer to the left edge of the lane the vehicles track.

0768 Control of Adverse Wind Near Buildings

Parvis Merati, (Grad. Asst., Armour Coll. of Engrg., Dept. of Mech. and Mechanical and Aerospace Engrg., Illinois Inst. of Tech., Chicago, Ill. 60616), **Roald Wigeland**, (Asst. Prof., Armour Coll. of Engrg., Dept. of Mech. and Mechanical and Aerospace Engrg., Illinois Inst. of Tech., Chicago, Ill. 60616) and **Hassan Nagib**, Member, ASCE, (Prof., Armour Coll. of Engrg., Dept. of Mech. and Mechanical and Aerospace Engrg., Illinois Inst. of Tech., Chicago, Ill. 60616)

Transportation Engineering Journal, Vol. 108, No. TE5, September, 1982, pp. 509-521

The characteristics of basic "flow modules" associated with an isolated building, a fence and building combination, and a building with various canopies in a simulated atmospheric boundary layer are documented. The rolled-up horseshoe-shaped vortex at the windward base of the building, the shear layer originating at the leading edges of the roof and sides of the building, the stagnation region on the face of the building, and the tornado-like vortices in the recirculating region in the lee of the building are several of the flow modules that exist around an isolated building. The installation of a porous fence in open areas produces effective protection for pedestrians, while its presence in front of the building worsens the condition there. The fence solidity does not have any appreciable effects on the flow field. The presence of a canopy generally improves the severe conditions existing near the ground in front of a single building. The best position for attaching the canopy is at one-quarter height of the building.

0769 Flexible Pavement Evaluation and Overlay Design

Sudipta S. Bandyopadhyay, Member, ASCE, (Chf., Advanced Geotechnology, NFS/National Soil Services, Houston, Tex.)

Transportation Engineering Journal, Vol. 108, No. TE6, November, 1982, pp. 523-539

A rational method for flexible pavement evaluation and overlay design using the Dynaflect deflections is presented. Evaluation charts and design curves are based primarily on the results of a comprehensive two-year study during which the Dynaflect deflections were systematically monitored on selected flexible pavements located strategically throughout Kansas. Various factors, including the pavement thickness, volume of traffic, subgrade soil type, temperature and average rainfall are taken into consideration. The adequacy of the sample size of the deflection measurements is tested statistically throughout the study. A pavement evaluation scheme consisting of a visual condition survey and a structural diagnosis chart based on the Dynaflect measurements is presented. The data from the study are used to develop temperature adjustment factors and seasonal adjustment factors. Charts are presented whereby representative values of DMD and SCI can be used to compute the thickness of the overlay required for the pavement to sustain the design traffic. The implementation of the developed subsystem in a pavement management system is illustrated by a flow chart.

0770 Protection of Pipelines Subject to Surcharge Loads

Eugene M. Kovach, Member, ASCE, (Project Engr., Sanitation Dist. of Los Angeles County, Whittier, Calif.)

Transportation Engineering Journal, Vol. 108, No. TE6, November, 1982, pp. 540-549

Surcharge loads from new fills, railroads, and structures often require design of reinforcement to increase the field supporting strength of existing sewers and storm drains. Pipelines located in easement and rights of way, or designed for shallow earth cover, are most vulnerable. Because the pipe length is often short, the unit cost for reinforcement is high, often exceeding unit costs for new sewers. Load analysis, assumptions, and design details are presented. The concrete arch encasement and its load factors are reviewed. Its use and misuse are considered. References for the designer are also cited. Alternatives to increasing pipe strength are shown by examples of reinforced concrete arch and slab designs. Recommendations are made for design and construction of new pipelines and for preventing future surcharge problems. The necessity of good inspection reporting pertaining to trench width and soil description is stressed.

0771 Optimization Methodology for Bus Garage Locations

T. H. Maze, Assoc. Member, ASCE, (Asst. Prof., School of Civ. Engrg. and Environmental Sci., Univ. of Oklahoma, Norman, Okla. 73019), **Snehamay Khasnabis**, Assoc. Member, ASCE, (Assoc. Prof., Dept. of Civ. Engrg., Wayne State Univ., Detroit, Mich. 48202.) and **Mehmet D. Kutsal**, Student Member, ASCE, (Research Asst., Dept. of Civ. Engrg., Wayne State Univ., Detroit, Mich. 48202)

Transportation Engineering Journal, Vol. 108, No. TE6, November, 1982, pp. 550-569

An optimization model is developed for determining both the optimal bus garage locations and optimal assignment of buses to those garages. The model uses a list of candidate garage sites, the transit system's service and operational characteristics and the characteristics of the regional highway system as input. As output, it produces the optimal number, sizes and locations of bus garage additions or changes. The model considers garage operating and garage construction cost and the costs of deadheading and driver relief under separate operating schedules for morning, all day, and evening route assignments on weekdays, Sundays and Saturdays. In the methodology development, the problem is first formulated as a discrete solution-space fixed-charge facility location problem. The fixed charge problem is formulated as a mixed integer program. However, because of the difficulty involved in solving a large-scale problem with a general purpose mixed-integer program, the problem is solved by a special purpose branch-and-bound process. This process utilizes an efficient sequence of classical transportation problem (a quick solving, special class of linear program). The methodology is demonstrated with a small-scale sample problem.

0772 Traffic Analysis for Large Construction Project

Robert G. Spencer, Assoc. Member, ASCE, (Civ. Engr., Engrg. Design, Tennessee Valley Authority, W3A19, 400 W. Summit Hill Dr., Knoxville, Tenn. 37902), **Arun Chatterjee**, Member, ASCE, (Prof. of Civ. Engrg., Univ. of Tennessee, Knoxville, Tenn. 37996-2010) and **Jack B. Humphreys**, Fellow, ASCE, (Prof. of Civ. Engrg., Univ. of Tennessee, Knoxville, Tenn. 37996-2010)

Transportation Engineering Journal, Vol. 108, No. TE6, November, 1982, pp. 570-584

The paper presents various characteristics of commuter-traffic generated during the construction of Hartsville Nuclear Plant in a rural area in Tennessee, and a procedure for estimating such traffic for similar projects. The variations in daily traffic volumes and peaking characteristics are examined for locations nearby the construction site. For the commuter traffic generated by the site, automobile occupancy rates and the use of vans and buses are examined. The study includes the spatial distribution of commuters and it is found that the relative distribution of employees among surrounding counties can be predicted with a simple formulation of gravity model. A step by step procedure for predicting the traffic generated by large construction sites is presented.

0773 Mudslide Effects on Offshore Pipelines

Richard C. Swanson, (Sr. Research Engrg., Shell Development Co., Houston, Tex.) and **Warren T. Jones**, (Staff Research Engr., Shell Development Co., Houston, Tex.)

Transportation Engineering Journal, Vol. 108, No. TE6, November, 1982, pp. 585-600

In regions of unstable soils such as portions of the Mississippi Delta, occasional pipeline failures during periods of severe weather have been attributed to mudslides. The effects of such downslope soil movements on the deflection and resulting stress in a pipeline have been studied in an attempt to reveal methods of routing or design which would reduce the risk of failure. Results show that the chances of surviving a slide are increased if the pipeline outer diameter is reduced to as small a value as possible, if the pipe wall thickness is increased, and if some slack is available in the line. The chance of survival is greatest if the slide occurs in a direction perpendicular to the pipeline route since the failure mode is primarily one of tension. When the slide crosses the pipeline at other angles of incidence, the chance of survival is significantly lessened.

0774 Natural Gas Transmission System Optimization

Festus O. Olorunniwo, (Asst. Prof., Dept. of Industrial and Systems Engrg., Univ. of Michigan-Dearborn, 4901 Evergreen Rd., Dearborn, Mich. 48128) and **Paul A. Jensen**, (Prof., Operations Research Group, Mech. Engrg. Dept., Univ. of Texas, Austin, Tex. 78712)

Transportation Engineering Journal, Vol. 108, No. TE6, November, 1982, pp. 601-617

This paper develops a general model and advances a solution methodology for the optimum design of natural gas transmission networks. The complete static design problem addressed is that of determining the optimum number, locations, sizes, and operating characteristics of the components of the system by minimizing, over a finite time horizon, T, the present worth of costs such that no pressure or technological specifications are violated. Most published investigations of this problem have resulted in suboptimal solutions because they fixed a subset of the decision variables while optimizing the others. This work differs from that of previous attempts at the complete problem in that it includes the option of expanding an existing network; it considers more general cost functions and the solution times have been so reduced that larger problems and extensive sensitivity analyses can be investigated.

0775 Soil Pressures Due to Aircraft Loans

Glenn T. Baird, Member, ASCE, (Sr. Research Engr., New Mexico Engrg. Research Inst., Univ. of New Mexico, Albuquerque, N.M.) and **Phillip Nash**, (Task Officer, Air Force Engrg. and Service Center, Tyndall Air Force Base, Fla.)

Transportation Engineering Journal, Vol. 108, No. TE6, November, 1982, pp. 618-631

Soil pressure gages were installed in a crushed limestone backfill of a bomb damaged runway. Measurements were made due to static and dynamic aircraft, C-130 and F-4E, wheel loads. Static pressures are compared to pressure predictions from a nonlinear, finite element computer code and indicate a good correlation. Dynamic pressures correlated with static pressures indicate 25%-50% lower pressures.

0776 Hot-Mix Paving in Marginal Weather

David C. Colony, Fellow, ASCE, (Prof. of Civ. Engrg., Univ. of Toledo, Toledo, Ohio), **R. Kenneth Wolfe**, (Prof. of Industrial Engrg., Univ. of Toledo, Toledo, Ohio) and **George L. Heath**, (Prof. of Mechanical Engrg., Univ. of Toledo, Toledo, Ohio)

Transportation Engineering Journal, Vol. 108, No. TE6, November, 1982, pp. 632-649

Extension of the construction season for bituminous layers could result in benefits to the public and to the construction industry. Weather conditions in Ohio were studied during the months of March, April, May, October, and November. Application of those weather data to a previously developed thermophysical model yielded the number of days during those months which were acceptable for construction of hot-laid bituminous mats. The criterion of acceptability was an average mat temperature at least 175°F (79°C) 10 minutes after placement. Probabilities of accpetable construction days were calculated by weeks for mat thicknesses between 1.25 in. (32 mm) and 3 in. (76 mm), using the thermophysical model and the current specification of the Ohio Department of Transportation (ODOT). Application of the thermophysical model to mats at least 1.5 in. (38 mm) indicated that construction during March and November could be significantly more frequent under the proposed thermally based specification, as compared to the current ODOT specification. Favorable conditions for 1.25 in. (32-mm) mats were seldom found during the five months studied. Air temperature alone does not adequately characterize the construction environment for thin mats.

0777 Marketing Implications of Perceptions of Transit

Wilfred W. Recker, (Prof. of Engrg., School of Engrg. and Inst. of Transportation Studies, Univ. of California, Irvine, Calif.) and **Harry J. Schuler**, (Riverside County Transportation Comm., Riverside, Calif.)

Transportation Engineering Journal, Vol. 108, No. TE6, November, 1982, pp. 650-661

Perceptions of Public Transit and other, psychographic, information are used to develop a segmentation-based approach to marketing public transit. Factor analysis is used to identify differences between the perception structures of male and female respondents. Logit analysis of intention to use public transit is employed to identify the salient dimensions of this decision process. The respondents are segmented into groups that are homogeneous with respect to these dimensions. The results of the analysis indicate directions for marketing approaches to the various subgroups. Empirical results reveal differences between male and female samples both in perception as well as in decision process.

0778 Pavement Roughness and Rideability: A State-of-the-Art Review

Michael S. Janoff, (Mgr., Transportation Programs, Ketron, Inc., Philadelphia, Pa. 19107)

Transportation Engineering Journal, Vol. 108, No. TE6, November, 1982, pp. 662-675

This paper reports on the results of a review of the literature and ongoing research concerned with pavement roughness and rideabiity. The material covers definitions and concepts, subjective rating methods, objective (physical) measurements, statistical methods, ongoing research and uses of roughness, profile, and rideability data. Much of the past work in pavement roughness and rideability has stressed the physical aspects of the road surface, and the analysis of such physical data, rather than the public's perception of pavement rideability defined in terms of the human response to roughness. Only a few researchers have considered the psychological issues of scaling and the data analysis necessary to define subjective ratings and relate these ratings to physical measures of roughness. Ongoing research by both the Pennsylvania Department of Transportation and the National Cooperative Highway Research Program will attempt to address these issues.

0779 Driver Gap Acceptance Phenomena

Olusegun Adebisi, (Lect., Dept. of Civ. Engrg., Ahmadu Bello Univ., Zaria, Nigeria)

Transportation Engineering Journal, Vol. 108, No. TE6, November, 1982, pp. 676-688

Results from a field study of gap acceptance characteristics of drivers executed merging maneuvers from stopped positions on side roads to main roads are presented. Gap acceptance parametrs are first estimated from the field data in an aggregated form and then for individual traffic flows observed on the main roads. It is found that flow on the main roads substantially influences the gap acceptance characteristics of drivers on the side roads. The practical implications of this influence on the accurate computations of average delays and intersection capacity, normally based on gap acceptance parameters, are also investigated. It is found that when these values are derived from gap acceptance parameters estimated from the aggregated data; they differ by as much as 20% from those derived from gap acceptance parameters estimated from the disaggregated data.

0780 Wind Tunnel Modeling of Building Internal Pressure

Henry Liu, Member, ASCE, (Prof. of Civ. Engrg., Univ. of Missouri-Columbia, Columbia, Mo. 65211)

Transportation Engineering Journal, Vol. 108, No. TE6, November, 1982, pp. 691-696

An inspectional analysis is made on the differential equations governing the fluctuations of the building internal pressure generated by wind for both single and multi-room buildings. The result is used to find the similarity parameters required for wind tunnel modeling of the internal pressure. How the results are to be used in practice is analyzed.

0781 Anatomy of Environmental Regulation

Rae Zimmerman, (Assoc. Prof. of Planning, Grad. School of Public Administration, New York Univ., New York, N.Y.)

Journal of the Urban Planning and Development Division, Vol. 108, No. UP1, May, 1982, pp. 1-10

Environmental permits have proliferated over the past decade as a device for implementing environmental law and policy. This paper describes the phenomenon of proliferation, the state response to it, and two methods for further changes in the administration of permits to mitigate the complexity of the permitting process. Proliferation of permits can be understood in terms of the number of types of permits, permits issued within each type, agencies administering them, and variations in procedures from permit to permit. Two administrative models could be used to achieve greater administrative simplicity in permitting. The first is permit sequencing which involves ordering permits during the review process from the more comprehensive in scope (in terms of information and environmental analysis requirements) to the more specific, which would avoid repetitive analyses. The second approach is to use a high volume permit as a "pivotal" which would act as a screen for the review of other permits. Thus, these two techniques would in effect achieve the objective of stemming the proliferation in permits in a more comprehensive way than the approaches currently being implemented.

0782 Citizen Participation in New York City Government

Bernard Haber, Fellow, ASCE, (Partner, Hardesty & Hanover, New York, N.Y.)

Journal of the Urban Planning and Development Division, Vol. 108, No. UP1, May, 1982, pp. 11-14

Following the example of London, England, New York City developed the first stage of Governmental decentralization in 1969 by forming 62 Community Planning Boards. In 1977, by Charter Revision, 59 Community Boards were established with expanded powers. These Boards operate in districts varying between 100,000 to 250,000 population. District Offices provided day to day monitoring of delivery of services through a small permanent community board staff headed by a District Manager. The Community Board members oversee the District Manager and the District Office. The Boards advise on all land use matters, the capital and expense budgets of the City of New York and provide planning for the welfare and orderly development of its district. After four years of operation in this final stage of decentralization with coterminus service delivery districts, the system is on the road to success.

0783 Conjoint Analysis of Residential Preferences

Irene Findikaki-Tsamourtzi, Member, ASCE, (Doctoral Candidate, Dept. of Civ. Engrg., Stanford Univ., Terman Engrg. Center, Stanford, Calif. 94305) and **Jarir S. Dajani**, (Assoc. Prof., Dept. of Civ. Engrg., Stanford Univ., Terman Engrg. Center, Stanford, Calif. 94305)

Journal of the Urban Planning and Development Division, Vol. 108, No. UP1, May, 1982, pp. 15-33

A methodology for studying the effect of different locational factors on individual residential preferences is developed. Observations on the overall preferences of individuals are used to construct profiles of prospective residents. The Conjoint Analysis (CA) method is used for this purpose. Preferences for alternative development proposals are measured with a multidimensional composite index of Residential Location Utility (RLU). Criteria for the selection of the dimensions of the index and the form of the compositional rule are postulated. A vector of the RLU is estimated by the use of ordinal regression in conjunction with linear programming. The methodology was implemented in the study of residential preferences of the population of Assistant Professors of Stanford University. The data collection technique is described and results of the analysis are presented. The members of the population are classified according to their demonstrated behavior in selecting a residential place. An informal validation of the results is made by comparison with direct statements by the respondents on their preferences.

0784 Civil Engineering Planning Guide: Part I—Framework for Planning

Task Committee to Prepare a Planning Guide of the Urban Planning and Development Division, Richard S. Howe, chmn.

Journal of the Urban Planning and Development Division, Vol. 108, No. UP1, May, 1982, pp. 34-69

A Task Committee of ASCE's Urban Planning and Development Division has prepared Part I of an eventual three part planning guide for field or classroom use by civil engineers. Part I of the *Planning Guide*, "Framework for Planning," is an overview of the planning process. The role of engineers in the planning process and the various types of planning are discussed. Geographical and temporal contexts, coordination, and institutional framework are all revealed to be integral elements of the process. The steps in the planning process are ranged as follows: awareness of need, data collection and analysis, development of goals and objectives, clarification and diagnosis of the problem, identification of alternative solutions, analysis of alternatives, evaluation and recommendation of actions, development of implementation program, and surveillance and monitoring. Public involvement is critical to the success of this planning process. Planning tools used for any of these steps can be simple or complex. Various tools are assessed in each of these areas.

0785 Civil Engineering Planning Guide: Part II — Transportation Planning

Task Committee to Prepare a Planning Guide of the Urban Planning and Development Division, Richard S. Howe, chmn.

Journal of the Urban Planning and Development Division, Vol. 108, No. UP1, May, 1982, pp. 70-100

The urban transportation planning process has changed dramatically during the past 25 years. This paper describes the important characteristics of such a planning process, including the key characteristics of urban travel, the short- and long-range planning processes, and the project programming process. Both the planning methodology and the types of actions considered in the transportation planning process are discussed. It is concluded that even given fiscal constraints, government cutbacks, and increasing reliance on local government commitments, there will be a need for a planning process such as that described in this paper.

0786 TVA Hydro Scheduling Model: Practical Aspects

Richard M. Shane, (Civ. Engr., Water Systems Development Branch, Tennessee Valley Authority, Norris, Tenn. 37828) and **Kenneth C. Gilbert**, (Mathematician, Water Systems Development Branch, Tennessee Valley Authority, Norris, Tenn. 37828)

Journal of the Water Resources Planning and Management Division, Vol. 108, No. WR1, March, 1982, pp. 1-19

Computer programs and mathematical methods have been developed to model long term week-to-week variations in water level, discharge and electrical generation for the 42 reservoirs operated by TVA. The model has an interactive structure with CRT graphics for easy use by TVA reservoir management personnel to evaluate the impact of new operating requirements on established objectives, check reservoir system status to warn of possible future problems, forecast reservoir systems operation and develop new long range operation guides.

Notable features of the model include: (1) the observation of important system operating constraints and priorities, (2) extreme event analysis capability, (3) a special linear programming structure that provides the flexibility to modify operating objectives and experiment with new types of long range efficiency guides, and (4) interactive CRT graphical output analysis.

0787 TVA Hydro Scheduling Model: Theoretical Aspects

Kenneth C. Gilbert, (Staff Member, TVA Water Resource Management Methods, Tennessee Valley Authority, Engrg. Lab., Norris, Tenn. 37828) and **Richard M. Shane**, (Staff Member, TVA Water Resource Management Methods, Tennessee Valley Authority Engrg. Lab., Norris, Tenn. 37828)

Journal of the Water Resources Planning and Management Division, Vol. 108, No. WR1, March, 1982, pp. 21-36

The optimization procedures used in the Tennessee Valley Authority (TVA) HYDROSIM model are described. Scheduling of the TVA operated reservoir system poses a stochastic, nonlinear, multireservoir problem with both nonviolable (physical) and violable (operating) constraints. A linear programming procedure for implementing the preemptive priority constraints and a nonlinear search procedure for minimizing the power cost function are described. A guide development program based on stochastic dynamic programming is also described.

0788 U.S. Geological Survey Federal-State Program

Thomas J. Buchanan, Fellow, ASCE, (Asst. Chf. Hydro., U.S. Geological Survey, 441 National Center, Reston, Va.) and **Bruce K. Gilbert**, Member, ASCE, (Deputy Asst. Chf. Hydro., U.S. Geological Survey, 441 National Center, Reston, Va.)

Journal of the Water Resources Planning and Management Division, Vol. 108, No. WR1, March, 1982, pp. 37-45

The U.S. Geological Survey Federal-State Cooperative Water Resource Program is a partnership between the Geological Survey and State and local agencies for the collection of the hydrologic information needed for the continuing determination and evaluation of the quantity, quality, and use of the Nation's water resources. The first Cooperative Program was started in 1895; the Congress gave formal recognition to the partnership in 1928 and authorized Federal funding of not more than 50 percent for cooperative programs with the Geological Survey, with total funding over $80 million. The process of project selection in the Cooperative Water Resource Program is a mutual effort in which Geological Survey represents national interests, including the needs of other Federal agencies, and the cooperator represents State and local interests. The result is a balanced program that involves careful evaluation of needs, priorities, and resources. A number of typical examples of projects within the Cooperative Program are presented.

0789 Reservoir Management in Potomac River Basin

Richard N. Palmer, Assoc. Member, ASCE, (Asst. Prof., Dept. of Civ. Engrg., Univ. of Washington, Seattle, Wash.), **James A. Smith**, (Hydrologist, Interstate Commission on the Potomac River Basin, Rockville, MD), **Jared L. Cohon**, Member, ASCE, (Prof., Dept. of Geography and Environmental Engrg., Johns Hopkins Univ., Baltimore, Md.) and **Charles S. Revelle**, (Prof., Dept. of Geography and Environmental Engrg. Johns Hopkins Univ., Baltimore, Md.)

Journal of the Water Resources Planning and Management Division, Vol. 108, No. WR1, March, 1982, pp. 47-66

Techniques for determining the yield of multireservoir water supply systems have been developed and applied to the system that serves the Washington, D.C. metropolitan area (WMA). This system is composed of the Potomac and Patuxent River Basins and five reservoirs which lie in these basins. The yield of each reservoir in the system was determined using simulation models. Simulation and linear programming models were developed to determine the yield of the reservoir system when operated jointly with the Potomac River. Linear programming models were developed which incorporated projected water demands and limitations on the distribution and treatment capacities of the agencies that supply water to the WMA. These models were used to determine the regional demand that can be met. A multiobjective analysis was made of the potential tradeoffs between conflicting water uses on the basin. The models indicate that the yield which results from the proper joint operation of the system is significantly greater than the yield of the individual components of the system.

0790 Investigation of Daily Flow Forecasting Models

Hasan Yazicigil, (Asst. Prof., Dept. of Geology, Southern Illinois Univ., Carbondale, Ill.), **A. Ramachandra Rao**, Member, ASCE, (Prof., School of Civ. Engrg., Purdue Univ., W. Lafayette, Ind.) and **G. H. Toebes**, Member, ASCE, (Prof., School of Civ. Engrg., Purdue Univ., W. Lafayette, Ind.)

Journal of the Water Resources Planning and Management Division, Vol. 108, No. WR1, March, 1982, pp. 67-85

Two types of models for forecasting daily flows from a basin are investigated. These model types are a segmented model whose components are multi-input linear models of the basin between gaging stations, and an unsegmented or fully lumped stochastic basin model of the ARMA (p,q) family. Three approaches to segmented model construction are investigated. The results show that a constrained linear systems estimation method gave models which produced forecasts with a smaller bias than the ordinary least-squares method. The addition of an error model further improved the forecasting performance of the segmented models. A fully lumped stochastic model of the basin is developed and its forecasts are compared with those given by the segmented models. The results show that the segmented models provide better forecasts but exhibited more bias when compared to the unsegmented model of the basin. The results are obtained by using the data from the Green River Basin in Kentucky.

0791 Evaluation of Environmental Assessment Methods

Robert Nichols, (Research Assoc., Inst. for Environmental Studies, Univ. of North Carolina, 311 Pittsboro St. 256H, Chapel Hill, N.C. 27514) and **Eric Hyman**, (Research Intern, East-West Center, Environment and Policy Inst., 1777 East-West Road, Honolulu, Hawaii 96848)

Journal of the Water Resources Planning and Management Division, Vol. 108, No. WR1, March, 1982, pp. 87-105

Twelve representative methods for environmental assessment are described, evaluated, and compared in terms the following seven evaluation criteria: treatment of the probabilistic nature of environmental quality, incorporation of indirect and feedback effects, dynamic characteristics, multiple-objectives approach to social welfare, clear separation of facts and values, facilitation of participation, efficiency in resource and time requirements.

0792 Water Law Primer

William E. Cox, Assoc. Member, ASCE, (Assoc. Prof., Dept. of Civ. Engrg., Virginia Polytechnic Inst. and State Univ., Blacksburg, Va. 24061)

Journal of the Water Resources Planning and Management Division, Vol. 108, No. WR1, March, 1982, pp. 107-122

An overview of legal principles controlling water resources use and development is provided. The body of legal principles is a combination of federal and state law, with federal law generally having a greater role with respect to water quality protection than with respect to allocation which primarily has been a state function. The basic pattern of water law evolution is given, with separate consideration given to watercourses, ground water, and diffused surface water. Within each of these areas, common law doctrines and legislative developments are discussed. Current water management problems and dissatisfactions with water law which likely will result in continuing change are considered.

0793 An Alternative to the Peripheral Canal

Gerald T. Orlop, Fellow, ASCE, (Prof. of Civ. Engrg., Univ. of California at Davis, Davis, Calif.)

Journal of the Water Resources Planning and Management Division, Vol. 108, No. WR1, March, 1982, pp. 123-141

The present mode of water transfer through the Sacramento—San Joaquin Delta is not adequate to meet both future export requirements and maintain desired water quality. A 42-mile long Peripheral Canal that would convey project water around the estuary is favored by state agencies and water contractors but opposed by Northern California water interests who prefer an "open" water transfer alternative. Water quality control effectiveness was evaluated by means of mathematical models for the present system, the Peripheral Canal, and the alternative. Results of simulations indicated that both the Peripheral Canal and the alternative can control salinity intrusion in the western Delta. The Peripheral Canal was shown to maintain net seaward flow in most channels, while the alternative, because of its through-Delta nature, caused "upstream" flow in channels leading to export pumps. The alternative was shown to be more effective in maintaining water quality in the central and southern Delta. It was found to be capable of assuring water quality at the export location adequate for intended uses.

0794 Estimation of Small-Scale Hydropower Potential

Duane A. Poppe, (Grad. Student, School of Engrg. and Applied Sci., Univ. of California, Los Angeles, Calif.) and **William W-G. Yeh**, Member, ASCE, (Prof., School of Engrg. and Applied Sci., Univ. of California, Los Angeles, Calif.)

Journal of the Water Resources Planning and Management Division, Vol. 108, No. WR1, March, 1982, pp. 145-151

A statistical method of determining installed capacity and energy yield of potential small-scale hydropower sites is explored. Thirty watersheds with existing reservoirs were selected for this study from several hydrologic regions. A hydropower estimation model is developed. It is shown that by relating hydropower potential to watershed characteristics other than streamflow, estimates can be made with reasonable accuracy using a stepwise multiple regression model. Potential applications of the model are identified.

0795 Reservoir Systems Optimization

Otto J. Helweg, Member, ASCE, (Assoc. Prof., Dept. of Civ. Engrg., Univ. of California, Davis, Calif.), **Robert W. Hinks**, (Visiting Asst. Prof., Dept. of Civ. Engrg., Univ. of California, Davis, Calif.) and **David T. Ford**, (Hydr. Engr., Hydrologic Engrg. Center, U.S. Army Corps of Engrs., Davis, Calif.)

Journal of the Water Resources Planning and Management Division, Vol. 108, No. WR2, June, 1982, pp. 169-179

Although systems analysis tools have been available for over a decade, no major reservoir system presently uses optimization models for real-time operation. Of the many existing reservoir systems, the Central Valley Project in California, and the Tennessee Valley Authority project in Tennessee were examined as representative systems. Though the TVA is actively engaged in developing optimization models, the reasons neither is presently using optimization models are summerized as lack of upper management support, lack of qualified personnel in the development and use, inflexibility of many optimization models, and numerous associated problems such as hardware deficiencies. The potential net benefits of optimization are, however, apparently quite large which should encourage such systems operators to utilize these tools.

0796 U. S. Grain Exports: Transportation and Pricing Systems

Lowell D. Hill, (L. J. Norton Prof. of Agricultural Marketing, Univ. of Illinois, Urbana-Champaign, Ill.)

Journal of the Water Resources Planning and Management Division, Vol. 108, No. WR2, June, 1982, pp. 180-190

Exports of grains from the U.S. in 1979 were 4.6 billion bushels (121.1 million tonnes) and are projected at 6 billion bushels (157.9 million tonnes) by 1985. Over half of these will flow through the Gulf Ports. The problems involved in moving these volumes through the ports are primarily logistical in matching origin and destination capacities and timing. Water transport is a major link in the market chain. Its further expansion is limited by physical capacity and the introduction of competitive rail rates. Logistical problems can be resolved by vertical integration by exporters into assembly and transportation, or by changing policies and regulations that presently reduce competition and efficiency. The actions most likely to reduce logistical problems and improve market performance include equalizing subsidies among modes, transferring the ownership of transportation right-of-ways to public bodies, and reducing regulations that restrict the responsiveness of a free market to changes in demand and supply.

0797 Optimal Risk-Based Hydraulic Design of Bridges

Yeou-Koung Tung, (Asst. Prof. of Civ. Engrg., Univ. of Nevada, Reno, Nev. 89557) and **Larry W. Mays**, Member, ASCE, (Assoc. Prof. of Civ. Engrg., Dept. of Civ. Engrg., Univ. of Texas, Austin, Tex. 78712)

Journal of the Water Resources Planning and Management Division, Vol. 108, No. WR2, June, 1982, pp. 191-203

The hydraulic design of highway bridges entails the evaluation of flood hazard associated with a proposed bridge structure and the determination of socio-environmental effects of the bridge project. A procedure for treating hydrologic uncertainties is incorporated into an optimization framework to develop the optimal risk-based design of the bridge openings and the embankment heights that minimizes the total expected cost. The optimization technique used is called Hooke-Jeeves pattern search. The procedure to determine the optimal bridge design is applied to Leaf River, Mississippi. The effect of using different procedures for hydrogic parameter estimation and the effect of using different hydrologic probability models on the optimal design are examined.

0798 Municipal and Industrial Water Demands of Western U.S.

Erat S. Joseph, (Prof. and Chmn., Dept. of Civ. Engrg., Southern Univ., Baton Rouge, La. 70813)

Journal of the Water Resources Planning and Management Division, Vol. 108, No. WR2, June, 1982, pp. 204-216

Municipal water requirements are directly related to population whereas industrial demands are related to the level and type of industrial activity. The emphasis on domestic energy production will exert a strong influence on future industrial water demands. The most critical water using regions for mining will be the western regions. Anticipated conservation measures will have a significant impact on water savings in the industrial sector where much conservation can be expected through process changes, recirculation, and several emerging technologies. Water utility planners who forecast the raw water storage volume need to consider the combined effects of anticipated population rise, industrial mix and its likely growth, and the probable savings due to conservation. The material presents some guidelines together with regional tables in an attempt to provide a framework to aid in the estimation of the present and future municipal and industrial water requirements in the western United States.

0799 Reregulation on an International River — A Case Study

John E. Peters, (Regional Chf., Water Resources Branch, Inland Waters Directorate, Environment Canada, Dartmouth, Nova Scotia, Canada) and **Lawrence J. Bergen**, (Deputy Chf., Planning Div., New England Div., U.S. Army Corps of Engrs., Waltham, Mass.)

Journal of the Water Resources Planning and Management Division, Vol. 108, No. WR2, June, 1982, pp. 217-233

The St. Croix River, the International Boundary between Mew Brunswick, Canada, and Maine, United States, was subject to water use conflicts in the early 1970's. At that time, the complaints of residents and commercial camp owners interested in water supply, recreation and shorefront properties, intensified. In April 1971, the International Joint Commission approved the International St. Croix River Board of Control's request to study two problem areas - water levels of Spednic Lake and flow fluctuations below the Vanceboro Dam - which were controlled by the Georgia Pacific Corporation, under an order of the Commission. The study was directed towards a better understanding of how and why the reservoirs were operated the way they were, and an attitude survey of local residents and camp owners in order to obtain an appreciation of their concerns. An operating plan for the reservoir system was developed, one which most nearly met the needs of all concerned.

— — — — — — — — —

0800 Water Distribution Systems: A Spatial and Cost Evaluation

Robert M. Clark, Member, ASCE, (Chf., Drinking Water Research Div., Municipal Environmental Research Lab., Environmental Protection Agency, Cincinnati, Ohio 45268), **Cheryl L. Stafford**, (Research Geographer, Drinking Water Research Div., Municipal Environmental Research Lab., Environmental Protection Agency, Cincinnati, Ohio 45268) and **James A. Goodrich**, (Research Geographer, Drinking Water Research Div., Municipal Environmental Research Lab., Environmental Protection Agency, Cincinnati, Ohio 45268)

Journal of the Water Resources Planning and Management Division, Vol. 108, No. WR3, October, 1982, pp. 243-256

Problems associated with maintaining and replacing water supply distribution systems are reviewed. Some of these problems are associated with public health, economic and spatial development of the community, and costs of repair and replacement of system components. A repair frequency analysis has been completed for distribution system maintenance events (leaks and breaks). The economic implication of various replacement strategies and the effect of water quality (corrosivity) on water loss and system cost are examined. This analysis is based on the data acquired from tne large (260 mgd; 11.39 m^3/s) and one smaller (20 mgd; 0.88 m^3/s) water utility. As this study shows, once a length of pipe begins to require maintenance, its maintenance rate increases exponentially. Maintenance costs soon exceed the costs of replacement. Therefore establishing a timely maintenance and replacement program is extremely important from an economic and public health viewpoint.

— — — — — — — — —

0801 Worth of Inflow Forecast for Reservoir Operation

William W-G. Yeh, Member, ASCE, (Prof., School of Engrg. and Applied Science, Univ. of California, Los Angeles, Calif.), **Leonard Becker**, (Asst. Research Engr., School of Engrg. and Applied Science, Univ. of California, Los Angeles, Calif.) and **Robert Zettlemoyer**, (Civ. Engr., Dept. of Water Resources, Sacramento, Calif.)

Journal of the Water Resources Planning and Management Division, Vol. 108, No. WR3, October, 1982, pp. 257-269

This study assesses improved benefits that might be gained from the use of long-range streamflow forecasts in the operation of a typical multipurpose reservoir. Long-range is defined as a period of time ranging from one month to one year. The Oroville-Thermalito reservoir system of the California State Water Project (SWP) is selected for study; major benefits to be derived from long-term streamflow prediction for this system are hydropower generation, water conservation for irrigation and/or other beneficial uses, and decreased seepage damage to crops. Flood control benefits are potentially high but those benefits are more amenable to short-term rather than long-term forecasting. Incremental benefits are determined as functions of forecast time and accuracy on the basis of general adherence to current operating rules. The reservoir is theoretically operated over the 1914-73 time period; forecasts and comparisons are based on the historical streamflows. Procedure failure is defined as inability to meet a minimum storage constraint at any time over the 60-year period. To provide an additional evaluation of benefits from forecasting, analyses are also made using the historical means as estimates of future streamflows.

0802 Model for Optimal Size and Location of Detention

Larry W. Mays, Member, ASCE, (Assoc. Prof., Dept. of Civ. Engrg., Univ. of Texas, Austin, Tex. 78712) and **Phillip B. Bedient**, Assoc. Member, ASCE, (Assoc. Prof., Dept. of Environmental Science and Engrg., Rice Univ., Houston, Tex. 77001)

Journal of the Water Resources Planning and Management Division, Vol. 108, No. WR3, October, 1982, pp. 270-285

An optimization model has been developed for determining the minimum cost, size, and location of detention basins for urban watersheds. This model is based on dynamic programming and can handle any arbitrary arrangement of candidate detention basin locations and sizes. The detention system representation uses the concept of iso-drainage lines. Flood routing procedures for channel routing and detention basin routing are incorporated into the dynamic programing model. An example application is given for illustration of the model.

0803 Water Resources and Growth Management in South Florida

Stanley W. Hole, Fellow, ASCE, (Pres., Hole, Montes and Assoc., Inc., Naples, Fla.) and **Peter B. Rhoads**, (Dir., Resource Planning Dept., South Florida Water Mgmt. Dist., West Palm Beach, Fla.)

Journal of the Water Resources Planning and Management Division, Vol. 108, No. WR3, October, 1982, pp. 286-295

The role of water resource planning and regulation in the management of growth in

South Florida during the past decade has been to provide for orderly, directed growth that is consistent with resource limitations and local government plans and decisions. Potentially large quantities of potable water are available, and drainage problems can usually be overcome with effective engineering. These factors limit the effectiveness of managing growth by water resource constaints alone. The South Florida Water Management District's authority to regulate water use and drainage has affected, to some degree, the magnitude, distribution and timing of urban development by requiring new development to meet reasonable standards of flood protection, resource impact, and environmental protection. The district's philosophy during the past decade has been to support local government growth management efforts by coordinating plans and regulatory actions and by providing a mechanism for local land use decision makers to understand the water management implications of their decision prior to making it.

0804 Economic Analysis of Rehabilitation of Water Mains

Thomas M. Walski, Assoc. Member, ASCE, (Research Civ. Engr., Water Resources Engrg. Grp., U. S. Army Engr. Waterways Experiment Station, Vicksburg, Miss.)

Journal of the Water Resources Planning and Management Division, Vol. 108, No. WR3, October, 1982, pp. 296-308

Cleaning and relining of water transmission lines is economical if the costs are less than the savings in energy and pumping capacity which occur because of the increased carrying capacity of the pipe. Criteria are developed to determine if it is economical to clean and reline a pipe for two cases: (1) Flow not significantly changed by rehabilitation of pipe; or (2) system is looped so that change in carrying capacity significantly changes flow. The decision depends on the cost to clean and reline the pipe, the price of energy, the incremental cost of pumping capacity, the peak and average flow in the pipe, the nominal diameter, the interest rate, the year in which the pumping eguipment will be upgraded or replaced, and the Hazen-Williams C factor before and after rehabilitation. A sensitivity analysis of several of the variables is presented.

0805 Columbia River Treaty: Environmental Impact

Nicholas A. Dodge, (Chf., Water Management Branch, U.S. Army Corps of Engrs., North Pacific Div., Engrg. Div., Portland, Oreg. 97208)

Journal of the Water Resources Planning and Management Division, Vol. 108, No. WR3, October, 1982, pp. 309-320

Several major environmental impacts results from implementation of the Columbia River Treaty. The first one stems from the conservation of hydroelectric resources through coordinated planning and operation between 16 Pacific Northwest utilities and the British Columbia Hydro and Power Authority. The regional power system meets the load with a minimum of resources and cost to the rate payer. This has been spin-off benefits in terms of minimizing fuel consumption and air pollution associated with thermal-electric plants. Additional socioeconomic benefit is derived from the ability of treaty projects to control major floods and increase human safety. Although the anadromous fishery of the Columbia River has sustained setbacks caused by numerous factors, such as ocean over-fishing, undesirable forest and agricultural practices, and elimination of headwater habitat dams on the Columbia River are perceived by many as the major contributor to the decline. Added mitigation measures are now focused on construction and operation of bypass facilities and increasing flows to shorten the out-migration times to the ocean. To assure such flows, a significant reregulation strategy will be required. The solution to this problem is still unresolved but may require replacement of hydropower resources by expensive fossil-fueled thermal plants.

0806 Groundwater Management in the Southeast

David H. Howells, Fellow, ASCE, (Prof. Emeritus, North Carolina State Univ. and Univ. of North Carolina, Raleigh and Chapel Hill, N.C.) and **James C. Warman**, (Dir. Water Resources Research Inst., Auburn Univ., Auburn, Ala.)

Journal of the Water Resources Planning and Management Division, Vol. 108, No. WR3, October, 1982, pp. 321-328

State groundwater management programs in the Southeast are reviewed and critiqued. Strengths and weaknesses of different programs are examined with attention focused on legal and administrative issues, groundwater pollution, and geologicaL diversity. Conclusions and recommendations are listed with respect to needed legislation, institutional arrangements, funding levels, research, technology transfer, and education.

0807 Transverse Forces on Vertical Tube Array in Waves

Subrata K. Chakrabarti, Fellow, ASCE, (Dir., Marine Research and Development, Chicago Bridge and Iron Co., Plainfield, Ill. 60544)

Journal of the Waterway Port Coastal and Ocean Division, Vol. 108, No. WW1, February, 1982, pp. 1-15

In a wave tank test with a vertical array of tubes the inline and transverse forces on small instrumented sections of the tubes were measured. The transverse forces on them and the total force profiles around them are presented here at different spacings for a range of Keulegan-Carpenter number covering the test range. The maximum transverse forces are nondimensionalized to obtain the lift coefficients. The eddy-shedding process and the associated lift forces on the tubes are shown to be affected by the spacings between the tubes. The lift force frequencies are also investigated. The results are presented for various tube spacings and at right angles to the wave direction.

0808 Simulation of Large-Scale Circulation in Harbors

Wen-Li Chiang, Assoc. Member, ASCE, (Sr. Engr., Tetra Tech., Inc., Pasadena, Calif. 91107) and **Jiin-Jen Lee**, Assoc. Member, ASCE, (Assoc. Prof. of Civ. Engrg., Univ. of Southern California, Los Angeles, Calif. 90007)

Journal of the Waterway Port Coastal and Ocean Division, Vol. 108, No. WW1, February, 1982, pp. 17-31

A numerical model capable of simulating vertically-integrated tidal circulations in a harbor area for a substantial long period of time has been applied to the Los Angeles-Long Beach Harbor. The computed results show that the large gyre structures, which have also been observed in the field, correlate closely with those found in a hydraulic model. In the difference equations, the non-linear advective terms and the local acceleration terms are arranged on the same time level in order to improve the numerical stability. The numerical scheme is efficient for the hydrodynamic computation of tidal motion in a complicated harbor geometry.

0809 Wave Spectrum Changes on Sloped Beach

Bernard Le Mehaute, Member, ASCE, (Prof., Dept. of Ocean Engrg., Univ. of Miami, Fla. 33149) and **John D. Wang**, Member, ASCE, (Assoc. Prof., Dept. of Ocean Engrg., Univ. of Miami, Miami, Fla. 33149)

Journal of the Waterway Port Coastal and Ocean Division, Vol. 108, No. WW1, February, 1982, pp. 33-47

A comparison of all the different approaches to analyze the problem of wave spectrum change on a sloped beach is presented. It is shown how the heuristic, theoretical and numerical formulations compare or differ. It is shown that the transformation of a deep water directional spectrum into a shallow water directional spectrum can be done with little more effort than computing a one-dimensional spectrum by development of an inverse direction function only.

0810 Numerical Modeling Study of Lake Pontchartrain

Glenn D. Hamilton, (Chf., Data Systems, NOAA Data Buoy Office, NSTL Station, Miss. 39529), **Cecil W. Soileau**, Member, ASCE, (Chf., Hydr. Branch, Engrg. Div., U.S. Army Corps of Engr., New Orleans Dist., New Orleans, LA 70160) and **Arthur Stroud**, (Numerical Analyst, Compass Systems, Inc., San Diego, Calif. 92109)

Journal of the Waterway Port Coastal and Ocean Division, Vol. 108, No. WW1, February, 1982, pp. 49-64

A Hansen-type hydronamical-numerical model is shown to adequately simulate actual wind circulation, water levels, and particulate matter dispersion in Lake Pontchartrain. The model is an explicit numerical difference scheme based on leap-frog integration through time of the two-dimensional Eulerian form of the hydrodynamical equations to obtain a dynamical boundary-valve solution of tidal order. The differential equations are derived by integration of both x and y velocity components through a layer of assumed uniform density to achieve vertically averaged mean components. Terms to account for rotation of the earth, wind and tidal forcing, and dissipative frictional effects are included. The diffusion and advection method used is a Monte Carlo stochastic process in which the model provides the advectional flow. Two simple applications demonstrate the ability of this model to accurately represent the movement of a lake surface under the influence of tides and wind. A third application demonstrates the ability to model the injection and plume dispersion of particulate matter in lakes.

0811 Short Line Change Estimates Neat Oarai, Japan

Kazumasa Mizumura, Ph.D., (Prof., Dept. of Civ. Engrg. Kanazawa Inst. of Tech., Kanazawa, Ishikawa Pref., Japan)

Journal of the Waterway Port Coastal and Ocean Division, Vol. 108, No. WW1, February, 1982, pp. 65-80

To estimate the shoreline changes near the Port of Oarai when the groin was built, the field observations were done and the characteristics of the coast were investigated. A present shoreline in nature was simulated by hydraulic model tests with obtained wave conditions and the time-scale between nature and the model was determined. Shoreline changes in the future were predicted using the results of hydraulic model tests and this time-scale.

0812 Shock Pressure of Breaking Waves on Vertical Walls

M. Salih Kirkgoz, (Asst. Prof., Dept. of Civ. Engrg., Univ. of Cukurova, Adana, Turkey)

Journal of the Waterway Port Coastal and Ocean Division, Vol. 108, No. WW1, February, 1982, pp. 81-95

Breaking wave impact on vertical walls is studied experimentally with beaches having four different slopes. It is found that a breaking wave having its front face parallel to the wall at the instant of impact produces the greatest shock pressures. Deepwater wave steepness and the bench slope are the two parameters governing the magnitude and location of the maximum dimensionless impact pressure from breaking waves. The greatest pressures are produced with a beach sloping at about one in ten. The length of virtual mass which participates in the impact process, is not constant but varies vertically.

0813 Prediction of Water Level in Tidal Inlet

Kazumasa Mizumura, (Prof., Dept. of Civ. Engrg., Kanazawa Inst. of Tech., 7-1, Ohgigaoka, Nonoichi-machi, Kanazawa, Ishikawa Pref., Japan 921)

Journal of the Waterway Port Coastal and Ocean Division, Vol. 108, No. WW1, February, 1982, pp. 97-106

A very important and difficult problem is operation of gates at the inlet of tidal lakes in order to control water quality and quantity. The gate operation must sufficiently satisfy both flood control and irrigation. As the first step to optimize a gate control, the water levels downstream of the gate are predicted by a combined model using regression theory and control theory. The prediction model consits of two parts. A regression equation is used to predict the water levels downstream of the gate by the sea levels. A Kalman filtering algorithm is employed to identify parameters used to construct the state equation in the second part. This method is applied to the prediction problems of water levels on the downstream side of the gate at the inlet of Lake Kahokufgata, Ishikawa, Japan. It is possible to predict the fluctuations of the water levels under the arbitrary gate operations with one or two hours ahead of time. The Kalman filtering algorithm with the first and second order Markov models are studied.

0814 Calculation of Trapped Reflective Wave

Fred E. Camfield, Member, ASCE, (Hydraulic Engr., U.S. Army Coastal Engrg. Research Center, Kingman Bldg., Ft. Belvoir, Va. 22060)

Journal of the Waterway Port Coastal and Ocean Division, Vol. 108, No. WW1, February, 1982, pp. 109-111

An increase in reflected wave energy may result where structures are built along a section of coastline with no beach fronting the structure. In certain instances the structures may combine with a steep bottom slope to form a wave guide which can trap wave energy, magnify wave heights, and increase erosion along an adjacent section of shoreline. Calculations are presented to determine the effects of such trapped wave energy.

0815 Node Factor Variations and Joint Occurences

James R. Houston, Member, ASCE, (Hydr. Engr., U.S. Army Corps of Engrs., Waterways Experiment Station, Vicksburg, Miss. 39180)

Journal of the Waterway Port Coastal and Ocean Division, Vol. 108, No. WW1, February, 1982, pp. 112-115

In constructing exceedance frequency distributions for extremal heights resulting from joint occurrences of tsunamis or storm surges and the astronomical tide, a period of time for the analysis must be selected if direct numerical approaches are used. It is shown that if a particular year is selected for the analysis, the affect of the temporal variation in the lunar node factor on the exceedence frequency distribution is negligible.

0816 Pipeline Stability Under Finite-Amplitude Waves

Min-Chih Huang, (Grad. Research Asst., Dept. of Civ. Engrg. and Ocean Engrg. Program, Oregon State Univ., Corvallis, Oreg. 97331) and **Robert T. Hudspeth**, Member, ASCE, (Assoc. Prof., Dept. of Civ. Engrg. and Ocean Engrg. Program, Oregon State Univ., Corvallis, Oreg. 97331)

Journal of the Waterway Port Coastal and Ocean Division, Vol. 108, No. WW2, May, 1982, pp. 125-145

The lateral stability of a bottom-laid pipeline under finite-amplitude waves which are shoaling on a mildly-sloping, rigid beach is considered. Incipient sliding motion conditions are assumed in which the wave-induced hydrodynamic pressure force is statically balanced by the Coulomb bottom friction force on a rigid, impermeable seabed. The Morison equation is employed to compute the wave-induced hydrodynamic pressure force. This static balance results in a dimensionless stability parameter that may be equated to a dimensionless wave force acting on the pipeline in order to determine lateral stability. A dimensionless wave force is computed using water particle kinematics computed numerically from the finite-amplitude Dean stream function wave theory. Design curves for predicting this dimensionless wave force acting on a pipeline under various design wave conditions are provided. Engineering applications are included which use the dimensionless wave forces to predict the maximum depth required to stabilize a bottom-laid pipeline on a rigid, impermeable beach under specified design wave conditions.

0817 Wave Interaction with Large Submerged Structures

Henry William Kidman Bird, (Assoc. Engr., Santa Fe Technical Services Co., Orange, Calif. 92668) and **Robin Shepherd**, Fellow, ASCE, (Prof. of Civ. Engrg., Univ. of California, Irvine, Calif. 92717)

Journal of the Waterway Port Coastal and Ocean Division, Vol. 108, No. WW2, May, 1982, pp. 146-162

The interaction of waves with large fixed two-dimensional structures submerged in water of finite depth is examined. The boundary-value problem for the velocity potential of the wave is formulated under the assumptions of linear diffraction theory. One of the most powerful tools used to analyze this boundary-value problem is the Green's function. To derive the Green's function that satisfies all or some of the imposed boundary conditions is, however, sometimes difficult or even impossible. This paper presents an alternative method of numerical solution in the form of the boundary element method. The boundary element technique is implemented by using Green's second identity and a fundamental solution to form a boundary integral equation which is then applied numerically over the full boundary of the problem. Wave forces and pressures are calculated from the velocity potential of the wave. Results presented for several basic objects of symmetry compare well with the known solutions.

0818 Field Comparisons of Cross-Shore Transport Models

Richard J. Seymour, (Lect., Scripps Inst. of Oceanography, La Jolla, Calif.) and **David B. King**, Member, ASCE, (Research Asst., Scripps Inst. of Oceanography, La Jolla, Calif.)

Journal of the Waterway Port Coastal and Ocean Division, Vol. 108, No. WW2, May, 1982, pp. 163-179

During the Nearshore Sediment Transport Study experiment at Torrey Pines Beach, CA, in November 1978, beach profiles were measured over a several week period concurrent with extensive daily measurements of wind, waves, and currents. These surf zone data were used to predict the cross-shore transport of sand using eight models available in the literature. Several models were found to have skill in predicting major changes, but none was capable of predicting more than a third of the total beach volume variability.

0819 Fixed and Floating Axisymmetric Structures in Waves

Michael de St. Q. Isaacson, Member, ASCE, (Assoc. Prof., Dept. of Civ. Engrg., Univ. of British Columbia, Vancouver, B.C., Canada)

Journal of the Waterway Port Coastal and Ocean Division, Vol. 108, No. WW2, May, 1982, pp. 180-199

The numerical calculation of wave exciting forces, added-mass and damping coefficients and body motions for a floating vertical axisymmetric body, as well as the wave forces and runup for a fixed axisymmetric body are described. An integral equation method for fixed bodies is amended and extended to the case of a floating body. The method utilizes the body's axisymmetry to reduce considerably the computational effort required from that for bodies of arbitrary shape. Comparisons are made with available solutions relating to various specific configurations. As an example of the method's application, new results are also presented for surface-piercing conical structures.

0820 Measurmment of Velocities in Solitary Waves

Jiin-Jen Lee, Member, ASCE, (Visiting Assoc. in Civ. Engrg., W. M. Keck Lab. of Hydr. & Water Resources, California Inst. of Tech., Pasadena, Calif. 91125), **James E. Skjelbreia**, Assoc. Member, ASCE, (Grad. Research Asst., W. M. Keck Lab. of Hydr. & Water Resources, California Inst. of Tech., Pasadena, Calif. 91125) and **Fredric Raichlen**, Member, ASCE, (Prof. of Civ. Engrg., W. M. Keck Lab. of Hydro. & Water Res., California Inst. of Tech., Pasadena, Calif. 91125)

Journal of the Waterway Port Coastal and Ocean Division, Vol. 108, No. WW2, May, 1982, pp. 200-218

The water surface profiles and corresponding water particle velocities of several solitary waves have been obtained. The measurements of the horizontal and vertical velocity components are conducted using a two-dimensional laser-Doppler velocimeter (LDV). Results are presented for three different wave height-to-depth ratios: ϵ = 0.11, 0.19 and 0.29. The experimental results are compared with existing theories which follow different orders of approximation, and the theories are found to agree well with the experiments. It has been convincingly demonstrated that the LDV measurement technique offers an unmatched advantage to determine the water particle velocities under water waves, especially in the region above the still water level.

0821 Laboratory Study of Oil Thickness on Water Waves

Eugene Kordyban, (Prof., Mech. Engrg. Dept., Univ. of Detroit, Detroit, Mich. 48221)

Journal of the Waterway Port Coastal and Ocean Division, Vol. 108, No. WW2, May, 1982, pp. 220-224

Using a system which follows the water surface, the local thickness of an oil layer on water waves has been measured in a laboratory wave channel for a range of wave lengths and oil layer thicknesses. A comparison with linear wave theory is not satisfactory with the experimental results producing much larger thickness variation than predicted.

0822 Hydromechanics of Tidal Jets

Prakash B. Joshi, Assoc. Member, ASCE, (Sr. Engr., Northrop Corp., Aircraft Div., One Northrop Ave., Hawthorne, Calif. 90250)

Journal of the Waterway Port Coastal and Ocean Division, Vol. 108, No. WW3, August, 1982, pp. 239-253

A unified theoretical investigation of the fully-developed region of a tidal jet is presented. The jet is treated as a vertically-averaged flow and the effects of lateral mixing, bottom friction, and variable bathymetry are included. It is shown, without assuming any specific form of the similarity function, that the cross-stream length scale and the centerline velocity of the jet must satisfy certain differential equations in order for the jet to be self-similar. Solutions of these equations are obtained for some simple forms of bottom topography. For a flat, frictional bottom, the jet exhibits exponential growth and decay of the centerline velocity which is in agreement with earlier investigations. An exact analytical solution for the similarity function is obtained upon introduction of Prandtl's eddy viscosity hypothesis. The theory contains two free constants which must be determined in future experimental investigations.

0823 Sediment Transport on Gentle Slope due to Waves

Nobuhisa Kobayashi, Assoc. Member, ASCE, (Asst. Prof., Dept. of Civ. Engrg., Univ. of Delaware, Newark, Del. 19711)

Journal of the Waterway Port Coastal and Ocean Division, Vol. 108, No. WW3, August, 1982, pp. 254-271

A criterion for initiation of sediment movement on a gentle slope is established by balancing the forces acting on a sediment particle lying on the slope. A relationship for the instantaneous rate of bedload transport on a gentle slope is then derived from a simple and quasi-steady analysis of the motion of sediment particles. A gentle slope produces net movement of sediment even in pure oscillatory flow. The theory is in reasonable agreement with experiments conducted on a horizontal bed, corresponding to zero slope. The relationship derived for the net rate of bedload transport is applied to the problem of infilling of relatively small and symmetric trenches caused by wave action. The analysis includes the influence of the disposal distance of dredge spoil on the infilling. The model qualitatively explains observed infilling of test pits for San Francisco's ocean outfall and smoothing of ice gouges due to summer storms in the Southern Beaufort Sea north of Alaska, for which limited data is available.

0824 Natural Wave Trains and Scattering Transform

Dr. Rodney J. Sobey, Member, ASCE, (Sr. Lect., Dept. of Civ. and Systems Engrg., James Cook Univ., Townsville, Australia) and **Errol J. Colman**, (Engr., Woodside Offshore Petroleum Party Limited, Perth, Australia)

Journal of the Waterway Port Coastal and Ocean Division, Vol. 108, No. WW3, August, 1982, pp. 272-290

The potential of the nonlinear Schroedinger equation and the Zakharov-Shabat scattering transformation as a "third generation" analysis technique for natural ocean wave trains is considered. Such an approach retains many of the advantages of the "second generation" Gaussian random wave model but accomodates such fundamentally nonlinear concepts as Benjamin-Feir instability and especially envelope solitons. Wave envelope solitons are naturally occurring and stable wave groups which exist quite separately from uniform wave train solutions. Envelope soliton and uniform wave train solutions interact nonlinearly in physical space and are difficult to separately identify from a wave record. Resolution is possible, however, in scattering transform space. The direct scattering transform leads to the scattering data, a broadly analogous summary of a wave record to the Fourier spectrum. The inverse scattering transform recreates the wave train at the same or a later time, in a similar manner to the inverse Fourier transform. The results of exploratory computations of both analysis and synthesis of wave trains are included, together with a discussion of the potential application of the method to natural wave trains.

0825 Columbia River Entrance Channel Ship Motion Study

Shen Wang, (Prof., Rosenthal School of Marine and Atmospheric Sci., Univ. of Miami, Miami, Fla.) and **Scott Noble**, (Project Engr., U.S. Army Corps of Engrs., Portland Dist., Portland, Oreg.)

Journal of the Waterway Port Coastal and Ocean Division, Vol. 108, No. WW3, August, 1982, pp. 291-305

This study is a part of the overall program for the Columbia River channel development, directed by the U.S. Army Corps of Engineers. The objective of the study was, through direct measurements of vessel motions in the channel, to provide pertinent information for analysis of channel project modification. A portable measuring system with instrumentation to record pitch, roll, heave, heading and vessel position was specifically developed for this purpose. Statistical analyses of the measured data in terms of both short term and long term distributions of the vessel motions are conducted. The results indicate that a deeper but narrower channel is a viable option among the several alternatives being considered in the current development.

0826 Nonlinear Diffraction by Eigenfunction Expansions

Min-Chu Chen, (Sr. Engr., SONAT Offshore Drilling, Inc., Houston, Tex.) and **Robert T. Hudspeth**, Member, ASCE, (Assoc. Prof., Dept. of Civ. Engrg. and Ocean Engrg. Programs, Oregon State Univ., Corvallis, Oreg.)

Journal of the Waterway Port Coastal and Ocean Division, Vol. 108, No. WW3, August, 1982, pp. 306-325

A nonlinear diffraction theory correct to second-order is presented which is based on an eigenfunction expansion of the Green's function for fixed axisymmetric bodies subjected to two-dimensional sinusoidal waves. The boundary value problem for the second-order scattered velocity potential is linearly decomposed into two separate boundary value problems, each having only one inhomogeneous boundary condition. Numerical results indicate that the second-order contributions to the total hydrodynamic pressure force from the inhomogeneous free surface boundary condition are much less than the second-order contributions from the inhomogeneous boundary condition on the axisymmetric body. Second-order contributions are also found to be greater in intermediate water depth conditions than for deep water wave conditions. Theoretical numerical results are also compared with experimental values measured on a fixed vertical circular cylinder.

0827 Movable-Bed Tidal Inlet Model

Subhash C. Jain, Member, ASCE, (Assoc. Prof., Energy Div. and Research Engrg., Inst. of Hydr. Research, Univ. of Iowa, Iowa City, Iowa 52242)

Journal of the Waterway Port Coastal and Ocean Division, Vol. 108, No. WW3, August, 1982, pp. 326-342

The effectiveness of movable-bed tidal inlet hydraulic models in predicting prototype behavior is evaluated by comparing the model predictions with the observations made in the prototype. The calibration of the model, as measured by bed topography changes, is evaluated by means of quantitative indicators, including correlation coefficients and root-mean-square error. Evaluation of data from the Galveston Harbor entrance model revealed that the shoaling rates and distribution along the navigation channel predicted by the model are not in good agreement with the prototype data. Disagreement between model and prototype is believed to have been due to scale effects introduced by nonsimilarity of the physical processes, insufficient prototype data for model calibration, oversimplification of the available prototype data for use in the model study, and experimental errors.

0828 Unsteady Flow in China

Pin-nam Lin, (Visiting Prof., Colorado State Univ., Fort Collins, Colo.), **Zheheng Dai**, (Chf. Engr. and Dir., Zhejiang Inst. of Estuarine and Coastal Engrg. Research, Hanzhou, China) and **Kuanbin Li**, (Deputy Chf. Engr., Zhejiang Inst. of Estuarine and Coastal Engrg. Research, Hanzhou, China)

Journal of the Waterway Port Coastal and Ocean Division, Vol. 108, No. WW3, August, 1982, pp. 343-360

This article summarizes some results of research on unsteady flow problems encountered in China. These results are in general little known to the world outside of China. The report is not intended to be general, although it does touch on considerable variety of problems, such as one- and two-dimensional tidal flow in estuaries, dam-break waves, routing of suspended load and prediction of salinity.

0829 Parametric Approach to Wave Forecasting

M. Aziz Tayfun, Assoc. Member, ASCE, (Assoc. Prof., Dept. of Civ. Engrg., Coll. of Engrg. and Petroleum, Kwait Univ., P.O. Box 5969, Kuwait)

Journal of the Waterway Port Coastal and Ocean Division, Vol. 108, No. WW3, August, 1982, pp. 361-375

The prediction of wave characteristics generated by variable winds under fetch- and duration-limited conditions is considered using a parametric approach and a simplified form of the radiative transport equation. The wind-generated sea surface is characterized in terms of an equilibrium range spectrum with a well defined low wave number cut-off. By relating the local wind velocity to the rate of increase in wave energy and incorporating the refractive and frictional effects of transitional water depths into the formulation, governing equations are derived to predict the development of a wind-wave field in terms of integral properties of the spectrum such as mean period, direction, and significant wave height. Under deep water conditions, governing equations admit simple closed form solutions which are shown to compare very favorably with observational data as well as various wind-wave prediction formulae currently available. Consequently, they are numerically solved first for the case of wind-waves in uniform water depth and then for a circular basin with variable bathymetry as an illustrative general case.

0830 Steady Flow in Alluvial Channels

Gerd Holtorff, (Hydro., Drainage Division, 18 Charlotte Street, Port of Spain, Trinidad and Tobago)

Journal of the Waterway Port Coastal and Ocean Division, Vol. 108, No. WW3, August, 1982, pp. 376-395

A theory is deveped to predict the formation and deformation of alluvial channels by relating the parameters of the transported substances (fluid and sediment) to those of the channel geometry. Possible channel forms are meandering and braided. Meandering channels are formed if the flow dynamics correspond with the channel morphology. Braided channels occur if flow dynamics and channel morphology are incompatible. Alluvial channels are unstable because the stability criteria for the channel bed and for the channel banks are different.

0831 New Method for Tidal Current Computation

Jean Pierre Benque, (Head, Research Branch, Laboratoire National d'Hydraulique, Chatou, France), **Jean A. Cunge**, Member, ASCE, (Head, Applied Math. Dept., SOGREAH, Grenoble, France), **Jacques Feuillet**, (Research Engr., Maritime Hydr. Branch, Laboratoire National d'Hydraulique, Chatou, France), **Alain Hauguel**, (Research Engr., Maritime Hydr. Branch, Laboratoire National d'Hydraulique, Chatou, France) and **Forrest M. Holly, Jr.**, Member, ASCE, (Engr., Applied Math. Dept., SOGREAH, Grenoble, France)

Journal of the Waterway Port Coastal and Ocean Division, Vol. 108, No. WW3, August, 1982, pp. 396-417

The computation of two-dimensional tidal currents using the Alternating Direction Implicit Method (ADI) can be subject to numerical attenuation, parasitic oscillations, and poor reproduction of wave propagation when large time steps are used. The new method described in the paper is designed to overcome these difficulties. It is based on a fractional step method in which momentum advection is calculated using the method of characteristics, horizontal momentum diffusion is calculated using an implicit finite difference scheme, and wave propagation is calculated using an iterative alternating direction implicit algorithm. The resulting method has been incorporated in the CYTHERE-ES1 modelling system, in which tidal flat flooding and drying as well as wind effects and Coriolis acceleration are taken into account. The basic principles of the method, as well as its application to four schematic test cases and two engineering studies, are described.

0832 Floating Tire Breakwater Design Comparison

Craig T. Bishop, (Coastal Engr., Hydr. Div., National Water Research Inst., P.O. Box 5050, Burlington, Ontario, Canada)

Journal of the Waterway Port Coastal and Ocean Division, Vol. 108, No. WW3, August, 1982, pp. 421-426

The design curves from two independent studies of the wave transmission and mooring force characteristics of Goodyear-type floating tire breakwaters are compared. The CERC wave transmission results from prototype scale tests and the mooring force results of Harms from model scale tests are recommended for use in design. Both wave transmission and mooring forces are seen to depend primarily on the ratio of wavelength to breakwater beam, and less importantly on the relative draft and the wave steepness. Differences between the results of the two studies are attributed to scale effects and to different methods of data analysis.

0833 Bragg Scattering by Pile-Supported Structures

Robert A. Dalrymple, Member, ASCE, (Assoc. Prof., Dept. of Civ. Engrg., Univ. of Delaware, Newark, Del. 19711) and **John E. Fowler**, (Grad. Student, Old Dominion Univ.)

Journal of the Waterway Port Coastal and Ocean Division, Vol. 108, No. WW3, August, 1982, pp. 426-429

A hydrodynamic analogy of Bragg's Law is introduced, which indicates that regularly spaced piling in a large pile supported structure can reflect, or scatter, water waves in a coherent manner. An experiment is described which illustrates the principle.

0834 Second-Order Wave Forces on Vertical CyLinder

John N. Hunt, (Prof., Dept. of Applied Mathematics, Univ. of Reading, Whiteknights, England) and **Rafik E. Baddour**, (Lect., Arab Maritime Transport Academy, Alexandria, Egypt)

Journal of the Waterway Port Coastal and Ocean Division, Vol. 108, No. WW3, August, 1982, pp. 430-434

Second-order diffraction effects are included in evaluating the maximum horizontal forces on a vertical cylinder under a second-order deep water progressive wave. For large cylinders, the values of the maximum horizontal force are significantly higher than are predicted by linear theory due to second-order terms. Current engineering practice for slender cylinders is to compute the maximum horizontal force using fifth-order Stokes theory with a mass coefficient taken from linear theory. It is shown that this gives an adequate approximation to second-order diffraction effects in the maximum horizontal force, to within 5% for the steepest wave. Further examination of these results is encouraged.

0835 Calculating Modulus *k* of Cnoidal Waves

V. Chris Lakhan, (Teaching Asst., Dept. of Geography, Sidney Smith Hall, University of Toronto, 100 St. George Street, Toronto, Ontario, Canada M5S 1A1)

Journal of the Waterway Port Coastal and Ocean Division, Vol. 108, No. WW3, August, 1982, pp. 435-438

To facilitate practical application of a previously proposed cnoidal wave theory, an easily programmed procedure is presented to calculate the modulus k of cnoidal waves of a known height (or length) and period in water of a known depth. A method using bisection is employed because it is ideally suited to high speed computers.

0836 Wave Force Coefficients for Vertical Cylinders

Subrata K. Chakrabarti, Fellow, ASCE, (Dir. Marine Research, Chicago Bridge and Iron Co., 1501 N. Division St., Plainfield, Ill. 60544)

Journal of the Waterway Port Coastal and Ocean Division, Vol. 108, No. WW4, November, 1982, pp. 445-455

Wave forces on small sections of vertical tubes were measured in a wave tank test. The tube sizes and the wave parameters were such that both the inertia and the drag dominated areas were covered in the test. The surface of the tubes was sand-roughened to simulate the marine growth on these tubes. The hydrodynamic coefficients from these forces were computed with the help of the Morison equation. The roughnesses of the tubes were varied to show their effects on the inertia, drag, and lift coefficients. Mean values of these coefficients versus the Keulegan-Carpenter number were used to compute total forces on the tubes which were compared with the measured total forces. The drag and lift coefficients increased with the increased roughness of the tubes, while the inertia coefficients were relatively unaffected.

0837 Effects of Uncertainties on Extreme Wave Heights

Marshall D. Earle, (Sr. Oceanographer, Marine Environments Co., 10629 Crestwood Drive, Manassas, Va.) and **Ledolph Baer**, (Mgr., Coastal Waves Program, Office of Oceanography, National Ocean Survey, National Oceanic and Atmospheric Administration, Rockville, Md. 20852)

Journal of the Waterway Port Coastal and Ocean Division, Vol. 108, No. WW4, November, 1982, pp. 456-478

The relative effects of uncertainties due to record lengths, natural variability, and measurement or hindcast errors on estimated extreme significant wave heights has been assessed objectively through simulation. The probability distributions describing extreme heights at a given return period have confidence intervals broader than those computed from typically used techniques and show substantial positive biases. Both the widths of the confidence intervals and the biases increase with the size of the total uncertainty which represents the combined effects of measurement or hindcast errors and natural variability. Tables and nomograms are provided for practical assessment of record lengths, natural variability and errors on extreme wave estimates.

0838 Wave Scattering by a Rigid Thin Barrier

Philip L-F. Liu, Assoc. Member, ASCE, (Assoc. Prof., School of Civ. and Environmental Engrg., Cornell Univ., Ithaca, N.Y.) and **Madjid Abbaspour**, (Grad. Student, School of Civ. and Environmental Engrg., Corell Univ., Ithaca, N.Y.)

Journal of the Waterway Port Coastal and Ocean Division, Vol. 108, No. WW4, November, 1982, pp. 479-491

The boundary integral equation method (BIEM) is applied to examine the effectiveness of both the vertical and the inclined thin breakwater. The BIEM has been used widely for free surface flow problems in both ground-water flows and water waves. For the present problems, modifications of the method are made to treat the velocity singularity at the tip of the breakwater and the zero thickness of the breakwater. The computational domain is truncated by introducing two auxiliary boundaries at some distance away from the break-water. Analytical solutions, which satisfy the governing equations and the radiation boundary condition, with unknown coefficients are used along the auxiliary boundaries. The accuracy of the numerical technique is demonstrated by comparing numerical results with analytical solutions for a vertical breakwater in a deep water. Numerical results in terms of both reflection coefficient and transmission coefficient are presented for both vertical and inclined thin breakwater.

0839 Long Wave Problems in Laboratory Models

Stig E. Sand, (Ph.D. Research Engr., Inst. of Hydrodynamics and Hydraulics Engrg., Technical Univ. of Denmark, DK-2800 Lyngby, Denmark)

Journal of the Waterway Port Coastal and Ocean Division, Vol. 108, No. WW4, November, 1982, pp. 492-503

For the design of harbors, offshore terminals, floating platforms, etc. it is of great importance that the long waves in nature are correctly represented in the physical model, because of the dominant influence of the long waves on mooring forces and slow-drift oscillations. Mostly, the long waves are a function of the grouping of wind waves. The present second order wave generation analysis reveals several long wave terms of practical interest. Various terms dominate in different frequency ranges, and this influences the control of the wave generators. For practical purposes a simple procedure for estimating the group-induced long waves and the necessary amplitude of the generator piston is proposed. These quantities are obviously important for the choice of model scale, since the practical limits of piston stroke, resolution, water depth, etc. must be observed. The long wave errors in traditional model tests with only first order wave generation are investigated. In shallow water model tests the long wave effect tends to be underestimated, while it is exaggerated in deep water tests.

0840 Evaluating Hazard of Landslide-Induced Water Waves

Rudy Slingerland, (Asst. Prof. of Geology, Dept. of Geosciences, Pennsylvania State Univ., University Park, Pa. 16802) and **Barry Voight**, (Prof. of Geology, Dept. of Geosciences, Pennsylvania State Univ., University Park, Pa. 16802)

Journal of the Waterway Port Coastal and Ocean Division, Vol. 108, No. WW4, November, 1982, pp. 504-512

Dimensional analysis of the landslide parameters which control dimensionless maximum wave amplitude, η_m/d, when slides impact with a water body, shows that dimensionless slide kinetic energy, E_k, is most important. Data from two site-specific hydraulic studies combine to produce a consistent data set and give a statistically significant linear regression equation of $\log(\eta_m/d) = -1.25 + 0.71 \log(E_k)$, in which $E_k = \rho_s\, l w h V^2 (2 \rho g d^4)^{-1}$, l, w, h, ρ_s, and V are slide length width, thickness, density, and maximum velocity, ρ is water density, g is gravitational acceleration, η_m = maximum wave amplitude at a standard distance $r/d \approx 4$ directly in front of a slide, d = water depth at that site, and r = radial horizontal distance from the point of slide impact. The hydraulic model studies had scale factors of 1:120 and 1:300, with $0.4 < h/d < 0.8$. Landslides were simulated by 1/18- 1/3 cu ft (0.002-$0.005 m^3$) triangular or tabular bags of metal or gravel, with toes initially above water level. η_m/d, back calculated from observed waves and wave run-ups for the 1958 Gilbert Inlet slide in Lituya Bay, Alaska, and the 1905 glacier fall in Disenchantment Bay, Alaska, are accurately predicted by the regression equation in the first case and conservatively predicted in the second.

0841 Pass Cavallo, Texas: Case Study of Tidal-Prism Capture

George H. Ward, Jr., (Assoc., Espey, Huston, & Assocs., Inc. P.O. Box 519, Austin, Tex. 78767)

Journal of the Waterway Port Coastal and Ocean Division, Vol. 108, No. WW4, November, 1982, pp. 513-525

Pass Cavallo was until 1963 the only permanent inlet for the Matagorda Bay System, and has not been stabilized or protected in any way. In 1963, a deepdraft ship channel was constructed to provide access from the Gulf of Mexico, for which the Entrance Channel was dredged across Matagorda Penninsula 18,000 ft (5.5 km) up the coast from the Pass. This landcut proved to be scour-unstable, enlarging spontaneously to depths on the order of 50-60 ft (15-18 m) and widths on the order of 1,000 ft (300 m). Current measurements in Pass Cavallo prior to the creation of the landcut and approximately ten years after are analyzed to determine hydraulic characteristics These indicate that the Entrance Channel has captured approximately half of the tidal prism of Pass Cavallo. From both critical current and critical stress criteria, the reduction in current speed in Pass Cavallo associated with this loss of tidal prism is sufficient to render the Pass shoal-unstable, hence more susceptible to deposition.

0842 Maintaining Tidal Inlet Channel by Fluidization

Richard N. Weisman, Member, ASCE, (Assoc. Prof., Dept. of Civ. Engrg., Lehigh Univ., Bethlehem, Pa. 18015), **Anthony G. Collins**, (Asst. Prof., Dept. of Civ. and Environmental Engrg., Clarkson Coll. of Tech., Potsdam, N.Y. 13676) and **James M. Parks**, (Dir., Center for Marine and Environmental Studies, Lehigh Univ., Bethlehem, Pa. 18015)

Journal of the Waterway Port Coastal and Ocean Division, Vol. 108, No. WW4, November, 1982, pp. 526-538

Along the barrier islands off the east coast of the U.S. and at other areas of the world, there exist many tidal inlet channels which connect the ocean to back bay areas. An innovative technique is tested to maintain a stable, navigable channel. Water is pumped into a perforated pipe buried beneath the sand, and the water flowing through the perforations fluidizes the overlying sand. The fluidized sand along the pipe is then removed either by gravity flow, slurry pump, or entrainment in the overlying ebb tidal current. Laboratory experiments provided information necessary for the design of such a fluidization system, specifically, data on hole orientation, hole size, hole spacing, flow rate necessary for fluidization, and the effect of slurry removal on the channel shape. A small-scale field test, designed from the laboratory data and performed on a beach face, shows that the technique is technically feasible and that the design data is sound. In the field test, full fluidization along fluidization pipe was achieved and the fluidized sand was removed by pumping the slurry or, when the pipe was sloped, by gravity flow.

0843 Tidal Computation on Taiwan Strait

Fuh Yin, (Prof., Dept. of Oceanography, National Taiwan Coll. of Marine Sci. and Tech., Keelung, Taiwan, R. O. C.) and **Sen-Ho Chen**, (Research Asst., Dept. of Oceanography, National Taiwan Coll. of Marine Sci. and Tech., Keelung, Taiwan, R. O. C.)

Journal of the Waterway Port Coastal and Ocean Division, Vol. 108, No. WW4, November, 1982, pp. 539-553

The three-dimensional hydrodynamical numerical model for predicting tides is designed. It applies the one- and two-layer model to the Taiwan Strait, and solves the phenomena of M_2 and $M_2 + S_2 + K_1 + O_1$ tides. Tides computed by this model show good agreement with measured data. The co-tide lines, co-amplitude lines, tidal current chart, and tidal current ellipse are made. The open boundary points at the south and north sides of the Strait are established by the trial-and-error method using 21 measured tide fluctuation data along the coasts and islands of Peng-Hu.

0844 Dolos Breakwater Design Improvements

Tianzhong Fang, (Civ. Engr., The Bureau of Capital Construction, Ministry of Communications, The People's Republic of China)

Journal of the Waterway Port Coastal and Ocean Division, Vol. 108, No. WW4, November, 1982, pp. 554-568

There are two different failure modes in Dolos breakwaters: (1) Loss of stability; and (2) overstress. Bending tensile stress in armor units increases with design wave height, and a linear relation has been established for this phenomenon. A wave wall or adjacent vertical wall causes an increase in wave force and damage. Consequences due to settlement and consolidation should be considered in deep water breakwater design. The potential benefit of reinforcement of concrete armor units is proposed.

0845 Comparisons of Numerical Random Sea Simulations

Hang Tuah, (Grad. Student, Ocean Engrg. Program, Dept. of Civ. Engrg., Oregon State Univ., Corvallis, Oreg. 97331) and **Robert T. Hudspeth**, Member, ASCE, (Assoc. Prof., Ocean Engrg. Program, Dept. of Civ. Engrg., Oregon State Univ., Corvallis, Oreg. 97331)

Journal of the Waterway Port Coastal and Ocean Division, Vol. 108, No. WW4, November, 1982, pp. 569-584

Two numerical models previously developed by Rice for modeling Gaussian, white-noise, electronic signals are used to simulate in the frequency domain nonlinear random ocean waves in water of finite depth. The following two models were developed by Rice: (1) Nondeterministic Spectral Amplitude model (NSA), and (2) Deterministic Spectral Amplitude model (DSA). These two Rice models are used with finite Fourier transform (FFT) algorithm to simulate linear Gaussian random wave realizations by the process of filtering Gaussian white noise in the frequency domain. A nonlinear random wave interaction matrix then operates on the linear Gaussian wave spectra to generate a complex-valued nonlinear random wave spectrum that is correct to second-order in the Stoke's wave perturbation parameter. Both the lineear and the nonlinear realizations obtained from both of Rice's two simulation models are compared with a realization obtained from hurricane-generated real ocean waves measured in the Gulf of Mexico. Statistical comparisons indicate that the NSA Rice model is better for both linear and nonlinear random wave simulations when band-limited white noise is digitally filtered.

0846 Remodeled 42nd St. Bus Terminal Improves both Transit and the Neighborhood

Eugene J. Fasullo, Member, ASCE, (Asst. Chf. Engr. for Design, Research and Development, Port Authority of N.Y. and N.J., New York, N.Y.) and **Daniel M. Hahn**, Member, ASCE, (Engr. of Design, Tunnels, Bridges and Terminals, Port Authority of N.Y. and N.J., New York, N.Y.)

Civil Engineering--ASCE, Vol. 52, No. 1, January, 1982, pp. 41-43

A $160 million project to expand New York City's 42nd St. Bus Terminal was completed last year. The project is expected to improve commuter transit by increasing passenger facilities and bus loading positions by over 40%. It will also reduce midtown Manhattan traffic and pollution by providing an underground bus tunnel access and ventilation facility. The article describes the project's design and benefits, including construction of a facade truss and the decision to use composite construction, resulting from a value engineering study of the preliminary construction plans.

0847 Building Failures—Preventing Them, Learning from Them

Civil Engineering--ASCE, Vol. 52, No. 1, January, 1982, pp. 44-50

Ways to design and construct a structure so as to minimize chances of errors leading to failure are described. Held up as models of systematically investigating failures are the incident-study teams of the National Transportation Safety Board (airplane crashes) and the National Fire Protection Association (building fires). The incident-study mechanisms of both are profiled. Insurers of architect/engineer firms that design buildings are working with them to reduce errors and omissions. The activities of the top three insurers are profiled. A building developer whose approach is said to prevent failures is profiled.

0848 How a Structural Firm Transformed Itself into an E/A Firm Engineeriing

Eugene E. Dallaire, (Assoc. Ed., Civil Engineering Magazine, New York, NY)

Civil Engineering--ASCE, Vol. 52, No. 1, January, 1982, pp. 52-56

During the past decade, there have been a number of structural engineering firms that have broadened their scope to become engineer/architect firms. This article is an in-depth case history of one of those firms, Duchscherer Oberst Associates, of Buffalo, New York. At one time, a key part of the firm's work was structural engineering, done for architect firms designing schools. But when school work began to dwindle, DO Associates decided to build its own inhouse architectural capability and to seek out clients on its own. It sought architectural projecte where good engineering was of major concern. This approach led them into the design of industrial buildings, warehouses, small office buildings, and bus maintenance facilities. Case histories of these projects illustrate the current state of the art.

0849 How Well do P.E. Exams Relate to the Real World? NCEE Tries to Find Out

Civil Engineering--ASCE, Vol. 52, No. 1, January, 1982, pp. 62-64

The National Council of Engineering Examiners (NCEE), the group which furnishes a standard liscensing exam for professional engineers used by almost every U.S. jurisdiction, has surveyed some 5,000 registered engineers in all fields to find out what tasks they perform in their practice. The survey is part of NCEE's effort to affirm the validity of its engineering exams. Survey results are now being compared to the exam contents to make sure that the exams reflect all the requirements for minimum competence. The article describes the survey, how it was conducted and how it will be correlated with the exams. Tables present some of the data collected, including the list of 45 tasks identified as entry-level in the practice of civil engineering.

0850 Accelerated Population and Industrial Growth Spurs Completion of Water Project

Merja Lehtinen, (Production Ed., Civil Engineering Magazine, New York, N.Y.)

Civil Engineering--ASCE, Vol. 52, No. 1, January, 1982, pp. 65-66

The total federal government investment for the Southern Nevada Water Project is \$171,332,000. The project is one of the largest water resources solutions applied in the U.S. today and serves a population of over 400,000 in an arid region stricken by water depletion. Continued growth over the last few decades spurred investment and development of technical solutions to the water shortage.

0851 How to Create and Present Effective Slides

Merja Lehtinen, (Production Ed., Civil Engineering Magazine, New York, N.Y.) and **Robert A. Crist**, (Managing Dir., Publications and Technical Affairs, ASCE, New York, N.Y.)

Civil Engineering--ASCE, Vol. 52, No. 1, January, 1982, pp. 67-68

Speakers often devote hours of preparation to the main body of a prepared text, but fail to allow ample preparation time to creating their visual aids. Slides, visual aids which should enhance a speech if properly designed and presented, can make the difference between a lively presentation or one which is a dreadful bore. Here are some tips to make sure one does not rock one's audience to sleep.

0852 The Microcomputer Explosion in Civil Engineering Firms

Eugene E. Dallaire, (Assoc. Ed., Civil Engineering Magazine, New York, NY 10017)

Civil Engineering--ASCE, Vol. 52, No. 2, February, 1982, pp. 45-51

During the next five years, small and medium size consulting engineering firms will undergo many changes. The key changing force is the low cost ($2500 to $20,000) microcomputer. Already, an estimated 10 to 20 percent of these small and medium sized firms are using microcomputers. Some experts believe that the use of microcomputers in the civil engineering profession is on the verge of exponential growth. The areas being computerized include engineering calculations, accounting and other administrative matters, and word processing. Four in-depth case histories discuss the pros and cons of microcomputers and how the use of microcomputers can change a consulting engineering firm.

0853 Washington Office of the American Society of Civil Engineers

Louis M. Meier, (Counsel, Washington Office, ASCE, Suite 607-8, 1625 Eye St., N.W., Wash., D.C. 20006)

Civil Engineering--ASCE, Vol. 52, No. 2, February, 1982, pp. 51-53

The mission, organization, and major public policy activities of the American Society of Civil Engineers are described. Emphasis is given to the ASCE Washington Office and its relation to other organizations in the nation's capital. Federal issues confronting ASCE are also discussed.

0854 EPA's Nationwide Urban Runoff Program Designed to Produce Useful Results

Carl Myers, (Chf., Policy Evaluation Branch, Water Planning Div., Office of Water Program Operations, U.S. EPA, Wash., D.C.), **Dennis Athayde**, (Chf., Urban Nonpoint Source Section, Implementation Branch, Water Planning Div., Office of Water Program Operations, U.S. EPA, Wash., D.C.) and **Eugene Driscoll**, (E.D. Driscoll & Associates, Oakland, N.J.)

Civil Engineering--ASCE, Vol. 52, No. 2, February, 1982, pp. 54-55

EPA's Nationwide Urban Runoff Program (NURP) is composed of 28 projects which are obtaining data on the significance and control of urban runoff to aid federal, state, and local decision making. In a past article in *Civil Engineering* (Sept. 1981, p. 86), the usefulness of the results expected to come from NURP was challenged. Specifically, the lack of bioassays in the program was criticized . In this rebuttal, EPA engineers assert that the program was designed to produce definable, quantitative information about the kinds of problems which exist, their magnitude, prevalence, and geographical distribution. The article summarizes EPA's method of studying urban runoff, finding, among other conclusions, that bioassays are more appropriate after potential problem pollutants and areas are identified.

0855 Landmark Projects of Yesterday, Today, and Tomorrow

Civil Engineering--ASCE, Vol. 52, No. 2, February, 1982, pp. 56-59

Excerpts are discussed from major papers to be presented at the March 1982 Construction Equipment and Techniques for the Eighties Specialty Conference to be held at Purdue University. Topics examined include equipment and construction techniques for conventional and deep-water offshore platforms, construction of fixed, deep-water bridges, and the Hoover Dam project. Also included is an examination of what the construction industry will bring to outer space.

0856 Micro-Computer Program: Roof Purlin Design

John L. Cranmer, Jr., Member, ASCE, (Consultant, Van Horn, Tex.)

Civil Engineering--ASCE, Vol. 52, No. 2, February, 1982, pp. 60-61

A micro-computer program is explained in step-by-step manner as a shortcut for biaxial bending design with results in as little as 60 seconds. Biaxial bending requires the consideration of the effects of the loading in both the major and minor directions. Input can be made in degrees, roof slope in percent, or roof pitch in inches rise per foot run.

0857 Surveying and Photogrammetry Research Needs

Civil Engineering--ASCE, Vol. 52, No. 2, February, 1982, pp. 62-66

An ASCE-National Science Foundation committee listed and prioritized research needs in surveying and photogrammetry. Among them are the Department of Defense's Global Positioning System (satellite network), agricultural management and crop forecasting, measuring movement of the earth's crust along fault lines, thus hopefully to predict earthquakes; inertial surveying (using hardware descended from ICBM guidance systems); and developing standards, specifications and guidelines for digital data base systems (computer-made maps). Article briefly describes progress and problems in all five areas.

0858 Cofferdams Protecting New York Bridges from Ship Collisions

Daniel M. Hahn, Member, ASCE, (Engr. of Design, Tunnels, Bridges & Terminals Design Div., Port Authority of New York and New Jersey, One World Trade Center, New York, N.Y. 10048) and **Hugo E. Rama**, Member, ASCE, (Staff Engr., Tunnels, Bridges & Terminals Design Div., Port Authority of New York and New Jersey, New York, N.Y. 10048)

Civil Engineering--ASCE, Vol. 52, No. 2, February, 1982, pp. 67-68

During the past few decades, there have been a number of cases where ships have rammed bridge piers, triggering the superstructure's collapse. The most tragic case was the collapse of three spans of the Sunshine Skyway Bridge into Tampa Bay on May 9, 1980. Such cases underscore the crucial importance of protecting bridge piers from ramming by large ships. In the New York City area, the Port Authority has taken some pier-protection steps that have already fended off catastrophe. Specifically, the PA has constructed sand-filled cofferdams adjacent to piers on some of its bridges. These in effect serve as crash barriers, preventing mammouth vessels from ramming bridge piers.

0859 Problems of the Independent Consultant

James A. Christenson, Fellow, ASCE, (Christenson Consultants, Pittsburgh, Pa.)

Civil Engineering--ASCE, Vol. 52, No. 2, February, 1982, pp. 69-71

Problems of consultants who practice alone are considered in terms of the author's experience in developig a practice. Issues discussed include establishing a foothold, locating a market, marketing, and control over operations. Later decisions regarding whether or not to expand, reassessing goals, and assuring continuity of the practice are also covered. Four major stages in the evolution of a practice are identified: achieving independence, survival, establishing a foothold, and consolidation of the practice. Remarks on getting started are also included for the benefit of those considering going out on their own.

0859a The Booming Business in Wick Drains

Allen Morrison, (Asst. Editor, Civil Engineering Magazine, ASCE, New York, NY 10017)

Civil Engineering--ASCE, Vol. 52, No. 3, March, 1982, pp. 47-51

Prefabricated vertical wick drains, used to improve weak, saturated soils, have been used in Japan and Europe for 20 years but were introduced in the U.S. only four years ago. Since then they have replaced sand drains on most soil consolidation projects. Compared to sand drains, the plastic wicks offer quicker and less labor-intensive installation, higher reliability, equally or more rapid drainage, less environmental impact, and lower cost. Differences abound among the five wick drain systems currently available in this country. Some distributors claim the superiority of their products on the basis of such factors as installation method, filter strength and permeability, and discharge capacity. The article discusses these differences and their varying significance depending on site conditions. It also presents a brief history of vertical drainage, photos and diagrams of typical installations, and a list of drain suppliers.

0860 How to Reduce Construction Claims through Mediation

Donald R. Goodkind, (Sr. Vice Pres., Hill International, Inc., Los Angeles, Calif.)

Civil Engineering--ASCE, Vol. 52, No. 3, March, 1982, pp. 52-53

Ever-increasing costs of claims lead owners and contractors to turn to mediation as an alternative to litigation or arbitration. As a first step, recognize the risk of claims, and establish the cost and time objectives of a project at the outset. This way, each decision regarding disputes can be measured against the impact on the project. Set time limits on the design project and work for mutual agreements, instead of establishing guilt when there is a dispute. A mechanism to appoint a mediator acceptable to all parties involved on a project should be written into the contract. A case study is presented.

0861 Research and Development for Increased Productivity —Where We Must Go

Civil Engineering--ASCE, Vol. 52, No. 3, March, 1982, pp. 54-55

A well-focused research and development activity must be the essential foundation of any long-range success in maintaining high and increasing productivity in manufacturing, the efficient use of resources, a growing standard of living, and strength in the ultimate testing ground of international trade. This call for action to the engineering profession suggests the following areas for attention: (1) construction technology; (2) energy transfer; (3) biomedical processing; (4) office operations; (5) service industries; (6) chemical and metallurgical processing; (7) manufacturing technology; (8) transportation and (9) waste recovery.

1982 ASCE PUBLICATIONS

0862 Preventing Another Sunshine Skyway Bridge Disaster

Civil Engineering--ASCE, Vol. 52, No. 3, March, 1982, pp. 56-57

America will not soon forget the tragic collapse of the Sunshine Skyway Bridge into Tampa Bay almost two years ago. That accident was a costly lesson in the crucial importance of protecting bridge piers from straying vessels. The Florida DOT is in the process of planning some measures to prevent such accidents from occurring in the failure. They are leaning heavily to protecting bridge piers by constructing an embankment parallel to the bridge—one on the upstream side, another on the downstream side. Made of dredge spoil removed from the navigation channel, these embankments would stop a straying ship before it struck a bridge pier. Florida DOT is also looking into better navigational methods that would keep a ship from straying from the channel.

0863 Microcomputer Water Quality Simulation Model

Richard J. Hughto, (Sr. Water Resources Engr., Camp Dresser & McKee, Inc., 235 Wyman St., Waltham, Mass. 02154) and **Robert P. Schreibner**, (Water Resources Engr., Camp Dresser & McKee, Inc., 235 Wyman St., Waltham, Mass. 02154)

Civil Engineering--ASCE, Vol. 52, No. 3, March, 1982, pp. 58-59

Water quality models are widely used in the analysis of natural water systems to evaluate the impacts of wastewater handling alternatives and changes in land use. Linear, steady-state, first-order models are most commonly applied. This article presents a BASIC program for performing planning level analyses of stream and river systems. The program computes instream concentrations of DO, BOD, and NBOD.

0864 1981 Essay Contest Winners

Civil Engineering--ASCE, Vol. 52, No. 3, March, 1982, pp. 60-64

The fourth and fifth prize essays in Civil Engineering Magazines 1981 Essay Contest are Alan Mooney's "Have we abandoned our responsibility to nurture young engineers?" and Joel Rosenblatt's "Exploding some housing myths." Mooney contends that today's experienced engineers are much less willing to nurture the new engineer than was the case a decade or more

---

ago; and that this bodes ill for the future of the engineering profession. In his essay, Rosenblatt argues that Operation Breakthrough never really had any promise to substantially reduce housing costs. Reason: it focused on material and labor costs. The most significant cost item, though, are land costs, financing costs, administrative costs—none of which were addressed.

1982 ASCE TRANSACTIONS

0865 A Cost-effective Shallow Foundation Accommodates Three Feet of Settlement

Arthur H. Wu, Member, ASCE, (Specialist, Geotechnical Engrg., Naval Facilities Engineering Command, Headquarters, Alexandria, Va.) and **Dennis J. Scheessele**, Member, ASCE, (Supervisory Engr., Naval Facilities Engineering Command, Chesapeake Div., Wash., D.C.)

Civil Engineering--ASCE, Vol. 52, No. 3, March, 1982, pp. 65-67

An automotive maintenance facility to be built at the Anacostia Naval Station in Washington, D.C. had two major foundation problems: the probability of large settlement and the possibility of excessive foundation cost. In the past, pile foundations had been used to build in the base's filled land. After a subsoil investigation and foundation engineering analysis, Navy engineers designed a shallow foundation system to support the facility while allowing for settlements of 3 ft (0.9 m). The system consists of an engineered fill placed on a firm stone base to support spread footings designed as inverted "T"-shaped grade beams with enough rigidity to resist footing deflections. The superstructure was also reinforced and provided with control joints to accommodate the settlement better. Using this design, the risk of foundation failure has been minimized, and the foundation cost saving was about $335,000.

0866 World's Largest Installation of Fabric-Formed Pile Jackets

Eduardo W. Bindhoff, Member, ASCE, (Vice Pres., Intrusion-Prepakt , (Intl), Inc., Cleveland, Ohio) and **John C. King**, Fellow, ASCE, (Consulting Engr., Cleveland, Ohio)

Civil Engineering--ASCE, Vol. 52, No. 3, March, 1982, pp. 68-70

One major feature of fabric forming is permeability. One application of this technique is protective encasement of steel H and pipe piles at a major pier in Chile. The Huachipato Pier is used for unloading raw materials required for the manufacture of steel and for loading the steel products produced by *Compania de Acero del Pacifico S.A.*

0867 Engineer's Notebook: One-step Formula for R/C Tension Steel

Douglas Dong Woo Lee, Member, ASCE, (Consulting Structural Engr., Carter and Burgess, Inc., Ft. Worth, Texas)

Civil Engineering--ASCE, Vol. 52, No. 3, March, 1982, pp. 71

In the design of reinforced concrete beams, after loadings and materials are defined, the first step is to assume a beam depth. Then the required steel area and beam width are determined. A simple one-step formula is proposed to perform a preliminary design and strength check.

0868 **An Improved Design for Grain Storage Tanks Solves Problem of Brittle Failures**

German Gurfinkel, (Prof., Dept. of Civ. Engrg., Univ. of Ill., Urbana, Ill.)

Civil Engineering--ASCE, Vol. 52, No. 3, March, 1982, pp. 75-77

The welded tank features fracture-tough steel and tube stiffeners to overcome problems evident in several failures of major tanks in recent years. Used extensively for grain storage in the Midwest, these tanks are subject to grain-induced meridional compressive forces. A new design features fracture-tough steel, elimination of built- discontinuities, and prevention of excessive tension-stress fields.

0869 **World's First All-Rollcrete Dam**

Ernest K. Schrader, Member, ASCE, (Materials Engr., U.S. Army Corps of Engineers, Wella Walla, Wash.)

Civil Engineering--ASCE, Vol. 52, No. 4, April, 1982, pp. 45-48

Willow Creek Dam in Oregon will be the world's first concrete gravity dam designed specifically for and exclusively by roller compaction methods. The average in-place cost for all mass concrete in the structure is $18.93 per cubic yard (based on low bid unit prices) compared to an estimated $65 per cubic yard for conventional mass concrete. All test data to date show that, at any age and cement content, strengths for the roller compacted concrete are higher than would typically be achieved for conventional mass concrete of the same age and cement content. The article describes in detail the design and construction procedure project. A box discusses the definition of "rollcrete" and speculates on possible future applications.

0870 **Financing Water Projects: Count on Uncle Sam Less!**

William R. Gianelli, Fellow, ASCE, (Asst. Secretary of the Army, Civil Works, Washington, D.C.)

Civil Engineering--ASCE, Vol. 52, No. 4, April, 1982, pp. 49-52

The federal government has been the chief underwriter of many water resources projects in the U.S., but now changes in the Administrtion's policies have spurred new directions in planning and allocating costs. The emphasis is on greater participation of the other sectors involved such as local and state governments and private industry.

0871 **U.S. Response to Coal Demand: Port and Harbor Improvements**

Michael J. Gawinski, Member, ASCE, (Project Coordinator, Dravo Corp., Engineering Works Div., Pittsburgh, Pa.)

Civil Engineering--ASCE, Vol. 52, No. 4, April, 1982, pp. 53-57

The OPEC oil embargo, which caused long gasoline lines in 1973, caused incredibly long lines of coal ships along the East coast just last year. These ships came to bet coal to fuel overseas power plant boilers and found out ports unable to meet demand. U.S. coal loading is impaired because we have not kept pace with the world in port construction and harbor deepening efforts. These areas for improvement are now in the planning for virtually every U.S. seaport.

0872 A Primer on Microcomputers

Kenton H. Johnson, (Editor/Publisher, Engineering Computer Applications Newsletter, 5 Denver Tech. Center, P.O. Box 3109, Englewood, Colo. 80111)

Civil Engineering--ASCE, Vol. 52, No. 4, April, 1982, pp. 58-63

This article is a primer on microcomputers. It's intended for the many small and medium sized consultants and others in the civil engineering field who may be considering the purchase of a microcomputer. This article will help to remove the mystery from numerous computer buzz words—from input and memory to read-only memory and random-access memory; from machine language to mass storage media; from dot-matrix printer to operating system; from programming language compilers and interpreters to hard disk.

0873 Behind the Sines, Protugal Breakwater Failure

Civil Engineering--ASCE, Vol. 52, No. 4, April, 1982, pp. 64-67

This article highlights a report on the possible causes of failure of the massive, $178 million breakwater at Sines, Portugal. This was the largest breakwater of its kind in the world. On February 26, 1978, as construction neared completion, storm waves lashed the breakwater. By the time the storm had abated, two-thirds of the dolos units making up the armor layer of the breakwater were lost. A key reason for the breakwater failure, says this study by ASCE's Coastal Engineering Research Council was the fracturing of many concrete dolos units themselves during the storm. These dolos units are not as strong as model tests led investigators to believe.

0874 Cold Regions Engineering in Norway

Kaare Flaate, (Dir., Norwegian Road Research Lab., Oslo, Norway)

Civil Engineering--ASCE, Vol. 52, No. 4, April, 1982, pp. 68-69

The use of insulating materials is a key factor in cold regions engineering in Norway. It is applied in road construction in areas where frost heave and reduced bearing capacity would be severe. It is also used for protection of railroad tracks, highway tunnels, and for reinforcement of existing roads when other suitable materials are scarce. The article describes Norway's climate soil conditions, the various types of insulation, and the procedures for their installation.

0875 Causes and Preventions of Lamellar Tearing: Engineer's Notebook

Civil Engineering--ASCE, Vol. 52, No. 4, April, 1982, pp. 74

Within the past 10 years, a steel cracking problem, lamellar tearing, has arisen after welding of thicker sections and higher-strength steels. A separation underlying a weld joining plate or structural shape, it can occur in the construction of buildings, bridges and offshore platforms. Generally detected during fabrication, it has not usually resulted in structural failures, but its reparation can be costly. To combat it, the steel industry has funded a three-year research project at Lehigh University. The major findings of this study, on the causes and prevention of lamellar tearing, are reported in this article.

0876 Fiber-Reinforced Concrete Proves Worth for Airport Pavements

Robert D. Howell, (Vice Pres., J. H. Kleinfelder & Associates, Reno, Nevada)

Civil Engineering--ASCE, Vol. 52, No. 5, May, 1982, pp. 52-55

Laboratory and field tests indicate that fiber-reinforced concrete has higher flexural and tensile strength, higher impact resistance, and improved resistance to spalling and joint deterioration when compared to conventional Portland cement concrete. Two projects utilizing Portland cement concrete reinforced with steel fibers are described. The first is an apron overlay at Fallon Naval Air Station, the other is a new taxiway at Cannon International Airport in Reno. Both pavements have been in constant use since construction and are performing as designed.

0877 Energy Turns the World as a World's Fair Turns a City Around

Teresa Williams, (Asst. News Editor, *Civil Engineering Magazine*, ASCE, New York, N.Y. 10017)

Civil Engineering--ASCE, Vol. 52, No. 5, May, 1982, pp. 56-58

Energy Expo '82 opens May 1 and will run through October 31, 1982. Host City, Knoxville, Tennessee has taken advantage of the fair as a catalyst for growth and urban renewal for the city. Some $500 million in construction and investment has been generated in the area. Improvements include a massive restructuring of the interstate highway system and the redevelopment of a 72 acre industrial slum. An Energy Expo is appropriate for Knoxville which is considered the energy capital of the country - it is the home of the Tennessee Valley Authority, the world's largest public utility, and the University of Tennessee where extensive energy related research is conducted. Knoxville is also located near Oak Ridge, a center of nuclear research.

0878 Steel Cable-Stayed Bridges: Can They Compete in the U.S.?

Nils D. Olsson, Member, ASCE, (Dir. of Project Development, DRC Consultants, Inc., New York, N.Y.)

Civil Engineering--ASCE, Vol. 52, No. 5, May, 1982, pp. 59-61

During the past decade, concrete bridge designs have been winning more bids for contracts than steel cable-stayed bridges. Reasons for this trend are examined. A brief evolution of bridge technology is presented. Rapport between the designer and fabricator/constructor is discussed. The effect of the American steel industry is considered. It is argued that cable-stayed bridges permit greater design innovation.

0879 Nuclear Safety: A Problem of Technology or Ideology?

Richard L. Meehan, (Pres., Earth Sciences Assoc., Palo Alto, Calif.)

Civil Engineering--ASCE, Vol. 52, No. 5, May, 1982, pp. 60-61

The issue of nuclear safety is examined. It is debated whether the public reaction to the issue is emotional or rational and whether nuclear safety is a problem of ideology or technology. The development of nuclear technology and the public's reaction to it are compared to other historic technological developments.

0880 Natural Treatment Systems in Wastewater Treatment and Sludge Management

Robert K. Bastian, (Environmental Sci., Office of Water Program Operations, Environmental Protection Agency, Wash., D.C.)

Civil Engineering--ASCE, Vol. 52, No. 5, May, 1982, pp. 62-67

Criteria defining performance of natural wastewater treatment and sludge management systems are presented. Objectives of natural treatment systems are to help treat municipal wastewater and sludge in a cost-effective and energy-efficient manner and recycle and reuse them as resources. Land application of wastewater and sludge is examined. Problems with natural treatment systems are reviewed. Two widely accepted systems included slow-rate land treatment and rapid infiltration of wastewater through basins. Case histories are presented.

0881 Trouble Shooting ChemicaLly Resistant Masonry

Walter Lee Sheppard, Jr., (Pres., C.C.R.M., Inc., Havertown, Pa.)

Civil Engineering--ASCE, Vol. 52, No. 5, May, 1982, pp. 68-71

This paper is a series of case histories showing specifically how lack of understanding of the nature of the materials and their proper uses has caused failure in design, specification, inspection and installation of these structural materials. Examples range from ceramic tile paving on the sidewalk of a city street, brick (or tile) dairy floors, interior wall lining of a chemical/waste collection and holding tank, unsuitable concrete construction in several different tank exposures, improper concrete floor construction in an electrical equipment manufacturing plant, to problems with silicate monolithic floor topping in a steel mill.

0882 Computer-Aided Drafting in Civil Engineering

Michael F. Garrett, Member, ASCE, (Proj. Engr., Ellerbe Assoc., Inc., Bloomington, Minn.)

Civil Engineering--ASCE, Vol. 52, No. 5, May, 1982, pp. 72-78

Computer aided drafting (CAD) is defined as an interactive computer graphics system suitable for the generation of engineering documentation. The development of CAD is reviewed. The computer hardware of a CAD system is described. Applications of a CAD system are discussed. Advantages over traditional drafting are examined, although it is emphasized that CAD is a tool to be used along with existing methods and will probably not replace the drafting board. The economics of using a CAD system are shown.

0883 Will Reforms Improve Sewage Treatment Plant Performance?

Allen Morrison, (Asst. Ed., *Civil Engineering Magazine*, ASCE, New York, N.Y. 10017)

Civil Engineering--ASCE, Vol. 52, No. 5, May, 1982, pp. 79-81

After an exhaustive investigation, a Congressional subcommittee recommended more than 20 legislative and administrative reforms aimed at improving performance and economy of the wastewater treatment plants being built under the Clean Water Act. Some of the reforms were written into the 1981 Amendments to the Act which President Reagan signed last December. This article addresses key areas of reform, describing main features of the new law and discussing those areas in which further reforms may be forthcoming. Subjects included are reduction of federal funding level; liberalization of deadlines, definitions, and policies; accountability of municipalities and design engineers; turnkey projects; enforcement; and training of personnel.

0884 Minicomputers in Structural Optimum Design

Bogdan O. Kuzmanovic, Fellow, ASCE, (Sr. Struct. Engr., Geiswenger, Hoch and Assoc., North Miami Beach, Fla.) and **Nicholas Willems**, Member, ASCE, (Prof., Dept. of Civ. Engrg., Univ. of Kansas, Lawrence, Kan.)

Civil Engineering--ASCE, Vol. 52, No. 5, May, 1982, pp. 82-84

A problem dealing with the weight optimization of girders without transverse stiffeners is demonstrated. Constraints follow the 1978 AISC Specification. Actual programming is done in BASIC for an HP9830-A programmable desk calculator with a total amount of user memory of 3808 words (16 bits-two bytes per word). No analysis, but only synthesis, is performed. The theoretical proportions obtained do not necessarily conform to any particular set of available plate size, and may need adjustment to produce a practical design.

0885 The LOOP Project: Taking the Ship to the Port

James C. Webb, Member, ASCE, (Asst. Exec. Dir., Offshore Terminal Authority, Louisiana Dept. of Trans. and Dev., Baton Rouge, LA)

Civil Engineering--ASCE, Vol. 52, No. 6, June, 1982, pp. 43-46

(*1981 OCEA Award Winner.*) LOOP—the Louisiana Offshore Oil Port—is a $700 million, multi-faceted project for the unloading of oil directly from supertankers 19 miles offshore in the Gulf of Mexico to a national distribution system. The problem leading to the project's inception was the lack of a natural deepwater harbor capable of accommodating the deep drafts of crude-carrying supertankers. The article describes the project's major features: a marine terminal with two platforms, one for operations control and one for pumping; mooring buoys that can handle three supertankers simultaneously; pipelines and pumps for unloading oil; huge underground salt dome storage caverns; and a 25-million bbl brine storage reservoir, used in moving crude oil into and out of the storage caverns.

0886 Bloomington Lake—Potomac River Water Resources Project

James W. Peck, Member, ASCE, (Col., U.S. Army Corps of Engineers, Baltimore, Md.)

Civil Engineering--ASCE, Vol. 52, No. 6, June, 1982, pp. 47-49

The recently completed $174 million Bloomington Lake project has improved flood control and water quality to the benefit of millions of people downstream. The headwaters of the Potomac River are now controlled by engineering technology. Transportation in the area for the mining industry was also improved by use of tunnels and railway bridges. The project was a boon to local employment as well. This project was a 1982 OCEA Award of Merit.

0887 The San Diego Trolley

Maurice Carter, (Dir. of Planning and and Operations, Metropolitan Transit Dev. Board, San Diego, Calif.) and **Roger Clark**, (Dir. of Engrg. and Construction, Metropolitan Transit Dev. Board, San Diego, Calif.)

Civil Engineering--ASCE, Vol. 52, No. 6, June, 1982, pp. 50-53

(*1981 OCEA Special Achievement Award in Cost-Effective Public Transit.*) The 16-mile long San Diego Trolley system, a light-rail transit project providing tourists and commuters with a link between downtown San Diego and the Mexican Border, was built without federal assistance, or federal red tape. Partly due to the retention of local control, and partly due to other factors, the Trolley was the least expensive transit system per mile to be built in this country in the last 40 years. This article examines the design of the system in detail, and explores the areas in which the designers were able to save time and money. These areas include: the use of light-rail technology, which allowed planners to avoid costly grade separations; simple, standardized design for most stations on the route; the purchase of 13 miles of existing railway tracks; and the use of a self-service, barrier-free system of fare collection.

0888 Olmos Dam Modifications

Civil Engineering--ASCE, Vol. 52, No. 6, June, 1982, pp. 54-55

(*1981 OCEA Nominee.*) The 55-year-old Olmos Dam, five and one-half miles northeast of San Antonio, Texas, required major modifications to avoid a potential overtopping or structural failure. Such an event could occur from flash floods which sometimes sweep the San Antonio River Basin, resulting from unstable air from the Gulf of Mexico. One costly proposal would have added a layer of mass concrete to the entire downstream face of the dam; instead, designers found a way to bolster the non-overflow sections of the dam with prestressing tendons, avoiding the application of more concrete. Explosives were also used to remove the topmost portions of the dam to make way for new construction, a first in the annals of dam renovation.

0889 Water Pollution Control Plant Upgraded to Meet EPA Regs

Civil Engineering--ASCE, Vol. 52, No. 6, June, 1982, pp. 56-57

An Outstanding Civil Engineering Achievement Award Nominee, Philadelphia's Southwest Water Pollution Control Plant, has a $300 million addition which nearly doubled plant capacity to 210 mgd. Begun in 1975, the plant can remove 90% of the pollutants from raw sewage and industrial waste. The plant's control center is completely computerized and is one of the first to seriously incorporate computers into the plant operation. Plant operators were consulted during the design and their experiences were incorporated into the design. The plant has its own cryogenic fractional distillation oxygen unit which produces oxygen for the treatment process. Sludge gas is collected and utilized for sludge heating, space heating, scum incineration and grit incineration. A flow chart of the major processes of the plant is included.

0890 Spilled Petroleum Recovered from Atop Water Table

Gary M. Mathes, (Vice Pres. and Dir. of Engrg., John Mathes & Assoc., Inc., Columbia, Ill.)

Civil Engineering--ASCE, Vol. 52, No. 6, June, 1982, pp. 58-59

An undetected oil spill endangered the Mississippi Aquifer and the people who lived nearby. A thorough engineering analysis determined a solution. A containment procedure and recovery system regained the escaping petroleum products and guarded the water quality of the adjacent water table. This project was a 1982 OCEA Nominee.

0891 Orlando International: Airport of the 21st Century

Civil Engineering--ASCE, Vol. 52, No. 6, June, 1982, pp. 60-61

The latest technology in airport design from the apron overlays to the terminal itself was incorporated at Orlando International, the nation's fastest growing airport. Now it can handle 12 million passengers a year. In a few years, it will be able to handle 24 million. Modern architectural design of the main terminal is a noteworthy feature of this project which also is recognized for innovations in the apron configuration and the apron drainage design. This project was a 1982 OCEA Nominee.

0892 Denny Creek: The Toughest Bridge Washington State Ever Built

Civil Engineering--ASCE, Vol. 52, No. 6, June, 1982, pp. 63-65

The story of how highway planners and engineers in Washington State's Department of Transportation were stopped from building a highway and viaduct through a pristine wilderness is presented. An alternative plan which opponents would deem environmentally compatible was necessary. The basic issue: environmentalists didn't want a highway built through a pristine, wilderness area. The problem was resolved by redesigning the highway, making it a beautiful, elevated roadway and constructing it with minimal disruption to the forest. This project was a nominee for the Outstanding Engineering Achievement of the year.

0893 Sanitary Sewer Design

John M. Lorinc, (Corporate Projects Group, Pittsburgh-Des Moines Corp., Pittsburgh, Pa.)

Civil Engineering--ASCE, Vol. 52, No. 6, June, 1982, pp. 66-67

Good sanitary sewer design practice requires evaluation of alternative locations, slopes and pipe sizes. However, cost considerations generally keep alternative evaluations to a minimum. Use of the "Sanitary Sewer Design Program" gives the designer the opportunity to rapidly evaluate alternatives and choose the solution best suited to the clients' needs. Use of the program requires the designer to have a suitable topographic map with surveyed sewer line locations. The program is very interactive with the designer, requesting ground elevation, quantity of flow, and station value. The program gives the designer the option to design by minimum slope, fixed slope or invert, and drop manhole. Computer use allows the designer to perform error free calculations and to focus attention on the actual design work.

0894 Structural Safety

Carl Turkstra, (Dept. of Civ. Engrg., McGill Univ., Montreal, Canada)

Civil Engineering--ASCE, Vol. 52, No. 6, June, 1982, pp. 68-69

Future building codes will go far beyond the classical emphasis on dimensioning, calculations, and specification writing. Emphasis will shift from scientific thinking to strategic thinking, from micro design to macro design, leading to detailed analysis of the relationships between elements of the construction process. Legal, administrative, and social factors have a major influence on the outcome of a structural design and the incidence of failure. Until realistic models of the entire process are developed, structural safety analysis can only deal with a fraction of the problems in current practice.

0895 Some Proposals for Reducing Structural Failures

Peter Kocsis, Member, ASCE, (Struct. Engr., Bridge Div., City of Chicago, Chicago, Ill.)

Civil Engineering--ASCE, Vol. 52, No. 6, June, 1982, pp. 70-71

Illinois is one of the few states requiring a Structural Engineer's License: a P.E. alone is not enough to practice structural engineering there. Unfortunately, even in Illinois, the actual design is often done by an unlicensed person. A Structural Engineer's license should be required for every person who does structural design and he should be reguired to affix his seal to all design calculations and drawings. In addition, it should be required that all structural computations be checked by another registered structural engineer.

0896 Urban Creekside Design Features Parallel Streets

Lloyd C. Fowler, (Chf. Engr., Project Dev. Branch, Santa Clara Valley Water District, San Jose, Calif.) and **Stanley Wolfe**, (Civ. Engr., Project Dev. Branch, Santa Clara Valley Water District, San Jose, Calif.)

Civil Engineering--ASCE, Vol. 52, No. 6, June, 1982, pp. 72-74

Waterflow in cities—natural streams and rivers surrounded by wildlife, or concrete channels—must be controlled to prevent devastating flooding. Proper landuse restraints and creative designs can prevent flooding, yet preserve the natural habitat which forms along a waterway, even in an Urban environment. A project in Santa Clara County, California is described which illustrates this.

0897 Microcomputers: Do They Have a Place in Large Engineering Firms?

Martin W. Butalla, Member, ASCE, (Mgr., Computer Operations and Services, Ebasco Services, Inc., Atlanta Regional Office, Norcross, Ga.)

Civil Engineering--ASCE, Vol. 52, No. 6, June, 1982, pp. 80-81

During the past few years, microcomputers have really begun to take hold in small consulting engineering firms. Less well known is the fact that microcomputers are also beginning to take hold in engineering firms with large central computers. This article tells how one large firm, Ebasco Services, recently bought ten microcomputers. The move takes the load off the large central computer, freeing it for the big jobs. Meanwhile, computing power has been brought closer to the engineer-user, who can now use a locally available microcomputer for his smaller problems. Large firms are just beginning to realize the tremendous potential of microcomputers.

0898 Steel Structures Provide Flexibility

Merja Lehtinen, (Production Ed., Civil Engineering -- ASCE, New York, NY 10017)

Civil Engineering--ASCE, Vol. 52, No. 7, July, 1982, pp. 42-45

The steel structures of several major buildings are evaluated by the structural engineers responsible for the design and construction. Included are the Park Avenue Atrium in New York; the parking garage at Love Field in Dallas, Texas; the New York Stock Exchange; and Vanderbilt Stadium in Nashville, Tennessee, among others. Noted for flexibility and longevity, steel structures around America are featured.

0899 At Four Western Stadiums, Wood Comes Out on Top

Civil Engineering--ASCE, Vol. 52, No. 7, July, 1982, pp. 46-49

In recent years, several long-span timber roofs have been constructed atop large arenas and stadiums in the Western states. The choice of glued-laminated timber offers designers good economy, rugged appearance, and structural versatility. Four structures recently completed or under construction are depicted in words and photos in this article: the Tacoma Dome in Tacoma, Wash., the University of the Pacific Activities Center in Stockton, Cal., the renovated Portland Civic Stadium in Portland, Ore., and the University of Wyoming's Arena-Auditorium in Laramie, Wyoming. Structural systems for these roofs, varying from computer-designed domes to beam and truss systems, are discussed, and a box describes the design competition held for the Tacoma Dome.

0900 Engineering Schools Respond to Financial Crisis

Civil Engineering--ASCE, Vol. 52, No. 7, July, 1982, pp. 58-61

A mail survey of the 200-plus civil engineering departments at engineering schools asked each chairman for a case history of response to the financial and other crises facing the departments. The article reports on what six different schools have done. Types of responses to the crisis include: build links with construction industry; establish endowed chairs; limit undergraduate enrollment; use ABET accreditation visit as lever to get more money; introduce intern PhD and coop PhD; establish an Engineering Advisory Council made up of corporate presidents—people with enough clout attract big dollars from the state legislature and from industry.

0901 Inspecting Town Bridges in Connecticut

Surendra K. Ojha, Member, ASCE, (Chf. Bridge Engr., Macchi Engineers, Hartford, Conn.) and **Josef Womelsdorf**, (Proj. Mgr., Macchi Engineers, Hartford, Conn.)

Civil Engineering--ASCE, Vol. 52, No. 7, July, 1982, pp. 62-64

A federally financed inventory and inspection program for town bridges was carried out in Connecticut. This article describes how the authors' firm inspected 343 bridges of all types, assessed their condition and extent of deterioration, and reported their recommendations for rehabilitation. Forty-five percent were found to be either structurally deficient or functionally obsolete. Photos of interesting structures and some major cases of deterioration are included.

0902 Collapse of the Kansas City Hyatt Regency Walkways

Edward O. Pfrang, Fellow, ASCE, (Chf., Structures Div., National Bureau of Standards, U.S. Dept. of Commerce, Wash., D.C.) and **Richard Marshall**, Member, ASCE, (Center for Building Tech., National Bureau of Standards, U.S. Dept. of Commerce, Wash., D.C.)

Civil Engineering--ASCE, Vol. 52, No. 7, July, 1982, pp. 65-69

A summary of the National Bureau of Standards report on the collapse of the Kansas City Hyatt Regency walkways is presented. The report does not assign blame to any party; rather, it explains the technical factors leading to the collapse and rules out other, irrelevent factors. The article explains the objective and scope of the investigation, describes the walkways, and examines the dead and live loads of the walkways. Laboratory testing is described. Causes of the collapse are given.

0903 L.A.'s New Maintenance Management System: Key to Survival in an Era of Tight Budgets

John M. Tettemer, Member, ASCE, (Acting Chf. Deputy Engr., Los Angeles County Flood Control District, Los Angeles, Calif. 90051)

Civil Engineering--ASCE, Vol. 52, No. 7, July, 1982, pp. 69-70

Even in the wake of budget-cutting Proposition 13, the Los Angeles County Flood Control District was able to cope with their maintenance needs. Their secret was a major overhaul in their maintenance management system leading to cost savings of over 30%. The key idea behind the new system was the majority of work can be preplanned, prescheduled, preengineered, prefinanced, and preapproved. The District has 9,000 items, or routines, of preapproved work. Each routine has the schedule, personnel, equipment and material requirements. The completion of a routine brings a facility's condition up to the maintenance standard. The facility then deteriorates again from the maintenance standard to the acceptable maintenance condition, at which time the routine is repeated.

0904 Civil Engineers Have an Obligation to Become Active in Local Government

Joseph Covello, Member, ASCE, (Vice Pres., Jacobs Associates, San Francisco, Calif.)

Civil Engineering--ASCE, Vol. 52, No. 7, July, 1982, pp. 71-72

By education, training, experience, and a desire to accomplish tangible goals, today's civil engineer is more qualified than any other professional to serve the community. Yet, at a time when our volunteer participation is needed now more than ever, our profession has turned its back on society's needs. Today's communities can no longer afford the luxury of a civil engineer's non-participation. Our claim of being a "people-serving profession" is empty if we're not willing to work for the community as a volunteer.

0905 San Francisco Convention Center Goes Underground

T. Y. Lin, Member, ASCE, (Board Chmn., T. Y. LIN International, 315 Bay St., San Francisco, Calif. 94133) and **Jeppe Larsen**, (Vice Pres., Moscone Convention Ctr., San Francisco, Calif.)

Civil Engineering--ASCE, Vol. 52, No. 8, August, 1982, pp. 41-43

The Moscone Convention Center, built at a construction cost of –126 million, opened its doors to an admiring public on December 2, 1981. The Center's grand exhibit hall, 90 meters wide and 255 meters long, is roofed with eight pairs of post-tensioned concrete archess, creating an area of 23,000 sq. m, believed to be the world's largest column-free underground hall in the world. The arches are tied with post-tensioned cables encased in the 2.2-m thick concrete mat floor. These cables were utilized to move the ends of the arches inward by 9 cm, forcing a camber of 10 cm at the crown of the arches. The ties are curved vertically in the foundation mat thus exerting a downward force which helps to counteract the upward ground water pressure.

0906 Law of the Sea: U.S. Votes No

Merja Lehtinen, (Production Ed., Civil Engineering—ASCE, New York, NY 10017)

Civil Engineering--ASCE, Vol. 52, No. 8, August, 1982, pp. 44-46

The United States has voted against the international treaty which will regulate ocean and waterway passage, ocean mining, and fishing rights. Implications are far reaching as the precedents set in this law are thought to be applicable for future legislation on space. The U.S. disagreement with the legislation adopted by the United Nations General Assembly revolves mainly about the International Seabed Authority, the regulatory body which would govern seabed mining. defense industry minerals are involved in the seabed mining operations.

0907 Knoxville's Junction Functions

Hugh B. Carpenter, (Dir. of Operations, Tennessee Dept. of Trans., Nashville, Tenn.), **Billy D. Evans**, (Constr. Engr., Tennessee Dept. of Trans., Knoxville, Tenn.), **Terry S. Grubb**, (Traffic Engr., Tennessee Dept. of Trans., Knoxville, Tenn.) and **Robert A. Foy**, Member, ASCE, (Sr. Vice Pres. & Chf. Engr., Wilbur Smith and Assocs., Columbia, S.C.)

Civil Engineering--ASCE, Vol. 52, No. 8, August, 1982, pp. 47-49

The urban highway system in Knoxville was originally built as a local expressway in the early 1950's. It was designed to significantly lower standards than Interstate and the predicted future traffic was much less. When the downtown expressway system was incorporated into the Interstate 40 and 75 routes, a virtual automatic malfunction was created. In the fall of 1979, the Tennessee Dept. of Transportation, in cooperation with the Federal Highway Administration, began a $178 million "Phase I" construction program along the urban corridors of Interstates 40, 75, 640, and 275. This construction, which included replacement, widening, and addition of bridges, involved 28 separate construction contracts which were completed in April 1982. One of the most significant influences on design features was the need to accommodate massive volumes of traffic during construction.

0908 Urban Runoff Control Technology

George Zukovs, (Ontario Ministry of the Environment, Toronto, Ontario, Canada), **D. G. Weatherbe**, (Ontario Ministry of the Environment, Toronto, Ontario, Canada) and **J. Marsalek**, (National Water Research Inst., Canada Ctr. for Inland Waters, Burlington, Ontario, Canada)

Civil Engineering--ASCE, Vol. 52, No. 8, August, 1982, pp. 50-52

As urban drainage management has moved from basic research, interest has turned to technologies suitable for the control of drainage quantity and quality. This paper summarizes, through case histories, the experience of a number of Ontario municipalities in implementing runoff controls. Inlet controls, underground storage, stormwater management ponds and research experience in runoff quality controls are presented.

0909 R_x—Treated Lumber & Metal + Corrosion Ailments

Ronald F. Zollo, Member, ASCE, (Prof., Univ. of Miami, Coral Gables, Fla.) and **Edwin F. Heyer**, Fellow, ASCE, (Pres., Brill/Heyer Assoc., Engrs., Planners, Inc., Coral Gables, Fla.)

Civil Engineering--ASCE, Vol. 52, No. 8, August, 1982, pp. 56-57

The use of metal elements in contact with fire-treated lumber requires special precautions. It is a problem that has not received much attention in the literature directed to the design and construction industry. The corrosion of nailers is an example of the problem and illustrates how the combination of moisture and oxygen as well as the quality of the electrolyte affects the rate of corrosion.

0910 New Evapotranspiration Process: Key to Zero Discharge for Electroplaters

Civil Engineering--ASCE, Vol. 52, No. 8, August, 1982, pp. 58-59

The electroplating process and methods of handling electroplating wastes at the Naval Air Rework Facility in Pensacola, Florida are described. In the past, evapotranspiration as a treatment process for electroplating wastes was perceived as costly and energy intensive. It is shown here, however, that with vapor recompression evaporation, a pound of water can be evaporated with only 1/12th the energy it normally takes. The potential for the new evapotranspiration process is examined.

0911 What Some Deem Frivolous Research Sometimes Turns Out to be Extremely Important

Thomas M. Walski, Assoc. Member, ASCE, (U.S. Army Engineer Waterways Experiment Station, Vicksburg, MS 39180)

Civil Engineering--ASCE, Vol. 52, No. 8, August, 1982, pp. 60

The essay, "Stop the World, I Want to Get Off," in *Civil Engineering—ASCE,* December 1981, contends that many academic and research papers published in journals are irrelevant to the needs of the Society's practitioner majority. He says in many cases if you translate such papers into everyday English it becomes apparent they are of marginal or no value. In rebuttal, another ASCE member writes that he, too, often believes journal papers are nonsense on first reading; but when he digs deeper, he learns the meaning of words and concepts he didn't understand, often it turns out (sometimes after several years have elapsed) that academics' papers are highly valuable to those in practice. Commentary by CE's editor is included.

0912 Engineers Notebook: Checking Capacities of Eccentrically Loaded Columns on an Existing Bridge

Alex V. Tufano, (Assoc. Prof., Dept. of Civ./Mech. Technology, College of Staten Island, Staten Island, New York, N.Y.)

Civil Engineering--ASCE, Vol. 52, No. 8, August, 1982, pp. 61-63

This article focuses on the structural analysis of eccentrically loaded columns. A simple and novel graphical technique for determining the load-carrying capacity of the columns is presented. Specific attention is given to both the operating and inventory ratings of the columns. These ratings are then calculated for the as-built condition of the bridge and the as-inspected condition of the bridge.

0913 Chinese Engineer Completes Internship in U.S.

Civil Engineering--ASCE, Vol. 52, No. 8, August, 1982, pp. 66-67

As part of a growing dialog between the People's Republic of China and the United States—and following the visit of an ASCE delegation to China in 1979—it was proposed that a number of Chinese civil engineers spend a year in this country learning U.S. approaches to engineering practice. First to arrive was Shao-Ming Huang, who recently completed a one-year internship with a Montana testing and geotechnical firm. Here is what he learned as told by Huang and his employer Lee Walker of National Testing Laboratories.

0914 Chemical Spill Cleanup Named Project of the Year

Civil Engineering--ASCE, Vol. 52, No. 9, September, 1982, pp. 40-42

A 1978 hazardous-waste spill in Michigan has been cleaned up so successfully that the American Consulting Engineers Council named it the 1982 project of the year. Here is how the railroad involved, plus a cleanup contractor, an environmental consulting firm, and the state of Michigan did the job. Initial step was to remove 90 gondola cars of contaminated earth. However, even then considerable quantities of vinylidene chloride and phenol remained. That in the groundwater was removed by pumping through newly built wells and two-stage aeration in an aeration pond and via sprinklers depositing the water on the land.

0915 CE Interview: John Hernandez

Civil Engineering--ASCE, Vol. 52, No. 9, September, 1982, pp. 43-48

Since the Reagan administration took over the Environmental Protection Agency in 1981, there have been few decisions that have not stirred controversy. In this interview, Deputy Administrator John Hernandez, a Reagan appointee and second-in-command under Anne Gorsuch, discusses the regulatory changes he has helped make at EPA. Topics include: the reform of the Construction Grants program; the nation's progress toward meeting its clean water goals; water quality standards and public health; and the Administration's proposed legislative changes to the Clean Water Act.

0916 State Officials Sound Off on Most Pressing Water Quality Issues

Gene Dallaire, (Assoc. Ed., *Civil Engineering—ASCE*, New York, NY 10017)

Civil Engineering--ASCE, Vol. 52, No. 9, September, 1982, pp. 49-52

In the months ahead, Congress will undertake the awesome task of amending the nation's most important law regulating water pollution control: the Clean Water Act of 1977. The federal EPA has already forwarded proposed amendments to Congress. Yet, what changes do state water pollution control officials feel should be made in the law? What are the major issues? These questions are discussed in this article. Among the key points made: there is a need for long-term predictable funding from the federal government to the states; there is a need for a strong federal push to ensure that states develop effective pretreatment programs; the federal government should concern itself with the overall strategy of the nation's water pollution control program and not get bogged down in tactics and subtactics, which are best left to the states; constructing wastewater treatment plants should be a local responsibility—not a federal one; to become self-sufficient, municipal wastewater treatment facilities will have to increase their sewer charges; it is unwise to try and make all the streams in the U.S. "fishable and swimmable."

0917 Hazardous Waste Landfill: Some Lessons from New Jersey

Peter Montague, (Dept. of Chem. Engrg. and Center for Energy and Environmental Studies, Princeton Univ., Princeton, N.J.)

Civil Engineering--ASCE, Vol. 52, No. 9, September, 1982, pp. 53-56

There are four key elements in a secure, hazardous waste landfill: a bottom liner; a leachate collection system; a cover; and the natural hydrogeologic setting. Each of these elements is critical to success. New Jersey's experience with hazardous-waste landfills shows that synthetic liners are prone to problems—and that clay liners also have their troubles. Also, this experience shows that landfill covers are vulnerable to attack from erosion, vegetation, soil-dwelling mammals, sunlight, subsidence, and human activities. Four case histories are presented.

0918 Writing an EIS for an Emerging Technology: DOE's Coal Liquifaction Projects

Steven G. DeCicco, (Research Staff Member, Energy Division, Oak Ridge Nat. Lab., P.O. Box X, Oak Ridge, Tenn. 37830)

Civil Engineering--ASCE, Vol. 52, No. 9, September, 1982, pp. 57-61

The emerging coal-derived synthetic fuel industry may be unique in that it has received public review of potential environmental effects before the first commercial scale plant is built. The Department of Energy (DOE) Environmental Impact Statements established important precedents for the synfuel industry. An environmentally aware public wanted detailed analyses of a technology that had never been demonstrated on a commercial scale, of a plant that had not been fully designed, of health and environmental effects that were not completely understood, and of environmental regulations whose applicability were undefined. A strategy for dealing with technical uncertainty in an emerging technology was developed. The public had a definite influence on every aspect of these projects, from DOE policy to plant design. This early input should help facilitate development of the synfueld industry and enhance its benefit to society.

0919 Organic Chemical Fire in Illinois: Emergency Response and Cleanup

William H. Busch, (Mgr., Fire Operations Section, Illinois EPA, Springfield, Ill.) and **John Renkes**, (Mgr., Emergency Response Unit, Illinois EPA, Springfield, Ill.)

Civil Engineering--ASCE, Vol. 52, No. 9, September, 1982, pp. 62-65

In April 1980 at Hillsboro, Illinois, a fire in an agricultural chemical warehouse generated a hazardous smoke plume, and some 250,000 gallons of water used in fighting the fire were highly contaminated. Eighteen chemicals of varying toxicity were involved in the fire. The article describes how the owner, Local officials, the state EPA and a cleanup consultant evacuated area residents and contained and treated spilled materials. The large volume of fire-fighting water used led to potential contamination of surface and ground waters. Whether or not to try to extinguish such a fire, but rather let it burn, is discussed. The chemicals were largely eliminated as an environmental threat by use of soda ash to raise soil pH, and sprinkling powdered activated carbon over the ground and the polluted water treatment pond to soak up chemicals.

0920 Does Landfill Leachate Make Clay Liners More Permeable?

David C. Anderson, (K. W. Brown and Assoc., College Station, Tex.)

Civil Engineering--ASCE, Vol. 52, No. 9, September, 1982, pp. 66-69

Most environmental regulations require that clay liners exhibit permeabilities less than 1 x 10/u-^{7} cm/sec. While permeability is typically determined using pure water, the leachate to which clay liners are exposed are usually either highly contaminated water or organic liquids. Studies going back 40 years have found that clay can be as much as 1 million times more permeable to organic liquids than to water. Regulations governing the disposal of liquids in landfills have alternated between a complete ban and allowing the practice to continue unabated. If liquids are to be placed in landfills, a minimum requirement should be that the permeability of the liners be evaluated with whatever leachates to which the liners would be exposed.

0921 Controlling Productivity at a Hazardous Waste Site

John Zirschky, (Environmental Engr., Dept. of Agricultural Engrg., Clemson Univ., Clemson, S.C. 29631) and **Daniel Harris**, (Sanitary Engr., U.S. EPA, 25 Funston Rd., Kansas City, Kans. 66101)

Civil Engineering--ASCE, Vol. 52, No. 9, September, 1982, pp. 70-74

The hazards associated with waste disposal sites can make estimating the resources required for a remedial action difficult. During the recent clean-up of a dioxin disposal site, the manpower requirements for the various unit operations (i.e., cap removal, excavation) were monitored. Non-productive time, utilized primarily for rest breaks, accounted for approximately 40 percent of the manhours. Of the remaining manhours, approximately 50 percent can be attributed to the major unit operations. A total of approximately 1600 manhours were required for the on-site operations. Safety procedures are also briefly discussed.

0922 Getting Foreign Engineering Contracts

Merja Lehtinen, (Prod. Ed., *Civil Engineering—ASCE*, New York, NY 10017)

Civil Engineering--ASCE, Vol. 52, No. 9, September, 1982, pp. 75-77

How does a firm enter the world market for design engineering contracts? Based on interviews with Walter Hutchin and Chester Lucas of the Inter-American Development Bank and Sverdrup & Parcel, respectively, this article examines the first steps to be taken by the firm or individual engineers who plan to expand to compete internationally. Registration, credentials, and cultural subtleties are as important as marketing specific engineering skills.

0923 Replacing Corroded Cables on a Cable-Stayed Bridge

Juan Otaola, Jr., (Precomprimido, C.A., Caracas, Venezuela)

Civil Engineering--ASCE, Vol. 52, No. 9, September, 1982, pp. 78-80

The cable-stayed bridge across Venezuela's Lake Maracaibo is among the longest bridges of its type. But after years of service, corroded cables threatened to snap, which would have resulted in structural collapse. As the original design consisted of two layers of suspension cables passing over each of the six main towers, replacing the cables without damaging the structure seemed nearly impossible, given time and cost constraints. The solution, a design consisting of a new top saddle and the use of auxiliary cables, not only made changing the cables possible but also allowed traffic to continue during repairs. Design and execution took a total of two years. The new design has considerably increased the lifespan of the bridge, and future maintenance and cable changing has been greatly simplifed.

0924 The Infrastructure Problem and the Role of the Civil Engineer

Urban Growth Committee of the Urban Planning and Development Divison of the American Society of Civil Engineers, Marshall F. Reed, Jr., chmn.

Civil Engineering--ASCE, Vol. 52, No. 9, October, 1982, pp. 41-43

It is civil engineers who design, build, and maintain the nation's infrastructure. Along with public works officials who manage the urban infrastructure, the civil engineer has a responsibility to make known both the problems with the infrastructure and possible solutions to them. Engineers should focus on the technical aspects of urban infrastructure needs. In addition to the acknowledged great financial need, there is an equally important need to review design standards and performance levels. These standards and levels need to be changed where necessary to conform to severe financial constraints.

0925 Using "Helpers" Could Cut Construction Labor Costs 20%

Civil Engineering--ASCE, Vol. 52, No. 9, October, 1982, pp. 44-46

During the past decade, most growth in the United States' construction industry has been captured by open-shop contractors. The chief economic advantage enjoyed by open-shop contractors is their ability to use a high percentage of semi-skilled workmen, called subjourneymen, and to pay them accordingly. This article describes how subjourneymen are used by open-shop contractors and in union construction. The economic advantages of using subjourneymen are described. Recommendations for implementing subjourneymen clauses in union agreements are given.

0926 Improving Safety Helps Improve the Bottom Line

Civil Engineering--ASCE, Vol. 52, No. 9, October, 1982, pp. 47-50

A recent study by the Business Roundtable, drawing on research conducted by the civil engineering department of Stanford University, estimated that accidents cost the construction industry's industrial, utility, and commercial sectors nearly $9 billion a year. By making reasonable deductions in the frequency and severity of workplace accidents, the study concludes, the industry could save an annual $2.75 billion. This article presents a condensed version of the Business Roundtable report, detailing the economic impacts of construction accidents and the potential savings to owners and contractors if safety programs are established. The article also describes how owners may select safe contractors and use their influence to improve contractor safety.

0927 Hydropower Finds New Market in the Third World

Lori Ioannou, (Sr. Reporter, *Journal of Commerce*, New York N.Y.)

Civil Engineering--ASCE, Vol. 52, No. 9, October, 1982, pp. 51-55

While roughly half of the world's hydropower potential is in the developing countries, totaling about 1,200 gigawatts, only 10 percent has been tapped, a World Bank study indicates. The study forecasts that over the next decade about 100 gigawatts of hydro capacity will be added in about 60 developing countries. Developing countries in Asia, Latin America and North Africa are attracting the most attention from U.S. and Canadian hydro engineering firms. These nations are developing hydroelectric facilities as part of their long-term industrial development plans.

0928 Risk Analysis: Is EPA Changing the Rules?

Allen Morrison, (Asst. Ed., *Civil Engineering — ASCE*, New York, N.Y. 10017)

Civil Engineering--ASCE, Vol. 52, No. 9, October, 1982, pp. 60-63

Recent actions by EPA seem to indicate a departure from past practices in health risk assessment. In the past, findings that a substance is an animal carcinogen have usually led to regulatory action to control its environmental exposure. But in a recent decision on the widely used chemical formaldehyde, EPA said that proof of the chemical's carcinogenicity in animals was inadequate to support a decision to move quickly to control its use. This and other actions have led to widespread criticism of the agency by members of the scientific community. This article discusses the formaldehyde controversy and surveys the implications of EPA's actions for future risk assessment policy. Some of the generally accepted principles of risk assessment are reviewed.

0929 Firm's Management Approach Leads to 25% a Year Growth

Joseph C. Lawler, Fellow, ASCE, (Chmn. of the Board and Chf. Exec. Officer, Camp, Dresser & McKee, Inc., Boston, Mass.)

Civil Engineering--ASCE, Vol. 52, No. 9, October, 1982, pp. 64-66

The consulting engineering firm that would outlive its founders must grow to attract, develop and hold the professional and managerial talent necessary for survival. Growth must be based on integrity, professionalism and technical excellence if it is to contribute to corporate longevity, and it must be planned for. The firm that would grow must define its objectives, devise a practicable strategy for reaching them and reduce the whole to writing for implementation. Growth plans are not static but must be continually reviewed and updated to accommodate to changes in the world in which the firm operates. The step-by-step planning process followed by the author and his firm in developing the firm from a small local operation to a major international enterprise is described in detail.

0930 Engineering School Drives for National Leadership

George C. Beakley, (Assoc. Dean, College of Engrg. and Applied Sci., Arizona State Univ., Tempe, Ariz.) and **C. Roland Haden**, (Dean, College of Engrg. and Applied Sci., Arizona State Univ., Tempe, Ariz.)

Civil Engineering--ASCE, Vol. 52, No. 9, October, 1982, pp. 67-70

Engineering programs in many universities are undergoing severe problems of limited space, too few faculty, burgeoning student enrollments, out-of-date laboratory equipment and inadequate funding. Arizona State University has launched a three-way partnership of the industrial community, state government, and the university to remedy these problems. This account is a progress report of the first two years of this five year plan. Elements of the plan are discussed. These include higher academic standards, excellent research-oriented faculty, strong graduate student recruitment, continuing education, interactive instructional television programs, a research facility, establishment of research centers, state of the art research equipment, Prince Visiting Scholar Program, and privately funded faculty chairs. The changing economic and demographic conditions of the state of Arizona have been conducive to encouraging this type of innovative program. These changes are discussed.

0931 Slope Stability Program

John P. Cross, Member, ASCE, (Data Processing Mgr., Proj. Engr., STS Consultants, Northbrook, Ill.)

Civil Engineering--ASCE, Vol. 52, No. 9, October, 1982, pp. 71-74

This article includes an entire computer program, written in BASIC computer language. The program is a simplified version of programs which have been written for large computer systems. The program utilizes the modified Bishop method of slope stability analysis and can be used on small computers. This simplified or modified Bishop method is reasonably accurate for most purposes where the slope under analysis can be assumed to fail along a circular failure surface. This article is part of a series on CE Computing Basics.

0932 What Type of Sewer Pipe is Best? Life-cycle Cost Analysis Yields Answer

Thomas Moss, Member, ASCE, (Team Leader, Proj. Mgr., Camp, Dresser & McKee, New York, N.Y.) and **Edward J. Jankiewicz**, (Public Works Dir., City of Winchester, Winchester, Va.)

Civil Engineering--ASCE, Vol. 52, No. 9, October, 1982, pp. 75-76

In the past, many engineers have selected the type of sewer pipe for a particular project based mainly on a comparison of initial costs of concrete, steel, and other types of pipes. This approach fails to take into account that some sewer pipes don't last as long as others and will have to be replaced sooner. One American city and its consultants are now selecting sewer pipe on a life-cycle cost basis. Lifespans of various types of sewer pipes are examined and a strategy for comparing life-cycle costs is presented.

0933 Salvaging Cleveland

Allen Morrison, (Asst. Ed., *Civil Engineering — ASCE*, New York, N.Y. 10017)

Civil Engineering--ASCE, Vol. 52, No. 10, November, 1982, pp. 39-42

Cleveland, Ohio has suffered from shrinking inner-city wealth resulting in reduced budgets for its urban infrastructure. Now, led by the Greater Cleveland Growth Association, the city and suburban communities and special authorities have taken a cooperative look at where they stand. With help from the Urban Institute, they totaled infrastructure spending needs. They determined which expenditures could be postponed and which results could be achieved at lower costs. With revised priorities and costs, tactics for raising the necessary money were developed.

0934 Fiber Reinforced Concrete

Kneeland A. Godfrey, Jr., (Editor, *Civil Engineering — ASCE*, New York, N.Y. 10017)

Civil Engineering--ASCE, Vol. 52, No. 10, November, 1982, pp. 44-50

In the past 10 years or so, concrete reinforced with fibers one or two inches long has begun to be used, first on a demonstration and experimental basis, then in major everyday applications. This article cites landmark or pioneer case histories of use in these areas: dam spillways; tunnels; open rock slopes; airfield pavements; shore protection; refractories; building wall panels; structural and curtainwall; roofs of dome and barrel-arch shape; slabs on grade; and other. A few key documents providing design aids are cited, as are the names and addresses of five suppliers of fibers.

0935 Engineers Assess Aquaculture Systems for Wastewater Treatment—Part II

Sherwod C. Reed, (Cold Regions Research and Engrg. Lab., Hanover, N.H.), **Robert K. Bastian**, (Ofc. of Water Program Operations, EPA, Wash., D.C.) and **William J. Jewell**, (Cornell Univ., Ithaca, N.Y.)

Civil Engineering--ASCE, Vol. 52, No. 10, November, 1982, pp. 51-53

An overview of the status of aquaculture technologies for their routine use in municipal wastewater treatment is presented. The detailed assessments were conducted by six internationally recognized engineers and were based on presentations at a September 1979 Aquaculture Seminar at the University of California at Davis. Systems included wetlands, aquatic plants, fish culture and combinations of aquatic components. Most of the systems considered produce high quality effluent with lost cost and energy inputs. However, in most cases further development and/or demonstration will be needed before concepts are ready for routine use.

0936 What Future for America's Railroads?

Eugene E. Dallaire, (Assoc. Ed., Civil Engineering — ASCE, New York, N.Y. 10017)

Civil Engineering--ASCE, Vol. 52, No. 10, November, 1982, pp. 54-59

The public sees America's railroads as old, rundown, bankrupt and unreliable. However, after decades of decline, the rate of return on investments has begun to show a steady increase for the American freight-railroad industry. This is due to both the rising oil prices and deregulation on the railroad industry as a result of the Staggers Rail Act of 1976 and 1980. It is speculated that there will be a continuation of the railroad merger movement and a trend towards railroads merging with trucking, barge, and other transportation companies. Federal subsidies to operate the nation's intercity passenger railroad system will likely remain at their present modest level.

0937 Houston Moves to Unravel Traffic

Merja Lehtinen, (Production Ed., *Civil Engineering — ASCE*, New York, N.Y. 10017)

Civil Engineering--ASCE, Vol. 52, No. 10, November, 1982, pp. 60-63

Houston, Texas is growing and has more cash to invest in its infrastructure than most American cities. Roads are being repaired and improved, more highways built, and new management strategies applied to solve Houston's giant traffic problem. More than seven hours of congestion a day is the current norm. Here, the private sector is taking a lead in turning Houston around. ASCE members play key roles.

0938 The Fabricator/Designer Connection

J. C. Holesapple, (Sr. Vice Pres., Engrg. and Operations, Allied Structural Steel Co., Chicago Heights, Ill.)

Civil Engineering--ASCE, Vol. 52, No. 10, November, 1982, pp. 64-68

What can the designer-fabricator team do to improve the overall function of design details? One of the best ways is to begin with team work by bringing the fabricator in on the project at the outset. Since connections are one of the highest unit cost items in a structure, simplification will reduce the cost and generally improve the function of the structure.

0939 Jubail Industrial City: Engineering Monument

J. M. Engstrom, (Arabian Bechtel Co., Ltd., Jubail, Saudi Arabia) and **Jasim A. Hejji**, (Royal Commission for Jubail and Yanbu, Jubail, Saudi Arabia)

Civil Engineering--ASCE, Vol. 52, No. 10, November, 1982, pp. 69-73

The seawater cooling project is a major innovation in engineering accomplishments. Here the project engineers discuss the key civil engineering aspects of the Jubail, Saudi Arabia project with particular attention to the seawater cooling system.

0940 Energy Crisis Revisited: Conference Highlights Roles of Marketplace and Government

Allen Morrison, (Asst. ed., *Civil Engineering—ASCE*, New York, N.Y. 10017)

Civil Engineering--ASCE, Vol. 52, No. 12, December, 1982, pp. 32-35

At a recent conference on energy conservation co-sponsored by ASCE's Minnesota Section and the University of Minnesota, three speakers presented divergent views on the role of the marketplace in solving the world's energy problems. William Tucker, a freelance writer who writes about energy and the environment, suggested that the removal of price controls on oil has "essentially solved the energy crisis." Ellis Armstrong, chairman of ASCE's Energy Policy Committee, argued that within the next 10-15 years oil demand will begin to exceed the supply and that government leadership on an international scale is necessary to prepare for future energy needs. Bruce Anderson, editor of *Solar Age Magazine*, contended that the best way to prepare to cut our dependence on oil is through an agressive government-supported program of conservation and development of renewable resources. The article summarizes the presentations of each speaker and includes a box on an innovative solar home discussed at the conference.

0941 Shanghai Tunnel Projects Spur Construction Innovations

Wang Zhen Xin, (Deputy Chf. Engr., Shanghai Tunnel Construction Company, China)

Civil Engineering--ASCE, Vol. 52, No. 12, December, 1982, pp. 36-38

Since underground construction in soft ground is newly developed in Shanghai, there is little experience there as compared to what is known for specific sites in other countires. Well known tunneling techniques had to be adjusted to meet the needs of China's environment. In some cases, new methods had to be devised. Areas of tunnel construction for which adjustments or innovations were made are long distance pipe jacking, and the use of cast in-situ walls or precast concrete sheet piles. These techniques are described and illustrated.

0942 How Did These Firms Grow in Face of a Depressed Economy?

Kneeland Godfrey, Jr., Member, ASCE, (Ed., *Civil Engineering—ASCE*, New York, N.Y. 10017)

Civil Engineering--ASCE, Vol. 52, No. 12, December, 1982, pp. 39-42

Despite a decline in construction physical volume of about 25% in the past five years, some engineering firms have grown. Profiles of three such firms are presented. Management decisions that benefited these companies are described. They include: (1) Diversifying beyond housing related markets at a time when housing construction is down; (2) geographical diversification; (3) selling the firm's services during slack periods as well as during boom times; (4) use of market research to choose target areas; and (5) selective use of sophisticated engineering aids, such as computer aided drafting, to increase engineering productivity.

0943 Protecting Post-Tensioning Tendons in Concrete Structures

Morris Schupack, Fellow, ASCE, (Pres., Schupack Suarez Engineers, Inc., South Norwalk, Conn.)

Civil Engineering--ASCE, Vol. 52, No. 12, December, 1982, pp. 43-45

Patterns of corrosion problems related to tendon systems are analyzed based on the author's experience in evaluating over 50 corrosion related performance problems in prestressed concrete structures and in surveying prestressed concrete behavior. These relatively rare problems are often caused by very small initial savinggs in the overall cost of the structure or by poor execution of the works. Construction methods and details have been developed to minimize the recurrence of the types of problems which have occured in the past.

0944 CE Interview: John Hernandez (Part II)

Civil Engineering--ASCE, Vol. 52, No. 12, December, 1982, pp. 46-49

Since the Reagan administration took over the Environmental Protection Agency in 1981, there have been few decisions that have not stirred controversy. In the second part of an interview with Deputy Administrator John Hernandez more of the regulatory changes he has helped make at EPA are discussed. Topics include: Reform of the Construction Grants program; EPA's declining research and development spending; the agency's drinking water regulations; ocean dumping of sludge; and hazardous waste disposal. (The first part of this interview appears in the September, 1982 issue of *Civil Engineering—ASCE.*)

0945 Compression Rings Support Deep Cut

C. Lal Lalwani, Member, ASCE, (Principal, Barber & Hoffman, Inc., Cleveland, Ohio 44115)

Civil Engineering--ASCE, Vol. 52, No. 12, December, 1982, pp. 50-51

The plans for building the Cuyahoga Valley Interceptor Lift Station in Cleveland, Ohio, called for 126 ft. (38.4 m) by 105 ft. (32 m) construction, some 80 ft. (24.4 m) below grade. Bracing of this large and deep excavation was done by four reinforced concrete compression ring beams (largest of this kind in the United States). Clear and uncluttered working space within the excavated area was an expedient way to build the concrete framed Interceptor Lift Station. The complex design of the ring beams was done keeping in consideration the uneven loading patterns due to excavation procedures and due to non-uniform soil properties. A close coordination between design assumptions and the Contractor's operations was a must for the success of the project. Inclinometers were installed to monitor the mass movement and the stability of the soil.

0946 Are We Training Enough Civil Engineering Technicians?

Jerry A. Nathanson, (Instructor, Union County Technical Institute, Scotch Plains, N.J.)

Civil Engineering--ASCE, Vol. 52, No. 12, December, 1982, pp. 52-54

A basic difficulty in the world of engineering education today is that we may be training too many B.S. engineers and not enough technicians and technologists. Presently, the total number of B.S. level engineering graduates is considerably higher than the combined total of technicians and technologists graduates. If technicians are to provide "support" for engineers, just as nurses provide support for doctors, then we have far too few technicians. Thus, there may be a need to place specific limits on enrollments in B.S. engineering programs; and for a major push to increase technology enrollments.

0947 Stimulating Creativity in the Civil Engineering Profession

Morris Hochburg, (Mgr., Value Engrg., Greeley & Hansen, Chicago, Ill.)

Civil Engineering--ASCE, Vol. 52, No. 12, December, 1982, pp. 54-55

A major activity missing from both civil engineering education and practice are means for stimulating creativity. Ideas for boosting creativity in the profession and discussed here are: brainstorming sessions in several cities at once using satellite communications; awards for creative work by students; seminars on creativity in major cities; professional recognition to individuals and teams that generate creative ideas; subsidized lectures on creativity to appear before local ASCE chapters; compensation to engineers for creative ideas that are implemented; encouragement for experienced engineers to share their hard-won experience and creative ideas with others.

0948 Curbing High Turnover in Consulting Firms

David O'Donnell, (Assoc. Civ. Engr., City of Detroit, Mich.)

Civil Engineering--ASCE, Vol. 52, No. 12, December, 1982, pp. 55

Many consulting engineering firms suffer from high-employee turnover. This is bad for the firm, and bad for the individual engineer. Suggestions to reduce turnover are given. For example, when a firm has to let engineers go during slack periods, rehire the same engineers when work picks up; instead of laying off an engineer when work is slow, give him a second task to do—e.g. client contact or marketing of the form's services; instead of laying off an engineer, have him work fewer hours.

0949 Washington D.C. Suburban I66—Unique Compromise in Expressway Design

John N. Hummel, Member, ASCE, (Div. Chf. for Planning and Engrg., Arlington County Dept. of Public Works, Arlington, Va. 22201-2676)

Civil Engineering--ASCE, Vol. 52, No. 12, December, 1982, pp. 56-59

A controversial urban Interstate project with more than 40 years of history has reached the final stages of construction. Citizen participation in varying forms has led to the development of a unique facility using the latest in transportation technology and significant public amenities to fit into an urban residential environment. The process which led to this design shows that civic participation can exert positive influence on major public works projects.

0950 Hydrologic Applications for Electronic Spreadsheets

Stephen C. Hung, Member, ASCE, (U.S. Dept. of Trans., St. Lawrence Seaway Dev. Corp., Wash., D.C.)

Civil Engineering--ASCE, Vol. 52, No. 12, December, 1982, pp. 60-61

Electronic spreadsheets have many potential applications to civil engineering. They are ideal for the derivation of unit hydrographs and the construction of design storm hydrographs. They are also useful tools for backwater computations, using standard step method, for reservoir and channel routings and for flood frequency analyses and reservoir operation studies.

0951 Unified Soil Classification System

Jim Stevens, (Assoc. Prof., Civ. Engrg. & Environmental Sci., Univ. of Oklahoma, Norman, Okla.)

Civil Engineering--ASCE, Vol. 52, No. 12, December, 1982, pp. 61-62

The most common soil classification method used by foundation engineers in the United States is "A Standard Method for Classification of Soils for Engineering Purposes, ASTM D-2487," usually known as the "Unified System". The BASIC program shown in this article classifies soil in accordance with this system.

0952 Public Infrastructure—Are More Dollars Coming?

Civil Engineering--ASCE, Vol. 52, No. 12, December, 1982, pp. 63-65

The top subject at the October, 1982, New Orleans convention of the American Society of Civil Engineers was the infrastructure finance problem. Noted analysts of the situation, including Pat Choate of TRW Corp. and George Peterson of the Urban Institute, presented an analysis of the problem and approaches to various solutions. It was recommended that all public works functions (including highways, water supply, sewage disposal and treatment and refuse collection) should be fully reimbursed by user charges. Utilities are cited as an example of this method; their charges are high enough to pay for not just operating costs but for replacing old plant and equipment.

0953 1982 National Conference on Environmental Engineering Division Specialty Conference, Sheraton Ritz Hotel, Minneapolis, Minnesota, July 114-16, 1982

Walter K. Johnson, (editor) and **Dennis R. Martenson**, (editor)

New York: ASCE, 1982, 763pp.

The Proceedings of the 1982 ASCE National Conference on Environmental Engineering includes papers on a broad spectrum of environmental issues. Over one hundred papers were presented addressing environmental engineering planning, design, operation, and monitoring issues. The sponsoring group, the Environmental Engineering Division of ASCE, consists of four Management Groups: Air, Noise, and Radiation Management (ANRM); Solid Waste Management (SWM); Water Pollution Management (WPM); and Water Supply and Resources Management (WSRM), and papers were presented in each of these categories. Several topics selected by the management groups for particular emphasis at the conference, including some invited papers, were: Noise mitigation and a status report on nuclear concerns; hazardous waste management; wastewater treatment plant effluent disinfection; and acid rain and water treatment for the removal of specific pollutants. Other topics stressed included industrial and biological wastewater treatment, solid waste landfills, wastewater sludge processing, water quality and effluent standards, air pollution control and the land application of wastewaters. Several key papers on environmental policy and energy issues were also presented.

0954 **Application of Plasticity and Generalized Stress-strain in Geotechnical Engineering**

Proceedings of the Symposium on Limite Equilibrium, Plasticity and Generalized Stress-Station Applications in Geotechnical Engineering

E. T. Selig, (editor), Fellow, ASCE, (Prof. of Civ. Engrg., Univ. of Massachusetts, Amherst, Mass. 01003) and **R. N. Yong**, , (editor), Member, ASCE, (Prof. of Civ. Engrg. and Applied Mechanics, McGill Univ., Montreal, Quebec, Canada)

New York: ASCE, 1982, 355pp.

This volume on the application of plasticity and generalized stress-strain in geotechnical engineering is divided into two main sections, with an introduction on soil constitutive model assessment. The first section covers limit equilibrium plasticity applications in geotechnical engineering. A state of the art report on limit equilibrium plasticity in soil mechanics discusses failure conditions, plasticity theory, slope stability, active and passive earth pressures, and ultimate bearing capacity of foundations. Eight papers discuss these topics in depth. The second section on generalized stress-strain applications is also introduced with a state of the art report which covers use of constitutive models in finite element programs, theories of constitutive relations, and current non-linear relations. Seven papers that explore these topics include a deformation analysis for braced excavation in clay, characterization of the undrained anistropy of clays, and stress-strain influence on soil compactability. The papers included in this publication were presented at a symposium on geotechnical engineering. Resulting panel discussions are summarized. Graphs and charts illustrate many of the papers.

0955 **Application of Walls to Landslide Control Problems**

Proceedings of Two Sessions Sponsored by the Committee on Earth Retaining Structures of the Geotechnical Engineering Division of the American Society of Civil Engineers at the ASCE National Convention, Las Vegas, Nevada, April 29, 1982

R. B. Reeves, (editor)

New York: ASCE, 1982, 132pp.

The evolution of tieback restraint systems is presented. Case histories describe current practices for tieback design and construction and give examples of how the requirements were met for stabilizing specific slides. Topics include geologic aspects of landslide control using walls, analysis of wall supports to stabilize slopes, evaluation of landslide properties, and slide control by drilled cantilever pier walls and tangent pile walls. Site investigations, laboratory tests, design procedures, and construction methods are described.

0956 Bibliography on Ocean Energy

Waterway, Port, Coastal and Ocean Division Task Committee on Ocean Energy

New York: ASCE, 1982, 89pp.

With the prospect of an increasing shortage of energy resources, there has been a growing interest in renewable alternative sources of energy. An increasing effort is being directed towards the problem of extracting energy from the world's oceans since the oceans represent a vast potential resource of renewable energy. To help meet a response to national energy concerns, the Task Committee on Ocean Energy of the Waterway, Port, Coastal and Ocean Division of the American Society of Civil Engineers has compiled this bibliography of over 1300 entries on ocean energy. Subjects covered include wind-generated waves, tides, ocean temperature gradients, salinity gradients and ocean currents.

0957 Classic Papers in Hydraulics

Task Committee to Develop a Volume of Classics in Hydraulics of the Hydraulics Division of the American Society of Civil Engineers, John S. McNown, chmn.

New York: ASCE, 1982, 671pp.

Fourteen papers representing basic contributions which made possible rapid advances in analytical techniques in hydraulic engineering are presented. The papers reflect state-of-the-art- developments in hydraulics from 1936 to 1961, and original discussions of the papers are included. The topics covered include hydraulic jump, mechanics of fluid turbulence, bed load transport, transportation of suspended sediment, drawdown test to determine the effective radius of an artesian well, diffusion of submerged jets, mechanics of manifold flow, storage and the unit graph, and laboratory investigation of rubble-mound breakwaters. Three papers investigate open channels, including the mechanics of supercritical flow, turbulent flow through open channel constrictions, and studies of air entrainment in open channel flows. The papers were selcted on the basis of originality and lasting value.

0958 Cone Penetration Testing and Experience

Proceedings of a Session Sponsored by the Geotechnical Engineering Division at the ASCE National Convention, St. Louis, Missouri, October 26-30, 1981

G. M. Norris, (editor) and **R. D. Holtz**, (editor)

New York: ASCE, 1981, 479pp.

Experiences and advances in cone penetration testing are reviewed in these papers presented in a technical session at the National Convention of the American Society of Civil Engineers. Attention is focused on test techniques, applications, evaluation of data, case studies, and recent developments in equipment and procedures. Theory, correlations, and interpretation of cone penetration are covered in discussions of centrifugal models of the cone penetrometer, cone resistance, and soil classification. The piezometer cone and new equipment are reviewed in reports and case studies. Standards and applications are discussed including a comparison of the European standard cone penetration test (CPT) to ASTM standards. A review of offshore application of the cone penetrometer is also included. A state of the art report on penetrometer practice provides an overview of the value, equipment, evaluation, and special applications of the CPT as well as a discussion on the analytical use of CPT data.

0959 Consulting Engineering
A Guide for the Engagement of Engineering Services
(Manual and Report No. 45)

Committee on Standards of Practice

New York: ASCE, 1981, 54pp.

The purpose of this manual is to outline the functions of the consulting engineer in serving a client, the types of services usually offered, the various methods of determining compensation for engineering services, and the general ranges of remuneration that competent consultants receive for their services. Suggested procedures for interviewing and selecting a consulting engineer and guidance on contracts for engineering services are also provided. The manual is designed to serve the best interests of both the client and the consulting engineer and to foster better understanding between them.

0960 Design of Water Intake Structures for Fish Protection

Task Committee on Fish-Handling Capability of Intake Structures of the Committee on Hydraulic Structures of the American Society of Civil Engineers

New York: ASCE, 1981, 163pp.

This publication presents the results of a study on available technical information on the characteristics of water intake structures designed to eliminate or reduce damage to fish. Biological criteria and the reliability of hardware are assessed. Engineering factors influencing intake design are covered. Considerable attention is given to practical fish protection methods, including fish diversion methods, fish deterrence, passive physical exclusion methods, and design of traveling screens. Fine screening methods for small organisms are also discussed, including collection and removal methods and passive physical exclusion methods. The design of bypass, pipeline, and pump fish return systems are also covered. Many of the systems discussed in the book are illustrated. This publication has been prepared as an aid to the practicing engineer and current acceptable procedures using proven hardware are stressed.

0961 **Drilled Piers and Caissons**

Proceedings of a Session Sponsored by the Geotechnical Engineering Division at the ASCE National Convention, St. Louis, Missouri, Oct. 28, 1981

Michael W. O'Neill, (editor)

New York: ASCE, 1981, 153pp.

Drilled piers, also called drilled caissons, drilled shafts, bored piles, and cast-in-situ piles, have recently undergone a period of growth in their use as foundation bearing elements, uplift anchors, and lateral-load-carrying devices. Paralleling this growth has come an elevated level of understanding of the effects of various factors on the behavior of drilled piers. Such factors include in-situ soil conditions, type and direction of loading, pier geometry, and, very importantly, construction procedure. This publication was assembled by the Deep Foundations Committee of the Geotechnical Engineeering Division of ASCE to describe recent developments in the wide spectrum of analytical modeling and observed performance of drilled piers. Two papers address construction procedures, three address the axial load performance of drilled piers in a variety of soil profiles, two are concerned with lateral load behavior, and two are directed toward analytical modeling.

0962 **Emerging High Performance Structural Plastic Technology**

Proceedings of a Session Sponsored by the Structural Plastics Research Council of the Technical Council on Research of the American Society of Civil Engineers at the ASCE National Convention, Las Vegas, Nevada, April 26, 1982

Andrew Green, , (editor), Member, ASCE, (Pres., Composite Technology, Inc., Fort Worth, Tex.)

New York: ASCE, 1982, 84pp.

Structural plastics have forced their growth by better performance over conventional materials in hostile environments. Demand by industry for increased life of process equipment and structures has increased the use of the materials by a factor of 1-1/2 times that of conventional materials in the past five years. This growth is the result of engineers seeking out new engineering materials with greater strength and stiffness for use in corrosive applications. Structural plastics have been in use in the aerospace industry for over 30 years, however their practical application in civil engineering has been limited. Information and data has been hard to find, which in turn limits expanded use of these materials. The papers presented here discuss applications of structural plastics with some history behind them, modern quality control methods, and current information on fire resistance of these organic materials to help demonstrate their practical application. The information covers areas of vital interest to the designer: static load carrying ability, stress rupture, fire resistance, and quality evaluation of the products. The evidence presented here of successful application should give the designer more confidence in selected material for life cycle considerations.

0963 Engineering and Business: Converting Engineers to Businessmen

Proceedings of a Symposium Sponsored by the Engineering Management Division of the American Society of Civil Engineers

J. R. King, Jr., (editor)

New York: American Society of Civil Engineers, 1982, 74pp.

Business practices and the role of the engineer as manager are covered in the papers presented at a symposium on engineering management at the October, 1982 National Convention of the American Society of Civil Engineers. In a case history of the development, growth, and evolution of a consulting engineering firm formed in 1958, the setbacks and rewards of running a consulting engineering firm are presented. Considerations for the establishment of a new business within the engineering community are outlined in a subsequent paper. The development from engineer to businessman is reviewed from four different viewpoints including an assessment of desirable individual characteristics, an outline of fast track methods for developing managers, an overview of manager development within a matrix structure, and a review of manager development programs. Risk management from a contractor's viewpoint is presented and a basis for effective decisio-making in probabilistic risk assessment follows.

0964 Engineering and Construction in Tropical and Residual Soils

Proceedings of the ASCE Geotechnical Engineering Division Specialty Conference

Geotechnical Engineering Division

New York: ASCE, 1982, 735pp.

Thirty-one papers and six invited lectures presented at the specialty conference on engineering and construction in tropical and residual soils are included. The proceedings represent a world-wide cross section of soil problems. Geotechnical properties are discussed, including swell and creep properties of an African black clay, properties of laterite gravels, geologic classification of coralline deposits, and composition effect on suction of a residual soil. A section on soils investigation and field testing includes discussions of air foam sampling, full scale load testing, and design of drilled shafts in South Florida limestone. Landslides, slope failures in colluvium overlying weak residual soils, stability analysis of unsaturated soil slopes, and engineering problems of gravel deposits are considered in a section on slope stability. Four papers also cover aspects of excavation and dewatering. Six construction case histories are presented including the use of ground anchors, effects of tropical soils on airport design, dam construction problems in a tropical environment, and repair of an old earth dam using filter fabric. Many of the papers are illustrated with charts, graphs, photographs, and tables.

0965 Evaluation, Maintenance and Upgrading of Wood Structures

Subcommittee on Evaluation, Maintenace and Upgrading of Timber Structures of the Committee on Wood of the Structural Division of the American Society of Civil Engineers, Alan Freas, chmn.

New York: American Society of Civil Engineers, 1982, 428pp.

Any structure, regardless of the material from which it is made may be subject to review of its ability to perform a specific function or functions. This guide has been prepared to present information on technical aspects of inspection, evaluation, reinforcement, repair, and rehabilitation of timber structures. Factors which influence serviceability of wood structures include loadings, duration of loads, temperature, moisture and weathering. Effects of chemicals and fire, as well as insects, fungi, and other organisms that attack wood are also covered. Designing to avoid problems caused by these influences is discussed. Inspection techniques and equipment are described along with guidelines on where to look and what to look for. Criteria covered in a section of evaluation of wood structures include structural analysis, determination of loads, and estimating load carrying capacity.

0966 Failure of the Breakwater at Port Sines, Portugal

Port Sines Investigation Panel

New York: ASCE, 1982, 287pp.

The objectives of the Port Sines Investigation Panel were to collect perishable data on and to identify causes of the failure on February 26, 1978, of the rubble mound breakwater at Port Sines, on the Atlantic Coast of Portugal. This report is a result of that investigation. The structure is the largest of its kind in such an exposed environment, and is situated in a previously untried combination of unusually deep water and high-energy marine setting. Five simplified failure possibilities are evaluated: (1) that design criteria were exceeded by the February 1978 storm: (2) that the breakwater construction was faulty; (3) that the materials used for construction were substandard; (4) that the procedures followed during the design were incoplete or incorrect for this specific set of environmental conditions; (5) that the design was inadequate. The purpose of this investigation was not to find a crucial error or omission but to report on all possible problems, omissions, and errors that could have contributed to the failure. A description of the breakwater project is given, along with details of the design and construction of the project. The physical setting and the storm of February 26, 1978 are also described.

0967 Gravity Sanitary Sewer Design and Construction (Manual and Report No. 60)

Joint Task Force of the American Society of Civil Engineers and the Water Pollution Control Federation, Dan L. Glasgow, chmn.

New York: American Society of Civil Engineers, Wash., D.C.:Water Pollution Control Federation, 1982, 275pp.

This Manual provides both theoretical and pracctical guidelines for the design and construction of gravity sanitary sewers. The initial chapter introduces the organization and administrative phases of the sanitary swer project. Subsequent chapters are presented in a sequence detailing the parameters necessary to establish the design criteria, complete the design and award a construction contract. The Manual concludes with a discussion of the commonly used methods of sanitary sewer construction The manual is intended to be of practical use to the designer of a gravity sanitary sewer system and is based upon the experience of engineers in the field of sanitary sewer structural and hydraulic design. Charts, illustrations, and example problem solutions are used liberally throughout to reinforce the text.

0968 Guide to Right-of-Way Survey Practices

Task Committee on Right-of-Way Survey Practices of the Committee on Land Surveying of the Surveying and Mapping Division, William T. Wamback, chmn.

New York: ASCE, 1981, 13pp.

Guidelines are described for right of way surveys with regard to properly executed research, field surveys, monumentation, platting and recording of plats, and descriptions by trained personnel. This guide should be applied to surveys for new easements, new fee land acquisition, resurveys of existing easements, and resurveys of fee land holdings for transportation facilities and transmission facilities which cross one owner's property to serve others.

0969 A Guide to Urban Arterial Systems

Committee on Urban Arterial Systems of the Urban Transportation Division, C. Michael Walton, chmn.

New York: ASCE, 1981, 344pp.

The urban arterial system, a link between freeways and local streets, provides for the efficient collection, distribution, and transition of traffic among freeways, collector streets, local roadways, and property. This Guide discusses traffic engineering techniques and alternative modes of transportation as applicable treatments of common problems in arterial systems. This report is primarily interested in short-range strategies and the planning, implementation requirements, issues, consequences, and financial implications of each. These strategies are embodied in Transportation System Management (TSM), a concept of integrating transportation elements into a coordinated and efficient package to satisfy a range of operational issues with emphasis on low or moderate cost. In this Guide, two approaches to transportation management are considered: supply management and demand management. Under supply management, six components were investigated, including automobile facilities, high occupancy vehicle (HOV) facilities, light rail transit (LRT), non-motorized facilities, terminal facilities, and urban goods movement. Under demand managemtne, four components were investigated, including land use management, parking management, access management, and variable hours management.

0970 **Long Span Roof Structures**
Proceedings of a Symposium Held at the 1981 Annual Convention and Exhibit, St. Louis, Missouri, October 26-30, 1981

Committee on Special Structures of the Committee on Metals of the Structural Division, Charles Birnstiel, chmn.

New York: ASCE, 1981, 357pp.

The papers included in these proceedings treat a wide range of technical topics related to long span roof structures including new structural concepts, structural stability, the effects of snow loading, wind-structure interaction and advanced nonlinear dynamic analysis. Also discussed are the state of engineering design, construction industry practices, and the relationship of architects to engineers, as they relate to building safe long span roofs.

0971 **Managing Liability - The Individual's Challenge, the Organization's Challenge, the Project Manager's Challenge**
Proceedings of a Symposium Sponsored by the Engineering Management Division of the American Society of Civil Engineers at the ASCE National Convention, Las Vegas, Nevada, April 26-30, 1982

J. R. King, Jr., (editor), Fellow, ASCE, (Rady and Associates, Inc., Ft. Worth, Tex.)

New York: ASCE, 1982, 95pp.

These proceedings are the papers presented during the 1982 Spring Convention of the American Society of Civil Engineers to determine how engineers can live with, manage, and limit liability. Liability problems of the individual, the organization, and project management are presented and solutions suggested. Topics discussed include the engineer in the role of an expert witness, personal liability of the government engineer, risk allocation through contract clauses, and product liability. An introduction to bonds, liens, and insurance for construction companies is also included.

0972 **Planning and Environmental Criteria for Tall Buildings**
Planning and Design of Tall Buildings, Vol. PC

Council on Tall Buildings and Urban Habitat, D. Sfintesco, chmn.

New York: American Society of Civil Engineers, 1981, 1309pp.

The role of the tall building in the urban habitat and its interaction with urban systems is examined. It ranges over the full spectrum of engineering, architecture and social sciences. The entire volume is founded strongly on the systems approach, and upon the consideration of human factors in design—two factors that are of overriding importance if long-term success is to be achieved in high-rise projects. The results show that the most important decisions are probably not those that have to do with structure, mechanical apparatus, or architecture, but instead may be those that are made by investors, planning commissions, mayors, boards of supervisors, or governmental agencies. The volume considers the role of the social environment, interaction between public and private sectors, the intertwining economic considerations, architecture, urban design, and the integration with transit and associated parking requirements. The volume concludes with some chapters that relate to building management and ownership, energy conservation, and the case for more intensive application of project management and systems concepts.

0973 **Proceedings of the Conference Applying Research to Hydraulic Practice**

Peter E. Smith, (editor)

New York: ASCE, 1982, 743pp.

Approximately 75 of the papers presented at the 1982 ASCE Hydraulics Division Specialty Conference in Jackson, Mississippi are included in this publication. The papers examine many aspects of applying new approaches to solving hydraulic problems in hydraulic practice. Specific areas include drop structures, spillways, jets and mixing, tidal hydraulics, and tidal inlets. Particular attention is given to modeling in hydraulics including turbulence models, model studies, and verification of models of hydrologic transport and dispersion. Small hydro design innovations and applications are also considered and case histories of small hydro projects are presented. Design and performance of reversible flow trash racks are covered. The phenomena of sediment transport is covered including discussions on the energy/power approach to this phenomena. The control of water quality in reservoirs and releases is also examined.

0974 **Proceedings of the Conference on Construction of Power Generation Facilities**

Experience with the Implementation of Construction Practices, Codes, Standards and Regulations

Jack H. Willenbrock, (editor), (Dept. of Civ. Engrg., Pennsylvania State Univ., University Park, PA)

New York: ASCE, 1982, 623pp.

The impact of the present construction practice, codes, standards and regulations on the construction of power generation facilities is reviewed and examined in papers presented at the September, 1981 conference on construction of power generation facilities. The papers examine major problems related to codes and regulations, the difficulty of implementing some of the requirements, and the possible improvement recommendations that can be made. The major areas covered include: Nuclear standards, licensing, enforcement, and construction inspection programs; requirements and performance of concrete materials; quality assurance and quality control practices; concrete test requirements and process control practices; steel fabrication practices, concrete anchorage systems, plant modification and retrofit experience; concrete batch plant production transportation and field adjustment practices; concrete formwork, reinforcement and placement tolerances and practices; and concrete placement and post placement practices. The perspectives of the utilities industry, the architect-engineer, the government, and the concrete industry are also examined.

0975 Proceedings of the Conference on Directional Wave Spectra Applications

Robert L. Wiegel, Fellow, ASCE, (College of Engrg., Univ. of California, Berkeley, Calif. 94720)

New York: ASCE, 1982, 501pp.

These collected papers and their ensuing discussions presented at the September, 1981 symposium at the University of California on directional wave spectra applications examine the present possibilities, the difficulties, and the gaps that exist in obtaining directional wave spectra and utilizing it in the design, installation, and operation of ocean facilities and ocean vehicles. State of the art means of wave measurement technology are presented. Ship motions and directional analysis of vessel response are covered. Other topics examined include wave directionality inferred from stereo photos, directional spectra density from air- and spacebourne radar, and surface contour radar remote sensing of waves. The Maximum Likelihood Method (MLM) estimation of directional spectra is considered in three papers. Discussions on the use of directional spectra in design of offshore structures are included. Kinematics of directionally spread waves, long shore currents and bed shear stress, Monte Carlo simulation for shoreline processes, and evaluation of floating bridge behavior are among the other areas covered in the papers.

0976 Proceedings of the Conference on Grouting in Geotechnical Engineering

Wallace Hayward Baker, (editor)

New York: ASCE, 1982, 1018pp.

The 63 pages included in these proceedings came from 12 different countries and deal with the pertinent advances in grouting materials and technology that have occurred throughout the world over the past two decades. Materials for cement and mortar grouts, and chemical grout materials are discussed. Dam grouting technology and its applications are discussed. Along with design and control for dam grouting. Chemical grouting technology and applications, and behavior of chemically grouted soil are also covered. Papers are presented on grouting for tunnels, shafts, and mines. Alternative grouting technologies covered include the utilization of super-high pressure liquid jets, flash-setting grout, cement grouts in offshore steel structures, and pressure injection grouting of landfills. Applications of grouting technology and testing and control for grouting are also covered. Many of the papers are illustrated. These papers were presented at the Conference on Grouting in Geotechnical Engineering, sponsored by the Geotechnical Engineering Division of the American Society of Civil Engineers in cooperation with the ASCE/AIME Underground Technology Council.

0977 Proceedings of the Conference on Water and Energy Technical and Policy Issues

Fritz Kilpatrick, (editor), Member, ASCE, (U.S. Geological Survey, Reston, Va. 22092) and **Donald Matchett**, (editor), Member, ASCE, (Stone & Webster Engrg. Corp., Denver, Colo.)

New York: ASCE, 1982, 668pp.

Industrial, political, legal, hydrological, and environmental issues surrounding development of energy and associated water resources are examined in this collection of papers presented at the Eastern and Western Conferences on Water and Energy in Pittsburgh, Pa., (May 24-26, 1982) and Fort Collins, Colo., (June 28-30, 1982). Water use and requirements, and wastewater treatment are examined in respect to the synthetic fuel industry, coal slurry pipelines, surface mining, hydropower, and oil shale production. Water conservation and reuse are also examined. Trends in energy/water legislation, policy and public management, as well as citizen participation in energy/water resource problems are reviewed. Attention is given to new energy resources and impacts, and to water related research needs for energy development. Planning and management for energy and water resource development are discussed including a review of issues regarding water allocation for competing uses such as agriculture versus industry. Among the case studies included are a review of citizen participation in the EPA Chesapeake Bay Program, a municipal hydropower development project, and projects on water recovery and wastewater volume reduction in industry.

0978 **Proceedings of the National Workshop on Reservoir Systems Operations**

Gerrit H. Toebes, (editor), (School of Civ. Engrg., Purdue Univ., W. Lafayette, Ind.) and **Alice A. Shephard**, (editor), (Sigma Research, Inc., Richland, Wash.)

New York: ASCE, 1981, 595pp.

The National Workshop on Reservoir Systems Operations, which convened August 13-17, 1979, at the University of Colorado in Boulder, was jointly sponsored by the American Society of Civil Engineers and the Office of Water Research and Technology of the Department of the Interior. The 90 participants represented federal agencies and private organizations engaged in reservoir systems operation and academic institutions involved in systems analysis research pertaining to this field. Objectives of the workshop included enhancement of technology transfer through the establishment of two-way lines of communication between users and researchers during the proceedings and formulation of specific recommendations to ensure continuation of this effort. Among the major issues addressed were: the communications gap between practitioners and academicians; the form, the curriculum and methods of present educational offerings; and the validity of distinctions between simulation and optimization. Six case studies are presented to provide a spectrum of reservoir systems, including a variety of sizes, ages, purposes, and agencies. Each case study includes a discussion and group study report.

0979 **Proceedings of the Speciality Conference on Construction Equipment & Techniques for the Eighties**

Cliff J. Schnexnayder, (editor)

New York: ASCE, 1982, 393pp.

Thirty-three papers on construction equipment and techniques are presented. Construction equipment management is discussed, including criteria for equipment selection, utilization and maintenance. An evaluation of equipment financing alternatives is presented to facilitate decisions on whether to buy, lease, or rent equipment. Specific areas of construction that are covered include building construction techniques, marine construction techniques for both port and offshore structures, heavy-highway construction techniques, and underground construction technques. Construction involving prefabricated structures includes discussions on energy conservative housing and air inflated structures. Some case studies are presented on cold region construction practice. Other aspects affecting construction prctices and covered in this volume include environmental regulations and safety. One section is devoted to project cost control and forecasting. Many of the papers are accompanied with illustrations These papers were published to facilitate discussion at the Specialty Conference on Costruction Equipment and Techniques for the Eighties, Purdue University, March 28-31, 1982. The conference was sponsored by the Committee on Construction Equipment and Techniques of the ASCE Construction Division in cooperation with the School of Engineering, Purdue University, and the Construction Research Council of ASCE.

0980 **Proceedings of the Speciality Conference on Engineering and Construction Projects**
The Emerging Management Roles

Dale R. Kern, (editor)

New York: ASCE, 1982, 335pp.

The papers presented at the specialty conference on emerging management roles in engineering and construction projects focus attention on successful management approaches, and methods of keeping construction projects within budget and on time. Roadblocks to successful professional management goals are covered, including restrictive relationships and liability aspects. Engineering, construction, project, and program management approaches are defined and discussed. Relationships and responsibilities of the professional manager and the general contractor are explored. Application of various management systems and techniques to the scheduling of super projects, management for assembly-line construction, and evolution of integrated project control systems are examined. Alternative management approaches to construction projects are compared. Target markets for engineering services are identified. Suggestions are given for improving project productivity. A procedure is presented whereby performance of construction management may be measured. Case studies are presented for five different types of projects, each with special considerations in project management.

0981 **Proceedings of the Specialty Conference on Design, Construction and Rehabilitation of Public Transit Facilities**

Dan Jones, (editor)

New York: American Society of Civil Engineers, 1982, 403pp.

The 35 papers presented at the March, 1982 Specialty Conference on Design, Construction and Rehabilitation of Public Transit Facilities are published here. Some of the papers focus on the design issue, while others address innovations in bus usage, transit system management, and high-occupancy vehicle applications. However, rehabilitation and modernization are key issues and are given close examination. Nearly every facet of passenger and maintenance facilities is reviewed including maintenance and storage facilities for rail transit systems, bus garages, and passenger interface facilities. Downtown bus priority treatment is also covered. Cost considerations are addressed, including design and construction techniques to lower costs. Some papers discuss views on operations and maintenance costs and the life cycle cost philosophy. A major portion of these discussions are related to energy factors. Several case studies are presented on bus priority traffic signal control, installation and rehabilitation of light rail systems, and conversion of an existing freight line for carrying transit passengers.

0982 Proceedings of the Specialty Conference on Environmentally Sound Water and Soil Management

E. Gordon Kruse, (editor), **Chuck R. Burdick**, (editor) and **Yousef A. Yousef**, (editor)

New York: ASCE, 1982, 536pp.

These proceedings contain the papers and abstracts presented at the 1982 Irrigation and Drainage Division of the American Society of Civil Engineers Specialty Conference held in Orlando, Florida, July 21-23, 1982. The conference theme, "Environmentally Sound Water and Soil Management," is supported by presentations covering environmental issues related to irrigation and drainage. Specific topics of discussion include new techniques in stream and wetlands management, agriculture and watershed management, forestry and watershed management, the impact of agricultural practices on receiving waters, effects of irrigation on water quality, and nutrient leaching. Research needs and trends in irrigation and drainage are reviewed. Irrigation management in humid areas, irrigation water requirements, and use of wastewater for irrigation are also considered. Considerable attention is given to irrigation and drainage in the Southeastern United States including agricultural drainage, urban drainage management, and groundwater problems and outlook in the coastal plains of the region. Metrication in water resources is also covered.

0983 Proceedings of the Specialty Conference: Reducing Risk and Liability through Better Specifications and Inspection

New York: ASCE, 1982, 165pp.

Papers presented at the Conference on Reducing Risk and Liability through Better Specifications and Inspection, and transcripts of the ensuing discussions are included in these proceedings. The issues of risks and liabilities prevalent in construction and how they can be reduced by improving the quality of construction specifications and inspection practices are examined. Specific topics covered include management of submittals to reduce liability, improving interpretation of specifications, inspector training, reducing owner liability through paper inspection techniques, cost and quality control, construction scheduling, the role of general consultants, and the role of the resident engineer.

0984 Proceedings of the Symposium on Probabilistic Methods in Structural Engineering

Masanobu Shinozuka, (editor) and **James T. P. Yao**, (editor)

New York: ASCE, 1981, 409pp.

Papers are presented from a symposium on the application of probabilistic methods to structural engineering. The first of four sections deals with codes and specifications which are probability based. State of the art in probability-based design codes are discussed in relation to steel and concrete structures, design loads, advance code formats, and serviceability requirements. In the second section, methods of risk analysis are reviewed for various types of structures including aircraft, dam, offshore, nuclear, and LNG terminal structures, and water supply systems. The third section deals with modern analytical methods in structural engineering including random vibration, crossing rates, partial factors, active control, and fatigue reliability. Safety and reliability of existing structures are covered in the final section. Practicing engineers present case studies and state of the art in current practice. Results of relevant research projects are also presented. The symposium was co-sponsord by the Engineering Mechanics Division and the Structural Division of the American Society of Civil Engineers.

0985 Proceedings of the Workshop Addressing Societal Needs of the 1980's Through Civil Engineering Research

New York: ASCE, 1982, 335pp.

Critical issues facing society in ten subdisciplines of civil engineering are defined and the research required to provide solutions to these issues in each area is identified. The ten subdisciplines are categorized as construction engineering and management; enery; environmental (sanitary) engineering; geotechnical engineering; hydraulics, coastal, ocean, and irrigation engineering; structural engineering and structural mechanics; surveying and photogrammetry; and transportation engineering. Each subdiscipline is examined within the context of six societal needs including shelter; food, air and water; public safety; conservation; energy; and transportation. A total of 182 critical research needs is identified.

0986 Proceedings of the Workshop on Limit Equilibrium, Plasticity and General Stress-Strain in Geotechnical Engineering

Raymond K. Yong, (editor) and **Hon-Yim Ko**, (editor)

New York: ASCE, 1981, 871pp.

Papers and discussions from the workshop on limit equilibrium, plasticity, and generalized stress-strain in geotechnical engineering are presented. The papers address aspects of the problems of modeling, testing and evaluation, as well as examine the deficiencies, merits, and requirements for successful solution of the problem of specifications of proper and pertinent soil constitutive relationships and yield/failure criteria. The focus of the papers is on predictive models and formulations presently available for stress-strain and the plastic yielding of soils, and the requirements and procedures needed to provide a more accurate modeling of the stress-strain and yield performance of soils. A main concern is the accurate and proper formulation of stress-strain relations and yield theories applicable to soils subjected to various stress paths and stress histories. Several predictive models, reported in recent literature, are examined and studied. These models range from empirical/phenomenological models to forms of cap and state boundary models.

0987 Pure and Wholesome

A Collection of Papers on Water and Waste Treatment at the Turn of the Century

Committee on History and Heritage of ASCE, Neal FitzSimons, chmn.

New York: ASCE, 1981, 175pp.

Four papers published between 1899 and 1907 describe problems in sanitary engineering encountered during a time of rapid urbanization. In these papers and associated discussions, leading engineers discuss the most appropriate means of providing water of acceptable quality and the problems of pollution from sewage and solid wastes. The emphasis is on filtration, with discussions of filtration versus watershed protection to insure an uncontaminated source of water. Also, the question of the necessity of treating sewage, if in fact filtration is effective in producing a water that is "pure and wholesome," is debated. Attitudes and practices concerned with environmental quality were developed in this formative period which persist to this day. The introductory paper was prepared for the Centennial of the American Society of Civil Engineers in 1952 and provides an historical context in which to place the other papers.

0988 **Seismic Performance of Low Rise Buildings**
State-of-the-Art and Research Needs

Ajaya Kumar Gupta, (editor)

New York: ASCE, 1981, 299pp.

Papers and panel reports from a workshop on seismic performance of low rise buildings assess the state of the art, identify problem areas, and determine research needs associated with the design and structure of low rise buildings and the main reasons for their failure during earthquakes. These issues are approached according to classification of low rise buildings by material (concrete and masonry, wood, and light metal), and building use (residential, commercial, and industrial). Manufactured building components are also discussed. Implementation strategies for earthquake resistant design of low rise buildings and impediments to these strategies are included. Research needs are discussed in detail and prioritized in each of the panel reports. The workshop was sponsored by the National Science Foundation.

0989 **State-of-the-Art Report on Finite Element Analysis of Reinforced Concrete**

Task Committee on Finite Element Analysis of Reinforced Concrete Structures of the Structural Division Committee on Concrete and Masonry Structures, Arthur H. Nilson, chmn.

New York: ASCE, 1982, 553pp.

This comprehensive state-of-the-art report contains a summary of the special features of finite element analysis applied to reinforced concrete, including: constitutive relationships, failure theories, multi-axial stress theories, modeling of reinforcement, behavior on the interface between reinforcement and concrete, crack representation, mechanisms of shear transfer, cyclic and dynamic loading effects, and the time-dependent effects of creep, shrinkage, and temperature variation. Alternative methods for incorporating these features are reviewed. Examples of recent applications are given, and comparisons are made between analytical results and observed behavior where such information is available. Computer programs that are generally available are referenced in an appendix.

0990 Structural Optimization
Recent Developments and Applications

Ovadia E. Lev, (editor)

New York: ASCE, 1981, 214pp.

Current documentation of the state-of-the-art of structural optimization is provided. Recent developments in the field, between 1972 and 1980, are emphasized. Research in structural optimization is performed by two groups of scientists. One group emphasizes the analytical, conceptual aspect, the other emphasizes the algorithmical, numerical aspect. While this report concentrates on the numerical aspects and industrial applications of structural optimization, the analytical methods are also reviewed. Over 130 books and 1000 papers are listed as references and a section containing 22 optimization problems frequently referred to in the literature is also included.

0991 Urban Water Conservation
Increasing Efficiency-in-Use Residential Water Demand

J. Ernest Flack, (editor), (University of Colorado, Boulder, Colo.)

New York: ASCE, 1982, 99pp.

Water supply and wastewater flow problems have resulted in an increasing concern with urban water demand. This report is an assessment of various water conservation measures aimed at reducing residential water usage. Demand reduction alternatives are examined and those applicable to residential areas are identified. Structural means such as water meters, recycle systems, water saving devices and flow reduction devices are examined. System and household leakage reduction as well as water use restrictions are among the operational methods investigated. Social and economic methods of public education, building code modifications, horticultural changes, and pricing policy are also studied. The literature is reviewed to determine the water savings that each conservation method can accomplish. The amount of water savings and the return flow implications for each method are investigated. Estimates of the combined impact of several methods used together in a common program are postulated. An assessment, based on this available data, found that water demand reductions of as much as 35 to 40% are possible through implementation of a combination of water conservation methods without incurring undue hardship on most water users. It is concluded that demand reduction techniques could successfully be incorporated in a water utility's management program.

1982 ASCE MEMOIRS

0992 Samuel S. Baxter, Past President, 1970-1971, ASCE
1905-1982

Samuel S. Baxter was born in Philadelphia on February 6, 1905. He received his diploma in Municipal Engineering from Drexel Institute of Technology (now Drexel University) in 1925. He began working for the city of Philadelphia in 1923 as an employee of the old Philadelphia Department of Public Works. He remained with that department until 1942 and for the last two years of that period he served as assistant director of that department.

During World War II he served with the Army Corps of Engineers and attained the rank of major. His army service included three years on the Manhattan (atom bomb) Project at Oak Ridge, Tennessee.

From 1946 to 1952 he worked for the Philadelphia Bureau of Engineering, serving for the first four years as assistant chief engineer and as chief engineer from 1950 to 1952. He was chief engineer of Philadelphia's Walnut Lane Bridge, completed in 1951. It was the first prestressed concrete girder bridge in the United States. In 1952 he was appointed as the city of Philadelphia's first water commissioner after the post was created under the new city charter. His first and most urgent mission was to remove objectionable tastes and odors from the city's drinking water. He launced a $500 million modernization program for the city's water and sewage systems and one year after his appointment as water commissioner he announced that for the first time in ten years, Philadelphia's drinking water had received an unconditional certificate of approval from the U.S. Public Health Service. Baxter's work as water commissioner also included the construction of three large sewage treatment plants and the construction of three filter plants. He served as Commissioner and Chief Engineer of Philadelphia's Water Department until 1972 and during his tenure earned a reputation as one of the nation's leading civil engineers.

Samuel Baxter joined the American Society of Civil Engineers as a Junior Member in 1926, became an Associate Member in 1935, a Member in 1946 and a Fellow in 1959. He served as a director of ASCE from 1959 to 1962. From 1962 to 1965 he was chairman of the ASCE Water Resources Coordinating Committee, and in 1968 he was a member of the ASCE Committee on Public Water Policy and Legislation. From October 1970 to October 1971 he served as President of the Society. In his outgoing address to the Society he stressed that the main purpose of the Society was the advancement of the science and profession of engineering and that this could be achieved through hard work on professional work and in service to the community.

Through his achievements as a professional engineer and as a life-long resident of Philadelphia, Baxter did well to set himself as an example of that advice. He was a member of the American Public Works Association and served as its national president in 1947. He served as board chairman of that organization's Research Foundation in 1959. He was also president of the American Water Works Association from 1965 to 1966.

He was a Diplomate of the American Academy of Environmental Engineers, a member of the National Academy of Engineering, the National Society of Professional Engineers and the Society of American Military Engineers. He served as president of the Philadelphia Chapter of the American Society of Public Administration. In 1959 he was cited as Engineer of the Year by the Philadelphia Chapter of the Pennsylvania Society of Civil Engineers. In 1967 he received the honorary degree of Doctor of Engineering from Drexel University. He also served as a trustee of that University.

In addition to his professional activities he also participated in many civic organizations including trustee of the Methodist Hospital in Philadelphia, chairman of disaster services for the Southeast Pennsylvania Chapter of the American Red Cross, president of the Philadelphia Council of the Boy Scouts of America and chairman of the United Fund Torch

Drive.

He was considered a very professional person and it was this attribute that enabled him to survive effectively through many political changes in Philadelphia.

Samuel Baxter died on February 7, 1982. He is survived by his wife Norma; son, Dr. Richard A.; daughter Linda C. Downe; and four grandsons.

0993

William Robert Eaton, Life Member, ASCE[1]
1904-1982

William Robert Eaton, a resident of Chattanooga, Tennessee, died in Birmingham, Alabama, on January 25, 1982, following heart by-pass surgery. The son of William Robinson and Grace Hull Eaton, he was born November 10, 1904, in Lockwood, Missouri. He attended the public schools of Lockwood and Kemper Military Institute. He was graduated from the University of Missouri in 1928 with a Bachelor of Science degree in Civil Engineering. Following graduation, he worked briefly for the Texas Oil Company and later for the Illinois Highway Department at Olney, Illinois. In 1929 Bob was employed by the United States Geological Survey and was stationed in Chattanooga, Tennessee. He worked as a hydraulic engineer in the Chattanooga District Surface Water (SW) staff. He was transferred to the Middle Atlantic District Headquarters in Washington, D.C. in 1931, and in 1933 moved back to the Tennessee District. From 1943 to 1947 he was in charge of the Asheville, N.C. Subdistrict. Bob was designated District Chief (SW) for Louisiana in 1947 and in 1948 moved for the third time to Chattanooga, this time as Chief of the Tennessee District (SW). In 1958 he was designated as Branch Area District Engineer (SW) of the mid-continent area, but did not transfer to the area headquarters in St.Louis until 1961.

He retired in 1967 and returned to Chattanooga to live. In retirement he often enjoyed deer hunting and trout fishing. During this period, 1970-1980, he worked with the U.S. Coast Guard Auxilliary inspecting boats and teaching safe boating and seamanship at several locations in the Tennessee Valley area.

He was married to Pauline Scoggin on January 23, 1930, in Marion, Arkansas.

He became a Life Member of ASCE in 1969.

Survivors include his wife, Mrs. Pauline S. Eaton of Chattanooga; son, William Robert Eaton III of Birmingham, Alabama; daughter, Mrs. Sammie Eaton Skinner of Wilson, N.C.; three sisters, Mrs. Ruth Marsh of Lockwood, Missouri, Mrs. Kemper Reid and Mrs. Lionel Northup of Columbia, Missouri; four grandchildren, one great-grandchild; several nieces and nephews.

0994

Cecil Earl Eubanks, Life Member, ASCE[2]
1901-1982

Cecil Earl Eubanks died September 1982 in Knoxville, Tennessee. The son of John J. and Helen Sheppard Eubanks, he was born on May 11, 1901, in Williamsburg, Kentucky.

He received a Bachelor of Science Degree in electrical engineering from the University of Tennessee in 1924 and a Master of Science in civil engineering in 1959.

After graduation he was employed by the Engineering Department of the City of Knoxville until 1930. At this time he was transferred to the Knoxville Utilities Board and became the Assistant Superintendent of the water treatment plant. From 1940 until 1953 he continued to work in the engineering department of the Knoxville Utilities Board and from 1953 until 1957 he was Superintendent of the water treatment plant. In 1960 he retired from the Knoxville Utilities Board and was employed by the Mid South Engineering Company in Knoxville as an

[1]Memoir prepared by J. B. Perry, Member, ASCE.

[2]Memoir prepared by J. B. Perry, Member, ASCE.

environmental engineer. He continued to work in the employment of the Flynt Engineering Company until retiring a second time in 1979.

On December 26, 1940, he was married to Willow Deen Hatcher in Knoxville, Tennessee.

At the University, he was a member of Phi Kappa Phi and Tau Beta Pi fraternities. He was a Fellow and Life Member of ASCE, a Life Member of the National Society of Professional Engineers, a member and Past President of the Knoxville Technical Society, a member of the Water Pollution Control Foundation, the Tennessee Water and Waste Water Association and the American Water Works Association. He was a member of St. Luke's United Methodist Church. Member and Past Commander of Coeur-de-Lion Commandry No. 9, Knights Templar, Charles H. McKinney Lodge No. 433, Paxton Chapter Royal Arch Masons 184, Knoxville Counsel No. 75 and Acacia Eastern Star 362.

He is survived by his wife and one sister.

0995

Elvin Leslie Everett, Life Member, ASCE[1]
1910-1982

Elvin Leslie Everett died May 15, 1982, in Knoxville, Tennessee. The son of John Henry and Lethia Lankford Everett, he was born in Terre Haute, Indiana, on November 16, 1910. His father was a railway postal worker.

He attended secondary schools in Terre Haute, Indiana, graduating from high school there in 1929.

In 1934 he was graduated from the Rose Polytechnic Institute (today called Rose Hulman Institute of Technology) in Terre Haute with a Bachelor of Science Degree in Civil Engineering. During the school years he worked part time as a chainman for R. L. McCormick of Terre Haute and as a test hole driller on topographic surveys for R. E. Hutchins. After graduation he worked briefly for R. A. Koermer as a surveyors assistant and A. G. Morrison as a draftsman. Both these firms were located in Terre Haute. He began work in March 1935 as an apprentice draftsman with the newly created Tennessee Valley Authority at offices in Chattanooga, Tennessee.

He worked in the Engineering Design office of the TVA in what is today known as the Highway, Railroad, Site Development, & Bridge Group. During nearly thirty-five years with the TVA he distinguished himself in work with railroad design. He assisted in the design of numerous access railroads and access highways at the hydro, fossil, and nuclear generating plants of the TVA. He also assisted in the design of the civil features within the plants.

On March 26, 1941, he married Margaret Hughlett of Trappe, Maryland. The following year, 1942, they moved to Knoxville with the relocation of TVA design offices. He worked in Knoxville until his retirement in early 1970. One of his two brothers, Merle, also retired from the TVA and presently resides in Chattanooga.

Elvin became a life member of the American Society of Civil Engineers in 1977. He was a member of the American Railway Engineering Association. He held offices in the TVA Engineering Association and continued as an associate member after his retirement. He was also a member of the Fraternal Order of Eagles and a Mason. He was a Registered Professional Engineer.

After his retirement, he and Margaret traveled, mostly to Florida and mostly to get away from the cold East Tennessee winters. He also devoted as much time as possible to visiting his grandchildren and great-grandchildren.

Elvin Leslie Everett is survived by his wife; his daughter, Doris Atkins of Atlanta, Georgia; and two sons, John, stationed with the Army in Ft. Hood, Texas, and Charles, an employee of the TVA in Knoxville. He is also survived by seven grandchildren and seven great-grandchildren.

[1]Memoir prepared by J. B. Perry, Member, ASCE.

0996

Robert Dawson Henderson, Fellow, ASCE[1]

1919-1982

Robert Dawson Henderson died suddenly on January 1, 1982, in Knoxville, Tennessee. The son of Luke and Edith Alice Dawson Henderson, he was born on July 5, 1919, in Boston, Massachusetts. He attended the public schools in Sherborn, Massachusetts, and graduated from Northeastern University in Boston, Massachusetts, in June 1942 with a B.S. degree in Civil Engineering. Later, he attended Canal Zone Junior College at Balboa, Canal Zone, and did graduate work at the University of Tennessee.

Soon after graduation he was employed by TVA in Knoxville. He did not work long until he entered the U.S. Army Air Corps for 3 1/2 years during World War II. In service he served as an Advanced Link Trainer Operator Instructor. He served with the U.S. Air Force Reserve until 1955.

After being discharged, he worked for two years with the Panama Canal and Railroad, Balboa Heights, Canal Zone, as a structural engineer. Upon returning to TVA in 1950 he worked on some phase of structural design of each of the TVA fossil-fired steam generating plants, except one, and had some involvement in most of the hydro projects in the TVA system at increasing levels of civil engineering responsibility. His latest work as Senior Civil Engineer involved responsible technical and administrative supervision, coordination and review of work of a civil engineering design section of up to 48 employees on major structures of the Bellefonte Nuclear Power Plant, a 2-unit 2660 megawatt plant in northern Alabama.

He was a member of and an active participant in the National/Tennessee Society of Professional Engineers, American Society of Civil Engineers, American Institute of Steel Construction, American Welding Society, Knoxville Technical Society, National Management Association, Tennessee Joint Engineering Action Group, and the U.S. National Conference on Earthquake Engineering.

He was an active member of the Fountain City Presbyterian Church of Knoxville; Toastmaster's International, Fountain City Club; Great Smokey Mountain Council, Boy Scouts of America; Coach of all Knoxville Boys Club entries in Optimists Oratorical Contest; Old Smokey Railway Museum; National Railway Historical Society; B.P.O. Elks Lodge 160; and the East Tennessee Automobile Club. He has served in all of the offices of the Tennessee Valley Section, ASCE, and on numerous committees both section and national. He also served in most of the offices of the Knoxville Chapter, National/Tennessee Society of Professional Engineers and served on numerous committees.

He was awarded the Outstanding Service Award to Engineering for the Knoxville area in 1970; the Chi Epsilon Civil Engineering Honor Society, Chapter Honor Member, 1974; the Tennessee Society of Professional Engineers, Knoxville Chapter Outstanding Service Award to Profession, 1976; the Presidential Award of Merit, TVA National Management Chapter, 1976; and the ASCE, Knoxville Branch, Dedicated Service Award, 1980. As the 1974 TVA Combined Federal Campaign, Division of Engineering Design Co-Leader, he raised $115,000 or 105 percent of the goal.

On June 13, 1942, he was married to Beatrice Louise Berg Mason in Cochituate, Massachusetts.

He was a student chapter member while at the university and became a Junior Member of ASCE in 1942, Associate Member 1952, Member in 1959, and Fellow in 1964.

He was a registered engineer in Tennessee and Alabama.

He is survived by his wife, Beatrice L. Berg Henderson, and a son, Richard Douglas Henderson, of Knoxville, one sister and two brothers.

[1]Memoir prepared by J. B. Perry, Member, ASCE.

0997

Justus John Koy, Fellow, ASCE[1]
1908-1982

Justus John Koy, son of August Koy, Jr. and Emma Buchtein Koy, was born on August 19, 1908 in Bonus, Texas and died in Houston, Texas on February 6, 1982. He received the Degree of Bachelor of Science in Civil Engineering from Rice Institute, now called Rice University, on June 6, 1932.

Prior to graduation John had been employed by United Gas Pipe Line Company in its Houston, Texas and Shreveport, Louisiana Engineering Departments. After graduation from Rice, he entered full time employment with this company, serving in various capacities, and, except for a period of service in the armed forces during World War II, remained until his retirement in 1973.

In 1970, the Houston division in which John served was reorganized and became United Texas Transmission.

As District Engineer and later as Chief Engineer of United Texas Transmission, John's responsibilities included the supervision of engineering personnel performing surveying, mapping and recording of Right of Way for pipelines, well locations, fee land surveys and telephone transmission systems, as well as supervision of personnel preparing cost estimates of the company's construction projects and the mapping, inspecting and writing of completion reports for that construction.

During World War II from March 3, 1943 to August 8, 1945 John Koy served as a Master Sergeant in the 363rd Engineers, Special Texas Regiment, Company C, of the Persian Gulf Command under Generals Donald H. Connolly and Donald P. Booth. His duties took in supervision of pipelines, water filtration and sewer systems, and maintenance of supplies for utilities.

John joined ASCE as a Junior Member in 1932, became an Associate Member in 1942, a Member in 1959, a Fellow in 1961 and was made a Life Member in 1973.

In addition to John Koy's long period of service in ASCE, he was also an active member of the Texas and National Societies of Professional Engineers, Houston Engineering and Scientific Society, A & A Scottish Rite of Free Masonry, Persian Gulf Command Veteran's Organization and St. Mark's United Methodist Church.

There have been many who served ASCE, but few were as conscientious, responsible, dedicated and helpful to other members of the society as was John Koy. He epitomized what is required to be a truly professional engineer.

John Koy married Vivian Adah Grubbs on April 20, 1935 in Houston, Texas. In addition to his wife, John is survived by a sister, Mrs. Irene Koy Harigel.

0998

Hung-hsun Ling, Life Member, ASCE[2]
1894-1981

Hung-hsun Ling was born in Canton, China and received his early education there. He graduated in 1915 from the Government Institute of Technology (later called Nanyang College then changed to Chiaotung University) in Shanghai. Later, the Chinese government sent him to the United States where he attended Columbia University School of Engineering and worked for three years for the American Bridge Company in Philadelphia and New York. Early in his career in China, he was Chancellor of Chiaotung University for five years. In 1931 he became a member of the Chinese Engineering Society in Shanghai and also served as director of that organization.

[1]Memoir prepared by W. O. Swift, Jr., Fellow, ASCE.
[2]Abstract of a memoir prepared by Edith Chung-ying Kao and T. C. Tsao.

Mr. Ling was Director and Engineer-in-Chief of the Chuchow-Saochow Section of the Canton-Hankow Railway which was completed on the eve of the Chinese-Japanese war in 1937. He wrote an article on this project which appears in the August 1935 issue of *Civil Engineering—ASCE*. He worked on many of China's major railroad projects from 1929 to 1945.

He served as Deputy Minister of Communications in the Chiang Kai-Shek Government from 1945 until 1949. In 1949, when the communists took power on the mainland, he fled to Taiwan. In 1951 in Taipel, Taiwan he became Chairman of the Board of the Chinese Petroleum Corporation and held the energy post until his retirement in 1971.

In addition to being an engineer and administrator, Mr. Ling was also the author of many articles and books including the titles *Railway Engineering* and *A Comprehensive Survey of Railway Development in China*. He was also a member of the Chinese American Engineering Association in Peiping, China and President of the Chinese Institute of Engineers. He represented China at many international engineering conferences and was honored for his contributions to the Chinese government.

He was elected to ASCE on October 1918 as a Junior Member; became an Associate Member in 1922; a Member in 1931 and a Fellow in 1959. In 1963 he was awarded the status of Life Member.

Mr. Ling was married to Huei-iang Shih from 1920 until her death in 1934. He is survived by his second wife, Ruth Soo; his sons Chung-guang, Chung-yu, Chung-li, and Chung-mai; his daughters, Chung-hsiu Wong and Edith Chung-ying Kao; 15 grandchildren and one great-granddaughter.

0999 **Arnoldo Longoria, Member, ASCE**[1]

1924-1981

Arnolldo Longoria, son of Romulo and Francisca R. Longoria was born February 26, 1924 in Veracruz, Mexico. He attended the Pharr-San Juan-Alamo schools in Pharr, Texas where his parents had come when he was a small child. He graduated with honors. He went to Edinburg Jr. College (now Pan American University) where he was a member of the National Honor Society, Phi Beta Kappa and again graduated with honors. He transferred to the University of Texas where he received his B.S. degree in Civil Engineering in January, 1950.

Upon graduation he returned to the Lower Rio Grande Valley and worked with the Texas Highway Department in Rio Grande City for two years. After flipping a coin (one side for South America, the other for Houston), he and his wife Edna, whom he had married in August, 1950 came to Houston, in 1952.

Arnoldo worked with H. E. Bovay Engineers in Houston until 1973 on projects including the design of Kennedy Blvd., main entrance to the Houston Intercontinental Airport, the Kirby Lumber Mill and the Methodist Hospital Garage, for which he presented a paper at an A.S.C.E. Convention. He was head of the Structural Department for H. E. Bovay Engineers when he made the decision to enter private practice.

In 1973 he established Arnold Longoria, Consulting Engineer and Longoria Engineers, Inc. which he operated until his death.

Arnold was a registered professional engineer in the State of Texas and a member of N.S.P.E., T.S.P.E. A.S.C.E., The Concrete Institute, Civil Defense, University of Texas Exes, Friends of Alec (Univ. of Texas). As a member of the A.S.C.E. he helped with the International Conventions. He was instrumental in getting several students interested in engineering and always encouraged them. He worked with the Boy Scouts and served as a Little League Baseball coach.

Arnoldo served in the U. S. Air Force during World War II, at which time he became a naturalized citizen of the United States. He served in France, Italy, Africa and Corsica.

He was a member of Saint Francis de Sales Catholic Church in Southwest Houston.

[1]Memoir prepared by David L. Adair, Fellow, ASCE.

Arnoldo Longoria died on August 29, 1981 and is survived by his wife Edna; a son, David Longoria; daughters Norma (Mrs. John) Carrington; Guilda (Mrs. Randy) Adams; Margo Longoria and a grandson, Kyle David Carrington.

Arnoldo entered the A.S.C.E. in 1962 as a Member.

1000

Nathan Meyer, Fellow, ASCE[1]
1903-1982

Nathan Meyer, the son of Joseph and Ida (Conviser) Meyer was born in New York City on December 2, 1903, the sixth of seven children. His parents had emigrated to the United States from Eastern Europe in the late 19th Century, and his father was the owner of a wire and metal fabrication business. He attended New York public schools and graduated from the Polytechnic Institute of Brooklyn with a Bachelor of Science in Civil Engineering in June, 1925.

Mr. Meyer's professional career spanned 57 years of experience in both design and construction in the New York area, as described in the following chronology.

After graduation he worked as a structural engineer with Parker & Shaffer and then with Ivan O. Shaffer, performing design and field inspection on steel and reinforced concrete industrial loft and high-rise buildings in New York City. In 1928 he became Chief Engineer for the Ettinger Security Construction Corporation, a firm which specialized in high-rise reinforced concrete buildings, many of which were then being constructed in New York.

From 1936 to the start of World War II, he served as the principal of Dover and then Hillside Construction Companies which performed public building and heavy construction projects in the New York area. During the early war years of 1942-43, Mr. Meyer worked as a structural engineer for the MW Kellogg Co. designing petrochemical plant facilities needed for the nation's defense effort. From late 1943 to 1946 he worked for the William Ginsberg Company, an engineering firm specializing in projects for the printing industry.

In 1946 he organized the firm of Lenmar Construction Company to become involved in the increased volume of public construction work after the end of World War II. During the fifteen years of the firm's existence until 1961, about eighty heavy construction contracts were performed for various New York City agencies, including the Department of Parks, Public Works and Transportation. He continued to perform construction work as the principal of Reyem Associates, Inc. until 1968, being involved mainly with playground and sewage treatment facilities.

From 1969 until the day of his death, Mr. Meyer worked as an engineering consultant to the Rodolitz Realty Co., providing structural and mechanical design and construction supervision services on industrial and office developments in the New York area. In addition, from 1980 - early 1982, he did structural design for Baskam and Jurczyk, Structural Engineers, of Woodbury, N.Y.

Mr. Meyer was a registered Professional Engineer in New York. As a member of Madison Jewish Center in Brooklyn in the early 1950's, he directed the construciton of a new building for the congregation. As a young man, he was an early pioneer in "Ham Radio" and operated Amateur Radio Station W2AIV in Brooklyn beginning in 1923. His hobbies included boating, fishing and gardening and he was a master of all forms of "Do-It-Yourself" endeavors.

On October 1, 1937, he married Betty Figer. Mr.Meyer passed away on September 9, 1982 and was predeceased by his wife in 1977. He is survived by two sons, Michael J., of White PLains, N.Y., and Peter E. Meyer, of Montclair, N.J. and three grandchildren. His sons are both licensed Civil Engineers and present members of the Society.

Mr. Meyer was elected an Associate Member of the Society in 1931, a member in 1947 and accorded the grade of Fellow in 1959. He became a Life Member in 1968.

[1]Memoir prepared by Michael J. Meyer, Member, ASCE.

1001

Sheldon R. Prager, Member, ASCE[1]

1943-1981

Sheldon R. Prager was born in Newark, N.J. on March 11, 1943, son of Irving and Betty (Greenberg) Prager, and his elementary education was in that city.

He graduated from Drexel University (formerly Drexel Institute of Technology) in Philadelphia, with a Bachelor of Science degree in Civil Engineering in 1966. He continued at Drexel in teaching and research until 1968, when he received a Master of Science degree.

While at Drexel, he was employed with the Corps of Engineers in Philadelphia, and Site Engineers, Inc. of Moorestown, N.J. Following graduation, he spent a year with Converse, Davis and Associates of Pasadena, California and two years with Schnabel Engineering Associates of Bethesda, Md. From 1971 to 1975 he was employed as project engineer by Sanders & Thomas, Inc., Pottstown, Penna., on highway and soils as related to environmental projects in many eastern states, Florida and West Africa.

From 1975 to the time of his death, he was an associate in ERTEC (Earth Technology Corp.) Los Angeles, and acted as Manager of facilities engineering and engineering business development. He was also project engineer for geotechnical design for nuclear plant design in Puerto Rico and Iran.

He was a member of the American Society of Civil Engineers, Los Angeles Section; Structural Engineers Association of Southern California; and the Soil and Foundation Engineers Association, of which he was President in 1981.

Prager presented papers at several conventions and meetings of engineers on various phases of earth stabilization and structure siting, including one on "Foundation Investigations for the Site of a Nuclear Plan" at the Seventh European Conference on Soil Mechanics, at Brighton, England in 1979.

He performed stability analysis on numerous dams, including Castaic, San Joaquin, Lake Bard, and Whittier Narrows Dam in California; Blue Marsh Dam in Pennsylvania; Henderson, Cave Buttes, and Painted Rock Dams in Arizona.

In 1965 he married Paula Rozen. He is survived by his widow, a son, Marc, and a daughter, Jennifer.

1002

Walter Adolf Renner, Fellow, ASCE[2]

1915-1982

Walter Adolf Renner, son of Phillip and Catherine (Duy) Renner, was born June 17, 1915 in Aurora, Illinois, where his parents had settled after coming from Austria. Following high school in Aurora he became a student at the University of Illinois at Urbana where he received the degree of Bachelor of Science in Civil Engineering and the Master of Science degree. He worked as a research assistant at the University under Professor Nathan Newmark, performing analytical work on slab and girder highway bridges.

During vacations and after college he worked with Roberts & Schaefer Company, Engineers, Chicago as a designer, inspector, and as a resident engineer on concrete shell structure projects. From 1942 to 1946 he served as a lieutenant with the Hq. 820th Engineer Battalion in the battles and campaigns of Normandy, Ardennes, Rhineland and Central Europe where he received five decorations and citations.

After World War II Mr. Renner rejoined Roberts & Schaefer Company, serving as a designer, resident engineer, researcher, test supervisor for shell structures, assistant superintendent on industrial structures, structural manager, and ultimately as manager of their Los

[1]Memoir prepared by Jack Schoustra and John M. Server, Members, ASCE.

[2]Memoir prepared by Anton Tedesko, Honorary Member, ASCE.

Angeles office which developed the concepts and criteria for the ground installations of USAF's Minuteman Missile System. During the early 1960's Mr. Renner worked on the early concepts of what then was known as the Mobile Minuteman System, a concept which has come to life again 18 years later as the MX-System.

When Roberts & Schaefer Company closed its Los Angeles office in 1962, Mr. Renner and his family chose to remain in California. For several years he worked with the Rocketdyne Division of North American Aviation, then joined Agbabian-Jacobsen Associates of Los Angeles as an analyst. After 2 years Mr. Renner switched to the Ralph M. Parsons Company as head of their Civil-Structural Section. He served also as head of their Quality Control Section and was Parsons' chief engineer at their head office in Pasadena when he retired in 1981.

Mr. Renner was a member of ASCE's first folded-plate committee, and he was active in professional societies in areas of nuclear structures, structural safety, and strength evaluation of existing structures. Noted for his excellent theoretical background and sound practical experience, "Walt" Renner will be remembered for his honesty and his lovable, boyish personality

Mr. Renner died at his home in Pacific Palisades, California, April 13, 1982. On August 13, 1955 he was married to the former Jean Riley of Chicago who survives him. Surviving also are their four sons: Steven Renner of Grand Junction, Colorado, Mark Renner of Pacific Palisades, Brian Renner of Santa Barbara, and Paul Renner of Pacific Palisades, California.

1003

Everett Scroggie, Life Member, ASCE[1]
1904-1982

Everett Scoggied died May 23, 1982, in Knoxville, Tennessee. The son of James Brown and Janet Tate Micholson Scroggie, he was born in Rion, South Carolina on December 4, 1903. He attended the public schools of Columbia, South Carolina and graduated from high school there in 1922. He entered the University of South Carolina in Columbia and graduated with honors with a Bachelor of Science Degree in Civil Engineering in 1926.

While a student at the University, he worked part time as a draftsman, designer and did field surveys for the South Carolina State Highway Department in Columbia.

He entered the graduate school of Rensselaer Polytechnic Institute in Troy, New York, and graduated with a Master of Civil Engineering Degree in 1927.

He was married to Ruby Ethelyn Powell on August 7, 1926, in Spartanburg, South Carolina.

He was a member of two honorary fraternities; Phi Beta Kappa and Tau Beta Phi.

Upon graduation from Rensselaer, he continued to work for the South Carolina State Highway Department until January 1928. At this time, he accepted a position as Assistant Engineer with the Southern Railway Company in Charlotte, North Carolina. From September 1932 until November 1933 he was employed by the North Carolina State Highway Commission in Raleigh, North Carolina, in the design of bridges and doing bridge surveys. In November of 1933 he was employed by the Tennessee Valley Authority as a concrete designer.

He retired from TVA in 1968 after a distinguished career in the design of bridges. Upon retirement he was Head Civil Engineer of the Bridges and Development Staff. Numerous bridges that he designed for reservoirs throughout the Tennessee Valley area will be reminders of his outstanding talent and skill for many generations. Ten of his bridges received national recognition in the A.I.S.C. (American Institute of Steel Construction) prize bridge competition. Notable of these were the bridge over the Little Tennessee River below Fontana Dam, a three-span continuous bridge across an area of Wautauga Reservoir, a two-span continuous bridge across South Holston Reservoir, and the Tennessee State Highway bridge across Norris Lake.

[1]Memoir prepared by J. B. Perry, Member, ASCE, and Roy A. Dorsch, Fellow, ASCE.

In addition to being a member of AREA Committee 8 Masonry, he was an active member of Cedar Springs Presbyterian Church and active in Cub Scouting. He was an avid fisherman, hunter and coin collector. He was a past president of the Knoxville Technical Society in 1961 and a past director of the Knoxville chapter of the Tennessee Society of Professional Engineers.

He was a registered professional engineer in Tennessee, South Carolina, North Carolina and Georgia. He was a Life Member of ASCE.

He is survived by his wife Ethelyn, daughter, Dr. Lucy E. Scroggie who works for TVA in Knoxville, a son, Everett Scroggie, Jr., an audiologist at the University of Florida in Gainesville, three grandchildren and a sister, Mrs. B. R. Gibson, Columbia, South Carolina.

He was buried in Melrose Cemetery, Abbeville, South Carolina.

1004

Melvin K. Sparks, Member, ASCE[1]
1918-1980

Melvin K. Sparks was born on a ranch near Crystal City, Texas, May 10, 1918. His father was William Arthur Sparks; his mother was Hattie Zuercher Sparks. His father died when Melvin was four and one-half years old, and his mother married Fred L. Foster when Melvin was eight and the family moved to San Antonio for six years. They returned to Crystal City where Melvin finished high school in 1937.

While in high school Sparks worked at the local movie theater taking tickets. After graduation he was made assistant manager and then manager. During this time he joined the Chamber of Commerce and the Lions Club.

When war was declared in December, 1941, Sparks joined the U.S. Air Force. He served four years in the Reconnaissance Branch, including three and one-half years in England, attaining the rank of Staff Sergeant. After the war he returned to his job as manager of the Guild Theater in Crystal City.

In 1948 Sparks entered Texas A & M University and received a B.S.C.E. degree in 1951. After graduation he worked for Griffels and Valet Consulting Engineers for one year, then joined Brown and Root in Houston as a structural engineer.

During his long and very distinguished career with Brown and Root Melvin Sparks was assigned to many foreign construction engineering positions, beginning in 1955 and ending in 1980, including field engineer for a large dam and irrigation system in the Republic of Haiti, as field engineer for a major marine terminal, a large gas compression station, and the cathodic protection of 48 flow stations, all for Creole Petroleum, in Lake Maracaibo, Venezuela, field engineer and later, office engineer on a dam and power house project in Thailand, chief engineer on a large earth filled dam in Columbia, and many other domestic and foreign projects. It was on his last assignment for a large cement plant in Alabama that Mr. Sparks became suddenly ill and died on October 15, 1980.

Mr. Sparks is survived by his mother, who still lives in Crystal City; two sisters, Miss Lillian A. Sparks of Houston and Mrs. Bernice Prestage of Crystal City; a niece and a nephew.

Melvin Sparks had become a registered professional engineer in Texas and was active in the Boy Scouts, the Methodist Church, the Chamber of Commerce and the Lions Club of Crystal City, serving as secretary of that organization for several years. A man who loved the outdoors and actively participated in the sports of fishing, golfing, swimming and skiing, he was at the same time a modest and humble person, as generous in the giving of himself as he was of his means.

It would be impossible to list his many accomplishments as an engineer, but it must be pointed out that Melvin Sparks, who had first been a successful businessman, developed a piping layout for compressor stations that eliminated the need for many concrete pipe anchors.

[1]Memoir prepared by Anthony P. Wenzel, Fellow, ASCE.

Few engineers have been privileged to work on as many and varied projects as Melvin Sparks. His high personal accomplishments served not only as a tribute to him and what he believed in, but also to the profession of Civil Engineering as a distinguished career.

1005 Marvin Ellsworth Warner, Fellow, ASCE[1]

1918-1982

Marvin Ellsworth Warner, son of Abe and Hilda (Daum) Warner, was born April 1, 1918 in Minneapolis, Minnesota. A graduate of the University of Minnesota, he spent most of his professional life as a structural engineer with the Roberts and Schaefer Company in Washington, D.C., New York City, and Los Angeles. Following college he worked for them as an inspector, resident engineer, and designer on concrete shell structure projects for the U.S. Army and U.S. Navy. He was married on November 30, 1940 to the former Carol Lesher.

1944 to 1946 he served as a lieutenant in the U.S. Navy on a small aircraft carrier. After the war he worked for a short time as a designer in Joseph Faisant's office. In 1948 he was back with Roberts & Schaefer as designer, project engineer, and project manager on a great variety of projects: industrial and military structures, arenas, marine terminals, airplane hangars, shell structures, ballistic missile and space rocket launching facilities. As the Company's representative he was the URSAM Joint Venture's structural section head (1964-66) in the design of assembly and launch facilities for the manned moon landing program, including NASA's Vehicle Assembly Building, which received ASCE's Outstanding Civil Engineering Achievement Award.

Subsequently, he represented Seelye Stevenson Value & Knecht as the structural section head in the DUSAF Joint Venture in the design and supervision of the construction of the National Accelerator Laboratory at Batavia, Illinois, now known as the Fermi Lab.

In 1973 he worked with the Harmon Division of Research-Cottrell in Bound Brook, New Jersey, as chief structural engineer, involved in the design and construction of natural draft cooling towers. 1974 saw Mr. Warner in Colorado where he became an associate of industrial engineer H. V. Weeks. In 1976 he returned to Illinois to accept a permanent position as head of Fermi-Lab's engineering group. Thoroughly familiar with structures, facilities, equipment, etc., in the design of which he had been involved, he expected to hold this position with Fermi-Lab for many years to come. However, he died on May 16, 1982, while jogging near his home at Wheaton, Illinois.

"Marv" Warner was a hard worker whom his friends will remember for his practical competence, his boyish enthusiasm and his sincerity.

Mr. Warner is survived by his wife, Carol L. Warner, by his children Nancy, Bruce, Audrey, Kenneth and Richard, and by eight grandchildren.

1006 George Dewey Whitmore, Life member, ASCE[2]

1898-1981

George Dewey Whitmore died on February 9, 1981, in Woodstock, Illinois. The son of George Edward and Jennie Ames Whitmore, he was born on April 25, 1898, in Hillman, Michigan. He attended secondary schools in Jackson, Michigan, and completed high school in 1916. He began his career in surveying and mapping in 1917 when he became a rodman. In 1921 he was employed by the Village of Canfield, Ohio, as a designer engineer for water and sewer systems. During 1922 he worked for the City of Bluefield, West Virginia, in the preparation of a property atlas for the city. In 1923 he was employed by R. H. Randall and Company of Toledo, Ohio, as a district engineer for city surveys. From 1923 to 1926 he was employed by the City of Pittsburgh, Pennsylvania, as a resident engineer for geodetic and topographical surveys. During

[1]Memoir prepared by Anton Tedesko, Honorary Member, ASCE.
[2]Memoir prepared by J. B. Perry, Member, ASCE.

this time he received a civil engineering diploma from the International Correspondence Schools. From 1926 to 1932 he again worked for the R. H. Randall and Company as Vice President and Assistant Chief Engineer. From 1932 to 1933 he was in charge of geodetic and topographical surveys for the city planning commission of Toledo, Ohio. In November 1933 he began work as a cadastrial engineer for Engineering Services Division of the Tennessee Valley Authority in Knoxville, Tennessee.

In September 1936 he transferred to Chattanooga as Principal Civil Engineer and Head of the Engineering Surveys Section of the Engineering Services Division, later called the Maps and Surveys Division.

While in Chattanooga he was active in the formation of the American Congress on Surveying and Mapping and was elected its third President and served two terms. George will be remembered for his role as the dominant force in the establishment of the technical standards of the old Maps and Surveys Branch of TVA, which enabled it to be recognized as one of the foremost organizations of its kind.

George was endowed with a strong foresight, coupled with the willingness to listen, the technical ability to evaluate new ideas, and to organize practical application of them. He used these attributes to advocate and perfect methods for the first widely accepted use of aerial photographs in cadastral surveys and for the organization and development of procedures for the Valley Topographic Mapping Program using photogrammetry to produce 1:24,000-scale topography. Much of the success of the Valley Topographic Mapping Program can be attributed to his perseverance against resistance to the photogrammetric approach.

Early in World War II several thousand miles of strategic U.S. areas, considered to be potential invasion routes, were mapped by Maps and Surveys under George's guidance, and as the war progressed TVA's effort was shifted to the mapping of three-fourths of a million square miles of enemy held territory. During all this he found time to consult with the Brazilian Government on their mapping needs in the San Francisco Valley.

In September 1945 he transferred to the U.S. Geological Survey in Washington, D.C. as Technical Staff Chief of the Topographic Division.

Twelve years later he was named chief topographic engineer with overall responsibility for carrying out the U.S. National Topographic Mapping Program. From then until his retirement in 1968, Mr. Whitmore led a great advancement in the rate of map production and a continuing revolution in mapping techniques.

He was a past president of both the American Congress of Surveying and Mapping and the American Society of Photogrammetry. He also had served as chairman of the surveying and mapping division of the American Society of Civil Engineers.

He was the recipient of the American Society of Photogrammetry's Struck Award. He also earned a surveying and mapping award from the American Society of Civil Engineers, and the Interior Department's Distinguished Service Award. The Whitmore Mountains in Antarctica were named for him.

He was a member of the American Geophysical Union, the American Geological Institute, the Washington Society of Engineers, the American Polar Society and the Antarctician Society. He was an honorary member of the Canadian Institute of Surveying.

He was married to Helen O'Neill who preceded him in death in 1975.

He is survived by a son, Robert, of Naples, Florida; three daughters, Margaret Gallagher of Reisterstown, Maryland, and Patricia Tambone and Jane Whitmore, both of Woodstock, Illinois; 24 grandchildren, and four great-grandchildren.

— — — — — — — —

1007

Robert Zaborowski, Fellow, ASCE[1]
1917-1982

Robert Zaborowski, son of Bruno and Harriet (Klocko) Zaborowski, was born on May 30, 1917 in Cicero, Illinois. He attended Crane High School in Chicago, receiving advanced credits at the Central YMCA. In 1934 he became a student at the University of Illinois in Urbana and there received with highest honors the Bachelor of Science in Civil Engineering (1938) and also the Master of Science degree (1940). At the university he worked under Professor Wilbur Wilson on tests of structural steel tension members. On May 30, 1939 he married the former Julia Koch.

Mr. Zaborowski joined Roberts & Schaefer Company, Engineers, in Chicago in 1940 and was soon thereafter transferred to their Washington, D.C. offices. Mr. Zaborowski had a brilliant mind, became an expert in concrete shell design, and was involved in the design of shells for airplane hangars of the U.S. Army and the U.S. Navy, and for other military facilities. He was among the first to use electric strain gauges on full-scale structures. 1944-1945 he worked at Johns Hopkins University's Applied Physics Laboratory on the development of the proximity fuse and other R & D work in support of the war effort.

Back at Roberts & Schaefer in Chicago, he carried on research in the structural analysis of concrete frames leading to economies in long-span concrete shell structures. When the Company opened an office in New York City, he was appointed its manager and for years was there in charge of consulting engineering work on major industrial, institutional and government projects. His responsibility included administration, engineering and architectural engineering, design, cost estimating, and supervision of construction. Marine terminals, sports arenas, airplane hangars are examples of his work during that period.

Upon the merger of the Roberts & Schaefer Company into Thompson-Starrett Company, Engineers and Builders, he became vice-president and ultimately the president of Thompson-Starrett at their Chicago offices, engaged primarily in industrial structures, nationally and internationally.

Mr. Zaborowski entered private practice in 1960 as a partner in Dahlquist & Zaborowski, Architects and Engineers, at Glenview, Illinois. In 1968 he joined Wil-Freds, Inc., General Contractors, headquartered at Naperville, Illinois. He served as vice-president of this firm until his retirement in 1978.

Mr. Zaborowski was known as a most competent engineer of high intelligence, with a wide range of interests. He was a member of the first shell committee which wrote the 1964 report on Shell Design and Construction. He served also on other professional committees; he pioneered new designs and processes. His oral and written presentations were always outstanding.

Mr. Zaborowski died at his home in Naperville, Illinois, on May 1, 1982. He is survived by his wife, Julia K. Zaboroski, his sons Neil R. Zaborowski of DeKalb, Illinois, and Roy A. Zaborowski and his wife Barbara and daughter Elizabeth of Phoenix, Arizona.

[1]Memoir prepared by Anton Tedesko, Honorary Member, ASCE.

SUBJECT INDEX

AUTHOR INDEX